Curves and Fractal Dimension

Claude Tricot

Curves and Fractal Dimension

With a Foreword by Michel Mendès France

With 163 Illustrations

Springer-Verlag

New York Berlin Heidelberg London Paris
Tokyo Hong Kong Barcelona Budapest

Claude Tricot
Département de Mathématiques Appliquées
École Polytechnique de Montréal
Montréal, Québec, Canada H3C 3A7

Cover illustration: Disjoint balls of the same radius, adjacent to a boundary that is a rectifiable curve, and included in its interior. This appears as Fig. 17.2 in the text.

SCI
QA
567
.T75 13
1995
c. 2

AMS Subject Classification: 28A80

Library of Congress Cataloging-in-Publication Data
Tricot, Claude.
    Curves and fractal dimension / Claude Tricot.
      p.   cm.
    Includes bibliographical references and index.
    ISBN 0-387-94095-2 (New York). — ISBN 3-540-94095-2 (Berlin)
    1. Curves, Plane. 2. Fractals. I. Title.
QA567.T75 1993
516.3'52—dc20   93-4690

Printed on acid-free paper.

First French edition, *Courbes et Dimension Fractale*, © Springer-Verlag, Paris, 1993.

Production managed by Terry Kornak; manufacturing supervised by Genieve Shaw.
Photocomposed copy prepared using the author's TeX files.
Printed and bound by R. R. Donnelley & Sons, Harrisonburg, VA.
Printed in the United States of America.

9 8 7 6 5 4 3 2 1

ISBN 0-387-94095-2 Springer-Verlag New York Berlin Heidelberg
ISBN 3-540-94095-2 Springer-Verlag Berlin Heidelberg New York

Foreword

A mathematician, a real one, one for whom mathematical objects are abstract and exist only in his mind or in some remote Platonic universe, never "sees" a curve. A curve is infinitely narrow and invisible. Yet, we all have "seen" straight lines, circles, parabolas, etc. when many years ago (for some of us) we were taught elementary geometry at school.

E. Mach wanted to suppress from physics everything that could not be perceived: physics and metaphysics must not exist together. Many a scientist was deeply influenced by his philosophy. In his book Claude Tricot tells us that a curve has a non-vanishing width. Its width is that of the pencil or of the pen on the paper, or of the chalk on the blackboard. The abstract curve which cannot be seen and which does not really concern us here is the intersection of all those thick curves that contain it. For Claude Tricot it is only the thick curves that are pertinent. He describes in detail the way bumps, peaks, and irregularities appear on the curve as its width decreases.

This is not a new point of view. Indeed Hausdorff and Bouligand initiated the idea at the beginning of this century. However, Claude Tricot manages to refine the theory extensively and interestingly. His approach is both realistic and mathematically rigorous. Mathematicians who only feed on abstractions as well as engineers who tackle tangible problems will enjoy reading this book.

The real world is complex; real shapes are often complicated and real surfaces are often very uneven. There is a mathematical theory that describes such objects which is Mandelbrot's Fractal Geometry. In his seminal book "Les Objets Fractals" (1967), which is considerably extended in the English version, Mandelbrot shows how rivers, coastlines, clouds, trees, lungs, telegraphic messages etc. all obey fractal laws. Their dimensions are seldom integers.

Claude Tricot discusses at length the theory of fractals applied to curves. It transpires from this that curves may well have a fractal dimension which exceeds its natural dimension which is 1. We are reminded of Kandinsky, who makes the following observation in his famous book, "Point, Line and Plane" (1926): "A straight line is the total negation of the plane whereas a curved line is potentially the plane in that it contains the essence of the plane within itself." With his artist's eye he has understood that a line can mimic a plane. Claude Tricot with his mathematician's eye convinces us that erratic curves do indeed aspire to being planes.

Michel Mendès France

Introduction

This book contains ancient and modern mathematics. It has been written for those specialists in experimental sciences who have neither the time nor the taste to go through the ordinary styled pure mathematical books. In fact, this style, which is quite mysterious and hermetic, led many of the able to be discouraged and to proclaim that higher mathematics is a heaven reserved for a few fortunate ones. Also, there is another less common but more dangerous attitude of being led through the path of abstract methodology, which gives the impression of a perfect science where everything is flawless, but where everything is finally transformed into a pure rhetoric.

In the following pages, we made an effort to do good mathematics without falling into rhetoric. There is a risk with this: experimenters may view this as standing far away from real and concrete problems, and pure mathematicians may view this as being imprecise, without sufficient generality. In counterpart, we gain a book where serious scientists can find ideas and methods to assist them in their research.

In fact, nothing is more unknown than curves, especially their geometrical and analytical properties. Yet the subject is so wide that we felt the need to restrict ourselves to the notion of **simple curves** drawn in the **plane**. For a topologist, these curves have the same characteristics as a line segment, and thus there is not much left to say. But as soon as we leave the stratosphere of the *analysis situ*, we must realize that there are infinitely many varieties of curves that are defined in mathematical models of trajectories, boundaries of aggregates, geographical lines, and profiles of rough surfaces. Moreover, we need to know how to classify them and to characterize them. There are two families of curves: those of **finite length**, and those whose polygonal approximations can have as large a length as we wish as the precision gets better, these are said to be of **infinite length**.

A part of the analysis of curves is based on the study of their intersections with straight lines. These intersection sets can contain a finite number of elements or they can be infinite sets of Cantor's type. That is why we devote the first three chapters to real **sets of null measure**. This is a classical theme, dear to Cantor, Borel, and Hausdorff, and it allows the introduction of the Bouligand **dimension**, one of the most useful tools of analysis employed in this book.

The second part deals with curves of finite length, or **rectifiable curves**, their local properties, and the different methods that can be used to define and compute their lengths. This is also an ancient and forgotten theme; it has disappeared from the curriculum of graduate studies. It cannot be found except in some books on the theory of measure, where it is drowned and distorted by an abusive generalization. It is interesting and useful to rediscover all these theorems. However, not all is said on this subject; there still is a place for some new ideas and reflections.

Finally, the third part deals with curves of infinite lengths, or **nonrectifiable curves**, and above all those that are **fractal**. This is a more restricted problem than the study of fractal sets as introduced and popularized by Benoît Mandelbrot. Yet it is quite surprising to find, even in the case of curves, how scarce, hazy, and unmathematical are our notions on the subject. The less precise our thoughts are, the fuzzier our practical scientific path will be. How many still think that the west coast of Great Britain is a self-similar curve, or that the "compass method" allows us to estimate the fractal dimension, or even that the dimension of a curve can be simply computed by adding 1 to the dimension of its intersection with a straight line? Inasmuch as this book is concerned we made a considerable effort to give an absolutely precise definition to every employed term, then bit by bit, we deduce the notions and computation methods; some are old, some are new, but all are practically useful. This is why we do not mention the **Hausdorff dimension**, an excellent mathematical tool, but we believe that, as it stands, it has no practical application in the study of curves originated in other sciences: physics, biology, or engineering. For this we acknowledge the helpful discussions we had with other scientists, physicists and engineers, on the best characterization of experimental curves.

This book contains eighteen chapters and many illustrations in the hope that they will clarify the fine print. Almost all the chapters end with bibliographical notes. The reader finds a mathematical complement in the appendices: the most important is the one that deals with convexity, a notion used in the local analysis of a curve.

A word on the text:

— This text is not a classical succession of well-numbered definitions, propositions, and theorems; it is a departure from the usual mathematical style. We hope, like the books from the beginning of this century, that this will be a gain in clarity and interest. Almost all the mentioned results are proved, and the proofs, which are in small characters, are set between ▶ ◀, so that the reader who is not interested in the technical details may skip them. The symbol ◇ is used a lot; it replaces the word "Remark."

— A bibliographical note can be found at the end of almost every chapter. It is by no means exhaustive, especially in the "Fractal" part. Interested readers may consult the bibliographies of other works on this subject. This book is a complement on some points; it is not exhaustive. "Fractal" is a new science and we need to wait a half-century before the really good ideas can be filtered out.

— The typesetting was arranged by TEX. For the illustrations we used a combination of software: Mathematika, Illustrator, and the new Fractal Analysis (*Analyse Fractale*)—tailored by the members of my research team—which knows how to draw attractors and sausages and to calculate the dimensions.

It is a pleasure to mention the part taken by my research team: Stéphane Baldo, François Normand, and Salim Salem for the scientific part; Emmanuelle Goulet, Pierre Ferland, and Axel van de Walle for the computer processing; and Frederic Latreille, Uong Dinh Bich Chau, and Axel van de Walle for the illustrations. Their young and enthusiastic state of mind has been of considerable help when working on this book.

Claude Tricot

Contents

Part IV. Annexes, References and Index

1 Perfect Sets and Their Measure

1.1 Duality set—measure

It is nearly impossible to describe all types of subsets of the real line. There is a very large number of them, but few are useful in building scientific models. Fortunately, the needs of mathematical modeling are much more restricted than those of pure logic. Experience shows that the wider and more general a theory becomes, the fewer specialists it will address. Here we observe the paradox between generality and usefulness. In this book, we deliberately favor "usefulness." We consider sets of well-known types, not too sophisticated, but easily *constructible*. Exactly what are these sets?

On the real line, we first see the **intervals**; there are the open ones (denoted by $]a, b[$), the closed ones (denoted by $[a, b]$), and the semiopen or semiclosed ($]a, b]$, $[a, b[$). Other interesting sets are built from intervals.

The **open** sets are the finite or countable unions of open intervals, and the **closed** sets are the complements of the open sets. The following is a characterization of open and closed sets:

> *A set V is* **open** *if every element of V is the center of an interval (even a very small one) contained in V.*
>
> *A set E is* **closed** *if it contains all the limits of all convergent sequences of its elements.*

◇ The interval $]a, b[$ is open: for any x in the interval, one finds a small ϵ such that $]a, b[$ contains $]x-\epsilon, x+\epsilon[$. But it is not closed, because the sequence $(a+\frac{b-a}{n})$, whose elements are in $]a, b[$, converges to a, which does not belong to the set $]a, b[$.

Given a set E, add to it all the possible limits of convergent sequences of elements of E; we obtain a closed set $\overline{E}$, which we call the **closure** of E.

Given a family of open sets V_n, we may construct sets of the form

$$E = \bigcap_n V_n \; ;$$

these are the finite or countable intersections of open sets. The class of these sets strictly contains the class of all open and closed sets. The complements of these sets can be written as:

$$E = \bigcup_n F_n \; ;$$

these are the finite or countable unions of closed sets. The class of these sets also contains the open and closed sets. But it is not the same as the previous one.

Although these constructions suffice for the topological descriptions of the present work, let us note that by taking all intersections, unions, and complements of such sets, one defines the *Borel* sets (from E. Borel). In the language of set theory, the Borel family is the smallest class of sets that contains all the intervals and is *closed* under the intersection, union, and complement operations. In practical terms these are the only *constructible* mathematical sets. All the other sets are logical; we can only assume their existence.

The construction, by induction, of Borel sets has a very important advantage: the (Borel) *measure* of these sets is also determined. The measure of an interval is its length. Then we apply Borel simple rules (1898):

> "*The measure theory for sets only requires that: having defined the measure of an interval to be its length; we consider the following necessary conventions: the measure of a finite or countable union of a family of disjoint sets is the sum of the measures of its members, and the measure of the difference of two sets (the set from which we subtract must contain the subtracted set) is equal to the difference of their measures.*"

Thus there exists a duality between sets and their measures. In fact, constructing a set and calculating its measure follow the same procedure. Somehow this tends to confirm the famous scientific rule: "What we cannot measure does not exist." We shall find other applications of this rule throughout the second part of this book, concerning curves. Curves, as we shall see, cannot be practically studied unless they are defined in a correct manner, that is, unless they are *parameterized*. But every parameterization induces a measure on the curve.

◇ We can measure an object in many different ways. For example, to define measures other than the Borel measure, we may use time-related measures or probability measures. The Borel measure is founded on the notion of the interval length; it is often called *length*. To avoid confusion we speak of *sets of null length*, instead of null measure.

1.2 Closed sets and contiguous intervals

The length of an interval $I = [a, b]$ or $]a, b[$ is

$$L(I) = ba .$$

The length $L(V)$ of an open set V is no more difficult to compute. Such a set is either an interval or a disjoint union of open intervals, hence it suffices to sum the lengths of these intervals. We deduce the length of closed sets in the following manner: let the closed set F be contained in the interval $[a, b]$. When a is the smallest element of F and b is the largest, we say that $[a, b]$ is the *fundamental interval* of F. This is the smallest closed interval containing F. The complement of F in its fundamental interval is an open set. Thus it is a union of a countable family of disjoint open intervals $\mathbf{C}_1, \mathbf{C}_2, \dots$. Following A. Denjoy we call them the **contiguous** intervals of F. By Borel's rules we have:

$$L(F) = b - a - \sum_n L(\mathbf{C}_n) \, .$$

A closed set is of null length if the sum of the lengths of its contiguous intervals in $[a, b]$ is equal to $b - a$.

It is clear that the measure of a closed set does not depend on the positions of its contiguous intervals. But it is important to note that the *topological properties of F depend on these positions*. We shall give some typical examples that will spare us the heavy general description.

Example In the interval $[0, 1]$ consider

$$\mathbf{C}_n = \left] \frac{1}{n+1}, \frac{1}{n} \right[\, , \, n = 1, 2, \dots \, .$$

Once these intervals are taken away we are left with a closed set F whose elements are $\{\frac{1}{n}\}$ and $\{0\}$.

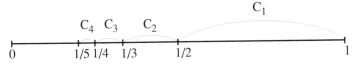

Fig. 1.1. *The closed set F, formed by the points 0 and $1/n$, $n \geq 1$. The contiguous open intervals $\mathbf{C}_1$, $\mathbf{C}_2, \dots$, have common endpoints (except point 1), hence F is countable.*

The points $\{\frac{1}{n}\}$ are called *isolated* points; each is the center of an interval containing no other points of F. We say that a set is *discrete* if all its elements are isolated. Here F is not a discrete set since it contains $\{0\}$, which is an *accumulation point*. This example suggests the following rule:

> When we arrange the contiguous intervals $\mathbf{C}_1$, $\mathbf{C}_2$, ... *from left to right or from right to left, according to the order of their indices, then the residual set F is formed by the endpoints of its fundamental interval, the isolated points, and eventually some closed intervals.*

But the family of closed sets is richer than the type just described. There are the *perfect* sets, which do not contain isolated points and are neither finite nor countable. Above all, there are the *perfect nowhere dense sets*, which contain no intervals. This is the topic of the next section.

1.3 Perfect sets

Example Cantor receives the credit for constructing the first closed uncountable set that contains no interval. Start with an interval $[0, 1]$ and take away the contiguous interval $\mathbf{C}_1 =]1/3, 2/3[$ of length $1/3$ centered at the middle of the initial interval. We are left with two intervals $[0, 1/3]$ and $[2/3, 1]$; repeat the same operation with the same ratio of $1/3$. At the nth iteration we get the union of 2^n disjoint intervals of length 3^{-n}, such that a contiguous interval of length 3^{-n-1} centered in the middle of each will be taken away. Since there is a countable number of steps, there is a countable number of contiguous intervals.

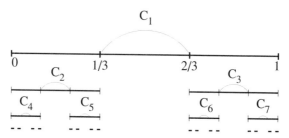

Fig. 1.2. *The construction of the Cantor triadic set by successive division. The intervals removed are one of length $1/3$, two of length $1/3^2$, four of length $1/3^3$, etc. No two of these can have a common endpoint.*

Evidently, the closed set F, the limit of the above construction, contains all the endpoints of the removed contiguous intervals; in fact, it contains many more elements as we shall see in §4. Between any two intervals there is a third. Thus no two intervals have a common endpoint. Moreover, F cannot contain an interval; in fact, if we take away the intervals up to the nth step, we are left with the closed intervals of length 3^{-n}. This length converges to 0 as n tends to infinity. Thus an interval contained in F must have length 0. Finally the length of F is 0 since there are 2^{n-1} contiguous intervals of length 3^{-n} and

$$1 - \sum_{n=1}^{\infty} 2^{n-1} 3^{-n} = 0 \,.$$

Another argument goes like this: at each stage we are left with two-thirds of the measure of the previous set; after n stages the measure of the left off set is $(2/3)^n$, which converges to 0 as $n \to \infty$. Thus the measure of the residual set is 0. This Cantor set can be totally described from the topological point of view as being a *perfect nowhere dense set*.

> A **perfect** *set is a closed set with no isolated points. A* **nowhere dense** *set is a set whose closure contains no interval.*

We can give some general rules about sets to construct such sets:

If the contiguous intervals $\mathbf{C}_1, \mathbf{C}_2, \ldots$ *are arranged so that:*
(i)   no two contiguous intervals have a common endpoint, then the residual set F is perfect.
(ii)   every point of F (hence, every point of $[0,1]$) belongs to the closure of $\cup_n \mathbf{C}_n$, then F is nowhere dense.

▶   The property (*i*) implies that F has no isolated points because any such point is a common endpoint of two contiguous intervals. The property (*ii*) implies that F contains no open interval because otherwise the middle of such an interval cannot be the limit of a sequence of elements of the contiguous intervals. Hence it will not belong to the closure of $\cup_n \mathbf{C}_n$.   ◀

◇   If we take away all isolated points from a closed set, we obtain a closed set. Repeating the same operation infinitely many times we are left with a perfect set (Cantor-Bendixon theorem). Taking away all open intervals from this set we obtain a perfect nowhere dense set. Cantor concluded in 1883 (replace "first power" with "countable"):

"Each closed set P of a power higher than the first can be divided in a unique manner into a set of first power and a perfect set."

Thus, the two sets described in the second and third sections are typical closed sets.

◇   A closed set of measure 0 cannot contain any open interval. Thus, it is nowhere dense. The converse is not true: it is easy to find perfect sets that are nowhere dense and whose measures are not 0. The division at each stage of the construction of a Cantor set by $1/3$ is the reason behind its null measure. This division is independent of the topological structure of F, which is due to the way in which we arrange the contiguous intervals. To give an example of a perfect nowhere dense set whose measure is not 0, we may simply take the contiguous intervals of a Cantor set away from the interval $[0,2]$: from $[0,2]$ take away the first contiguous $\mathbf{C}_1$, centered at 1 with length $1/3$; from each of the two intervals of length $5/6$ that are left, take away an interval of length $1/9$ centered at their middles; and so forth. At the nth stage we obtain a union of 2^n closed intervals of length $2^{-n} + 3^{-n}$. After infinitely many times we get a perfect nowhere dense set whose length is 1, since the total length of all the contiguous intervals is 1. A geometrical construction of this set is given in Chap. 6 (Fig. 6.4).

1.4 Binary trees and the power of perfect sets

As we have seen, the topological properties of the residual set F are solely determined by the position of its contiguous intervals. If they are taken away from left to right or right to left, then F is a discrete set, with a unique accumulation point. If they are arranged so that from between any two intervals we take away a third, then F is perfect. But here, the notation $\mathbf{C}_1$, $\mathbf{C}_2, \ldots$ of the contiguous intervals will not tell us the exact place of the contiguous $\mathbf{C}_n$ for any integer n with respect to the others. That is why we need another indexing scheme. The simplest is the dyadic scheme.

For example, let $\mathbf{C}_1$ be the largest interval (or one of the largest if there is more than one of the same maximal length), $\mathbf{C}_2$ the largest interval situated to the left of $\mathbf{C}_1$, and $\mathbf{C}_3$ the largest to the right of $\mathbf{C}_1$. Repeating the same procedure, the interval $\mathbf{C}_4$ will designate the largest interval situated to the left of $\mathbf{C}_2$, $\mathbf{C}_5$ between C_2 and C_1, C_6 between C_1 and C_3, C_7 to the right of C_3. This type of arrangement of the intervals can be symbolized by the tree in Fig. 1.3, where every nth node represents the interval $\mathbf{C}_n$. The fact that each chosen interval is always of maximal length assures us that none of the intervals will be left out.

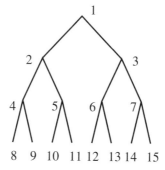

Fig. 1.3. *Arrangement of the contiguous intervals of a perfect set on the real line. Given $i \neq j$, $\mathbf{C}_i$ is on the left of $\mathbf{C}_j$ if i is on a branch eventually situated to the left of j.*

Consider the binary development :

$$
\begin{array}{cccc}
 & & & 8 \longmapsto 1000 \\
 & & & 9 \longmapsto 1001 \\
 & & 4 \longmapsto 100 & 10 \longmapsto 1010 \\
 & 2 \longmapsto 10 & 5 \longmapsto 101 & 11 \longmapsto 1011 \\
1 \longmapsto 1 & & & \\
 & 3 \longmapsto 11 & 6 \longmapsto 110 & 12 \longmapsto 1100 \\
 & & 7 \longmapsto 111 & 13 \longmapsto 1101 \\
 & & & 14 \longmapsto 1110 \\
 & & & 15 \longmapsto 1111
\end{array}
$$

The binary tree in Fig. 1.4 represents this development because it suffices to match each node n with the sequence of 0 and 1 of the path from the first node to the node in question.

Hence it is worthwhile to denote the intervals $\mathbf{C}(1)$, $\mathbf{C}(1,0)$, $\mathbf{C}(1,1)$, ..., $\mathbf{C}(1, i_1, \ldots, i_n)$, where i_k is either 0 or 1. A simple comparison of the sequences of 0s and 1s allows us to determine the exact place of the corresponding interval.

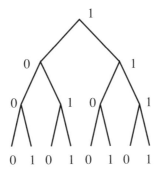

Fig. 1.4. *A simple method to find the binary representation of a natural number; each node is determined by its preceding sequence of 0 and 1; this sequence is the binary representation of the integer associated to the same node in Fig. 1.3.*

Moreover, this new notation has a very important consequence: *It is now possible to decide whether the perfect set F is countable.* In fact, if the perfect set contains an interval, then it is uncountable. If it does not, then it is nowhere dense; thus each point of it is the limit of a sequence of contiguous intervals whose lengths converge to 0. Conversely, to each $0-1$ sequence $i_1, i_2, \ldots$ there corresponds, according to the previous scheme, an infinite sequence of contiguous intervals $\mathbf{C}(1, i_1)$, $\mathbf{C}(1, i_1, i_2)$, $\ldots$ whose lengths tend to 0, and thus the sequence of contiguous intervals converges to a point of F. It is important to note that two different sequences lead to two different points of F. We conclude:

> *The binary arrangement of the contiguous interval defines a bijection between the set F and the set of all $0-1$ sequences.*

But the latter is uncountable; using the binary development of the real numbers we can establish a correspondence between the set of $0-1$ sequences onto the interval $[0,1]$. That's why we say (following Cantor) that this set has *the power of the continuum*. In conclusion:

> *A perfect set is uncountable; it has the power of the continuum.*

◇   Given a positive sequence (c_n) such that $\sum c_n = 1$, it is now possible to construct a perfect nowhere dense set whose fundamental interval is $[0,1]$, and the lengths of its contiguous intervals are precisely (c_n). Starting by renumbering the (c_n) according to the binary scheme $c(1)$, $c(1,0)$, $c(1,1)$,..., each $c(1, i_1, \ldots, i_n)$ is the length of the interval whose address is determined by the sequence $i_1, i_2, \ldots, i_n$ of 0s and 1s. Let $c(1)$ be the length of $C(1)$. To its left we must find all the contiguous intervals whose i_1 is 0. By adding all the lengths $c(1, 0, i_2, \ldots, i_n)$ we find the left endpoint of $C(1)$; the position of this point is entirely determined. Proceeding in this fashion we find the position of all other contiguous intervals.

1.5 Symmetrical perfect sets

Symmetrical perfect sets are a simple generalization of the Cantor set. The basic operation, labeled "dissection" (by Kahane and Salem), consists of taking away a centered segment of length l from a fundemental interval; we say that we have been applying a *dissection of length l*. Starting with the interval $[0, 1]$, for example, and a sequence of positive real numbers (l_n), apply a dissection of length l_1. On the two intervals that are left, apply a dissection of length l_2. At the nth stage, apply a dissection of length l_n on the 2^{n-1} intervals that are left. To be able to carry out this procedure, the l_n must satisfy:

$$\sum_{n=1}^{\infty} 2^{n-1} l_n \leq 1 .$$

Since between any two intervals we have placed a third, the residual set F is perfect. It is called *symmetrical* because each removed interval is centered at the middle of one of the leftover intervals. Thus, F is nowhere dense. Finally its length is given by:

$$L(F) = 1 - \sum_{n=1}^{\infty} 2^{n-1} l_n .$$

Particular case The Cantor set (also called the triadic Cantor set) corresponds to the case where $l_n = 3^n$.

A more general case is when the l_n form a *geometric progression* and the measure of F is 0. If $[0, 1]$ is the fundamental interval, l_n must be of the form:

$$l_n = (1 - 2\,a)\,a^{n-1} ,$$

where a is a positive real number $< 1/2$. Fig. 1.5 represents such a set for $a = 1/4$.

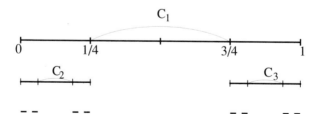

Fig. 1.5. *Symmetrical perfect set of ratio 1/4. Compare to the set in Fig. 1.2.*

Covering a symmetrical perfect set The nth stage of the construction of a symmetrical perfect set gives 2^n closed intervals of the same length that cover the set F. A. Denjoy called them the *isolating* intervals. If we note their union by F_n, then F can be written as:

$$F = \bigcap_{1}^{\infty} F_n .$$

Since the total length of F_n is $1 - \sum_{i=1}^{n} 2^{i-1} l_i$, the length of each of the isolating intervals is:

$$a_n = 2^{-n} \left(1 - \sum_{i=1}^{n} 2^{i-1} l_i \right) .$$

Thus, we can use a_n, instead of l_n, to define F, provided the a_n verify:

$$a_0 = 1, \; 2\, a_{n+1} < a_n .$$

The length l_n of the contiguous intervals of rank n will be such that :

$$l_{n+1} + 2\, a_{n+1} = a_n .$$

If F is a symmetrical perfect set such that (l_n) is a geometric progression of ratio a, then the lengths of the isolating intervals are also a geometric progression of ratio a, and we simply obtain:

$$a_n = a^n .$$

Here F is covered by two intervals of length a, or by four intervals of length $a^2, \ldots$, or by 2^n intervals of length a^n. In the case of the Cantor set, $a = \frac{1}{3}$.

1.6 Tree representation of perfect sets

Due to its precise symmetrical properties, symmetrical perfect sets do not suffice to exemplify the theory of all perfect nowhere dense sets. But describing these sets with words will make it rather difficult to comprehend. Thus, we opted for a finite tree representation, which we construct according to the following rules:

Let F be a perfect set whose fundamental interval is $[0, 1]$. For any integer n, divide $[0, 1]$ into 2^n binary closed intervals of length 2^{-n}. They are the intervals of rank n. Reject all those whose interior does not meet F. Each of the others has at least one point of F in its interior, hence infinitely many since F has no isolated points. We shall mention some properties that make these **white** intervals very useful in a graphical sense:

For any n, the white intervals of rank n cover F.

▶   This follows from the fact that F is perfect. If x is a point of F not in the interior of a white interval, then it is the endpoint of an interval of rank n. The point x is the limit of a sequence of elements of F. If this sequence contains a decreasing subsequence that converges to x, then the interval of rank n placed to the right of x is a white one. If the subsequence is increasing then the interval of rank n placed to the left of x is white. If the sequence has both increasing and decreasing subsequences that converge to x, then the intervals of rank n placed to the right and left of x are white. Thus x is the endpoint of a white interval, and since a white interval is closed, it follows that x belongs to a white interval of rank n. ◀

*Every white interval of rank n is contained in a white interval of rank n−1
and contains either one or two white intervals of rank n + 1.*

Thus white intervals of rank n can be represented as points, called *nodes*,
which are related to those of rank $n1$ by *segments* that symbolize the inclusion
relationship between them. We say that a node is an *embranchment* if it is the
endpoint of two segments. That is, the corresponding white interval contains two
others of succeeding rank. A *branch* is a set (finite or infinite) of related segments.

An infinite branch corresponds to a point of the perfect set.

▶   An infinite branch corresponds to a sequence of embedded binary intervals
that converges to a point of the perfect set.   ◀

An infinite branch contains infinitely many embranchments.

▶   Each white interval, represented by a node on this branch, contains at least
two distinct points of F, hence a sufficiently large white interval of rank n that
will be divided in two white intervals of rank $n + 1$.   ◀

In Fig. 1.6 we show the first subdivision of the simplest tree; where each
node is an embranchment, it represents the entire $[0, 1]$ interval. Generally, if for
every n we know all the white intervals of rank n, then we can determine the
associated perfect set. In fact there exists a bijection between the set of perfect
sets and the set of infinite binary trees that verifies the following property: all
their infinite branches contain infinitely many embranchments (this eliminates
the case of isolated points).

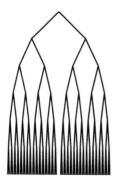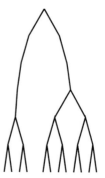

Fig. 1.6. *Binary tree representation up to rank 6 of (left to right) the interval
$[0, 1]$ (all nodes are embranchments); the interval $[0, 1/2]$; and the set $[0, 1/16] \cup
[7/8, 1]$.*

Examples Figs. 1.6, 1.7, and 1.8 represent the first subdivisions of different
trees representing either intervals or perfect nowhere dense sets. We could find
a logical notation that would theoretically determine these trees up to infinite
rank.

◇   Each segment issued from an embranchment can be marked by either 0 (if
it is to the left) or 1 (if it is to the right). Hence an infinite branch defines an

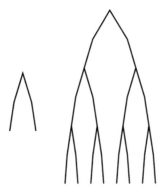

Fig. 1.7. *Up to rank 6, this figure represents the symmetrical perfect set of ratio* $1/4$ *in the interval* $[0, 1]$. *It is constructed according to the fundamental scheme shown on the left, which should be iterated infinitely many times.*

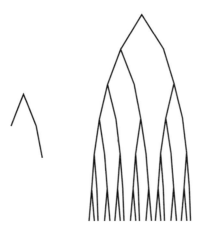

Fig. 1.8. *Up to rank 6, this figure represents a non symmetrical perfect set; it is constructed according to the fundamental scheme shown on the left. If* ω_n *is the number of nodes of rank* n, *then* $\omega_{n+2} = \omega_{n+1} + \omega_n$: *this is a Fibonacci sequence.*

infinite $0 - 1$ sequence. This is another way to show that a perfect set has the power of the continuum since each infinite branch contains an infinite number of embranchments.

1.7 Bibliographical notes

To verify certain notions of topology, one can consult any book on general topology. The recent books are all similar; a pioneer is [K. Kuratowski].

Modern theory widened Cantor's notions without changing them [G. Cantor]. An interesting translation of his main work was published in *Acta Mathematica*. Perfect sets were first defined by Cantor. For a detailed history, see [G. Dugac].

The passage of E. Borel cited in this chapter can be found in his collected work [Borel 1] and in his smaller book [E. Borel 3], which is very enjoyable to read—for a mathematical book. This scientific style has been lost unfortunately.

Perfect nowhere dense sets were behind the original work of [A. Denjoy 1, 2]. The construction of symmetrical perfect sets is reviewed in the first chapters of [J.-P. Kahane and R. Salem], which was for a long time one of the few French references on the geometry of sets of null measure before Mandelbrot's work. The tradition was more conserved in England due to the presence of A.S. Besicovitch, but it took a very technical aspect. The representation of perfect sets by trees comes from [C. Tricot 1].

2 Covers and Dimension

2.1 What is a null measure?

Throughout this chapter, and unless otherwise stated, the "measure" is the "length." The length of an interval u is $L(u)$. We have already seen that a Cantor set whose fundamental interval is $[a, b]$ is of null measure if the sum of the lengths of its contiguous intervals is $b-a$. But when a set is defined by **covers** of intervals, as in the case of perfect sets represented by binary trees, we are compelled to put the problem of measure in a rather different perspective. Moreover, what is necessary is to find a "good" definition for null measure that works for any type of set, not only for the closed ones. Here is a definition given by Borel:

A linear set E is said to be of null measure if for all $\epsilon > 0$ we can find a family of intervals $u_1, u_2, \ldots$ (at most countable) such that:

$$E \subset \bigcup u_n$$
$$\sum L(u_n) \leq \epsilon .$$

Such a family is a cover of E. Evidently, the cover gets finer when ϵ gets nearer to 0.

Example A countable set, as intuition will tell, is always of null measure. Let E be the sequence $(x_n)_{n \geq 1}$. For $\epsilon > 0$, consider the intervals:

$$u_n = \left[x_n - \frac{\epsilon}{2^n}, x_n + \frac{\epsilon}{2^n} \right] .$$

This is a cover of E, and the sum of the lengths of these intervals is equal to ϵ. Thus the measure of a discrete set is null as is the measure of all rationals that belong to $[0, 1]$, even though they are dense in $[0,1]$ in the sense that any real number can be approached by a sequence of rationals. This example, added to those given in Chapter 1, provokes the following remark: *The measure of a set does not determine its topological properties.* To study a set, it is necessary to find its measure, but this won't give all its realities.

The following equivalent condition, which does not use ϵ, is due to Borel:

The set E is of null measure if it can be covered by a family of intervals (u_n) such that:
(i) Every point of E belongs to infinitely many intervals u_n.

(ii) $\sum L(u_n) < \infty$.

▶   To show that the first definition implies the above, we consider a convergent series $\sum \epsilon_k$. For each ϵ_k, construct a cover $u_{n,k}$ of E such that $\sum_n L(u_{n,k}) \leq \epsilon_k$. The family of all the intervals $u_{k,n}$, once renumbered with a unique index, verifies *(i)* and *(ii)*. Thus, the first definition implies the second. Conversely: Let (u_n) be a sequence of intervals verifying *(i)* and *(ii)*. Then for any ϵ we can find an integer N such that $\sum_{n=N}^{\infty} L(u_n) \leq \epsilon$. Thus the family $(u_n)_{n \geq N}$ is a cover of E verifying the first definition.   ◀

The sequence (u_n) that verifies *(i)* and *(ii)* is a **Vitali cover** of E, a notion that will be used in Chapter 7. For more details see Appendix B.

We have already seen two types of Vitali covers for a perfect nowhere dense set F:

1. **Dyadic intervals** The family of dyadic intervals meeting F (*white* intervals) can be represented by a tree (Chap. 1, §6). Every infinite branch of this tree corresponds to a point of F. Thus the family of all these intervals forms a Vitali cover. Moreover, any cover of F, by these intervals, is determined by a family of nodes through which all the infinite branches pass. For example, the intervals of rank n constitute a cover of F. Assume that there are ω_n such intervals. Since the white intervals are embedded, it follows that $2^{-n} \omega_n$ is a decreasing sequence, hence it converges. If its limit is 0, then F is of null measure (according to the first definition of Borel). It turns out that this is not only a sufficient condition; it is also necessary:

Let ω_n be the number of white intervals of rank n. The set F is of null measure if and only if the sequence $2^{-n} \omega_n$ converges to 0.

▶   We first note that F is closed. Assume that F is of null measure; then for all $\epsilon > 0$, there is a cover (u_n) of F such that $\sum L(u_n) \leq \epsilon$. Each u_n is covered by either one or two white intervals of length less than $2\,L(u_n)$. Denoting these intervals by (v_m) and taking their union we obtain a new cover of F such that $\sum L(v_m) \leq 4\,\epsilon$.

We shall now construct a *finite* cover of F. Consider v_m and let v_m^* be the interval centered at the middle point of v_m with length $3\,L(v_m)$. Again (v_m^*) is a cover of F and $\sum L(v_m^*) \leq 12\,\epsilon$. Since F is closed and the (v_m^*) are open, by Borel's theorem we can find a finite subcover of F. Therefore there exists an integer M such that $F \subset \cup_1^M v_m^*$. Let N be any integer such that 3.2^{-N} is less than $L(v_m^*)$. Now, if we replace v_m^* by the white intervals of rank N contained in them, we obtain the ω_N white intervals covering F. Their total length is:

$$2^{-N} \omega_N \leq \sum_{1}^{M} L(v_m^*) \leq 12\,\epsilon\ .$$

We conclude that given any $\epsilon > 0$, for a sufficiently large N the above inequality holds. This suffices to prove that $\lim 2^{-n} \omega_n = 0$.   ◀

◇   To show the "nullity" of the measure of a set we generally use covers by intervals of different lengths. We have already seen an example at the beginning of this section: the cover of the rationals in [0,1]. This set is not closed, and to find a "small" cover we used intervals of different lengths.

2. **Isolating intervals** The notion of *isolating* intervals was mentioned in the discussion on symmetrical perfect sets (Chap. 1, §5). For such a set, the family of all isolating intervals is a Vitali cover of E, because every element of E is the limit of a sequence of embedded isolating intervals whose lengths converge to 0. The length of an isolating interval of rank n is a_n. An argument similar to the preceding one proves that:

A symmetrical perfect set is of null measure if and only if

$$\lim_{n \to \infty} 2^n a_n = 0 \ .$$

This is easily verified for isolating intervals of constant ratio a because $a_n = a^n$, and $a < 1/2$.

2.2 Hierarchy of sets of null measure

In the preceding analysis we entertained a very important idea, the relation between the null measure of sets and the convergence of some series. This naturally leads to a classification of these sets according to the "speed" of the convergence of their associated series. In 1913, Borel wrote (replace "exclusion intervals" with "covering intervals"):

"The sets of null measures play a fundamental role in the theory of functions. In fact it is always possible to embed the singularities of a bounded function in a set of either a null measure, or of a measure as small as we wish. [...] For these many reasons, the notion of sets of null measure seems to be a primary notion, but at the same time this is a general notion and there is no hope to fully understand this question, unless we thoroughly investigate this notion, that's to say we need not confuse these sets but classify them. It seems to me that the first step in this direction, which is imposed upon the analysts, is a classification based upon the asymptotical decrease of the exclusion intervals. Evidently, there are some transfinite problems, which we never hope that they will be entirely resolved, but they aren't any more difficult than those encountered in any question that deals with the general theory of 'growth' (like those of the theory of convergence of series of positive terms). But on the other hand, problems are, usually—if not always—presented independently from these difficulties.[...] Theoretically, the complexity of this classification is greater than this found in the endless study of convergent series of positive terms. Practically, a restricted number of simple classes suffices for the effective needs of analysis."

These reflections lead us to seek a hierarchy of the family of sets of null measure using the convergence of series of type $\sum L(u_n)$:

> *"We admit that: the faster this series converges the more rarefied the corresponding set of null measure is. We will later on encounter numerous cases, where it is possible for some sets to be covered by many families of intervals, such that the series of their lengths converges as fast as we please. Thus to define the rarefaction of a set of null measure, we should always consider the corresponding series whose convergence is fastest."*

Borel didn't pursue these ideas at this time. He resumed the study of *rarefaction* of sets much later, when he considered the sequence of the lengths of the contiguous intervals (Chap. 3). However, from the above paragraphs, in 1961 M. Frechet deduced a "qualitative" method to classify these sets. It goes like this (see Appendix A for more details on upper and lower limits):

Let E be a set of null measure, as in §1 we call a *majoring sequence* of intervals (u_n) any Vitali cover whose series of lengths converges. We say that E_1 is *more rarefied than* E_2, if there exists a majoring sequence $u_n^{(1)}$ of E_1, such that its series of lengths converges faster than the series of any majoring sequence $u_n^{(2)}$ of E_2. This can be translated as:

$$\limsup_{n\to\infty} \frac{\sum_{i=n}^{\infty} L(u_n^{(1)})}{\sum_{i=n}^{\infty} L(u_n^{(2)})} < 1 \ .$$

Moreover E_1 is *as rarefied as* E_2 if E_1 is not more rarefied than E_2 and E_2 is not more rarefied than E_1.

This attempt to classify sets of null measure was not pursued any further; it could be shown, with this qualitative definition in mind, that all closed sets of null measure (including the one-element sets) have the same order of rarefaction. Thus we do not obtain a fine classification. It is easy to see the reason: given a majoring sequence, it is always possible to extract a majoring subsequence of it whose series of lengths converges as fast as we please. Thus to classify the sets of null measure, it is essential to introduce new conditions on the envisaged covers. That's what G. Bouligand (1928), E. Borel (1948), and many others have done. The reminder of this chapter and Chapter 3 are devoted to different *quantitative* methods of classification that originated at this time and will lead to the notion of the (fractal) *dimension*.

2.3 Cantor–Minkowski measure

The notion of Borel measure is a fine one; it has been generalized by Caratheodory and Hausdorff. It has the properties needed by mathematicians. Most of all it is additive. However, the complexity of its computation goes hand in hand with the complexity of the topological structure of the set in question. There exists a "primitive" notion of measure due to Cantor (1884), which uses covers by equal–length intervals. Cantor attacked a more general problem, he sought

> *"A notion of volume or of size, of any subset P (continuous or not) of an n–dimensional space."*

The idea consists of replacing each point x of P with a ball $B_\epsilon(x)$ of radius ϵ and center x. By integration, it is easy to compute the n–dimensional volume of the union of these balls. Taking its limit when ϵ tends to 0, we obtain what we shall call the "volume" of P.

The union of the balls $B_\epsilon(x)$ is sometimes called the ϵ–dilation of the set P, but now it is known as the "Minkowski sausage of P" (a term coined by B. Mandelbrot) because it looks like a sausage and was first used by Minkowski in his study of curves. We denote it $P(\epsilon)$. Thus

$$P(\epsilon) = \bigcup_{x \in P} B_\epsilon(x) \quad \text{is the } \textit{Minkowski sausage} \text{ of } P.$$

On the real line the balls $B_\epsilon(x)$ are nothing but the intervals $[x-\epsilon, x+\epsilon]$, and the Minkowski sausage of P is formed by a finite number of intervals of length $\geq 2\,\epsilon$. Let us call L_c the Cantor measure; thus for a set P on the real line we have, by definition, that:

$$L_C(P) = \lim_{\epsilon \to 0} L(P(\epsilon)) \,.$$

Following are some remarks concerning L_c:

• The sausage $P(\epsilon)$ of a set P is identical to the sausage of its closure P; thus:

$$L_C(P) = L_C(\overline{P}) \,.$$

For example, the L_c measure of the set E_1 of rationals in $[0,1]$ has a value 1 that is the L_c measure of the interval itself.

• It follows that the only distinguishable sets, by the L_c measure, are the closed ones.

• The L_c measure of a closed set is equal to its Borel measure. A proof of this result is given at the end of this section.

• In general, the measure L_c is not additive. For example, let E_2 be the set of irrationals in $[0,1]$. Then $L_c(E_1) + L_c(E_2) = 2 \neq L_c(E_1 \cup E_2)$. Theorists regard this an inconvenience vis–à–vis Borel's measure.

• This inconvenience is compensated by an advantage: it can be efficiently computed. In "practice," we do not distinguish a set from its closure. The irrational numbers are a mathematical invention, and no doubt they are useful; Cantor's measure doesn't consider this subtlety. Better still we can say: practically, and due to our natural "shortsightedness," we can see only the Minkowski sausages at an ϵ–precision, where ϵ is some small positive number. We retain the following rule, which is suggested by many examples: *A measure that satisfies the "good" axioms, from the theorist's point of view, generally has no practical applications.*

- Cantor's idea was later used by Minkowski to compute the length of a curve. In Chapter 9, we see that this was behind G. Bouligand definition of *Cantor-Minkowski's order* or *dimension*, which will be discussed in §5.

▶   We show that if F is closed then $L_c(F) = L(F)$. It suffices to show that:

$$L(F) = \lim_{\epsilon \to 0} L(F(\epsilon)) \,.$$

Assume that F is not finite. In its fundamental interval $[a, b]$, F is defined by a sequence of contiguous intervals. We order them according to the decreasing order of their lengths:

$$c_1 \geq c_2 \geq \ldots \geq c_n \geq \ldots \,.$$

Let $\epsilon > 0$ and $n(\epsilon)$ be the integer n such that

$$c_n \leq 2\,\epsilon < c_{n-1} \,.$$

The Minkowski sausage of F covers all intervals of rank $\geq n(\epsilon)$ and a length ϵ from each endpoint of an interval of rank $< n(\epsilon)$. Using the additivity of lengths we have:

$$L(F(\epsilon)) = L(F) + 2\,\epsilon\,n(\epsilon) + \sum_{i=n(\epsilon)}^{\infty} c_i \,.$$

Let $\eta > 0$, η as small as we wish. Let M be an integer such that $\sum_{M}^{\infty} c_i \leq \eta$. Choosing ϵ so that $M\,\epsilon \leq \eta$ and $n(\epsilon) \geq M$, we obtain:

$$\epsilon\,n(\epsilon) = M\,\epsilon + (n(\epsilon) - M)\,\epsilon$$

$$\leq M\,\epsilon + \sum_{M}^{\infty} c_i \leq 2\,\eta \,.$$

Thus $\epsilon\,n(\epsilon)$ tends to 0. Since $\sum_{n(\epsilon)}^{\infty} c_i \leq \eta$ we find that:

$$L(F) \leq L(F(\epsilon)) \leq L(F) + 5\,\eta \,.$$

When η tends to 0, ϵ also tends to 0 so that we can deduce the desired limit.   ◄

    We note here that this proof could replace the one given in §1 for the binary intervals that cover F. Better still, in §1 it was useless to assume that F is of null measure. In fact, since the white intervals of rank n belong to the sausage $F(2^{-n})$, we have

$$L(F) \leq \omega_n\, 2^{-n} \leq L(F(2^{-n})) \,.$$

Immediately, one concludes that:

   If F is a closed set, then

$$L(F) = \lim_{n \to \infty} \omega_n\, 2^{-n} \,.$$

2.4 Space filling and the order of growth

With an ϵ–precision, the larger the portion of space filled by a set E, the greater the length of its Minkowski sausage $L(E(\epsilon))$. This length will tend to 0 with ϵ whenever E is a closed set of null measure. However, the speed of the convergence to 0 of this function constitutes an index of the "rarity" of the set. Thus we can define a qualitative degree of the *space filling* by comparing the orders of growth to 0 of the functions $L(E(\epsilon))$. Given two sets E_1, E_2 of null measure, we say:

> *The degree of the space filling by E_1 is higher than by E_2 if the convergence to 0 of $L(E_1(\epsilon))$ when ϵ tends to 0 is slower than the convergence to 0 of $L(E_2(\epsilon))$.*

$\diamond$ Let us state, once and for all, how we compare the convergence to 0 of two functions. Let $f(x)$ and $g(x)$ be two positive functions on $]0, b]$ $(b > 0)$. Assume that their limit when x tends to 0 is 0.

(i) We say that f converges to 0 faster than g if:

$$\lim_{x \to 0} \frac{f(x)}{g(x)} = 0 .$$

We sometimes say that the order of growth of f toward 0 is higher than that of g. We denote it by

$$f \succ g \ \ or \ g \prec f .$$

(ii) We say that f and g are equivalent, or they have the same order of growth, if there are two constants c_1, c_2, such that for all x,

$$0 < c_1 < \frac{f(x)}{g(x)} < c_2 .$$

We symbolize this equivalence relation by:

$$f \simeq g .$$

(iii) Finally, if there is a constant c such that for all x

$$\frac{f(x)}{g(x)} \le c ,$$

we write that

$$f \succeq g \ \ or \ g \preceq f .$$

For example, the two functions x and $2x + 3x^2$ are equivalent in the neighborhood of 0. But their order of growth is higher than the order of $\sqrt{x}$. If α and β are two positive real numbers, then $x^\alpha \succ x^\beta$ if and only if $\alpha > \beta$.

◇   The same terminology is employed when we compare the order of growth to infinity of two functions. Thus, the order of f is higher than the order of g if $f(x)/g(x)$ tends to infinity. Similarly if for all x, $f(x)/g(x)$ is bounded by two non-null constants, we say that f and g have the same order.

Examples
- Let E be a finite set. Then $L(E(\epsilon))$ is equivalent to ϵ. Thus all finite sets have the same degree of space filling.
- A symmetrical perfect set E of constant ratio $a < 1/2$ satisfies:

$$L(E(a^n)) \simeq (2a)^n$$

and

$$L(E(\epsilon)) \simeq \epsilon^{1 - \frac{\log 2}{|\log a|}} .$$

This tends to 0 faster than ϵ. Thus its degree of space filling is higher than the degree of a finite set. Furthermore the larger the a, the higher the degree. In fact, the convergence to 0 of the lengths of the isolating intervals is slower when a becomes larger. We conclude that the symmetrical perfect set of ratio $1/4$ is rarer, in this sense, than the triadic set of Cantor with ratio $a = 1/3$.
- Consider the set E whose points are $1/n$ where $n \geq 1$. Let

$$c_n = \frac{1}{n} - \frac{1}{n+1} \simeq \frac{1}{n^2} .$$

For all $\epsilon \simeq c_n$ we have:

$$L(E(\epsilon)) \simeq n\epsilon + \frac{1}{n} \simeq \frac{1}{n} .$$

Consequently,

$$L(E(\epsilon)) \simeq \epsilon^{1/2} .$$

Thus, E has the same degree of space filling as the symmetrical perfect set of ratio $1/4$.

2.5 Orders of growth and dimension

Is there a method to *quantify* the degree of space filling? The answer would be in the affirmative if we can provide a well-ordered hierarchy of the orders of growth. That is to say, given two orders of growth we can know whether the first is higher than, lower than, or equivalent to the second. But this is not always possible. Since we can always construct two functions whose ratio has no limit at 0, for example, an upper limit equals ∞ and a lower limit equals 0. Thus we are obliged to restrict ourselves to a reference family of comparable functions, giving rise to the notion of *scale of functions*.

> A **scale of functions** *in the neighborhood of* 0 *is a family $\mathcal{F}$ of functions defined in the neighborhood of* 0 *such that for any two functions f and g in $\mathcal{F}$ we have:*
> $f \simeq g$;

or $f \succ g$;

or $f \preceq g$.

To allow for the negative functions we shall, in general, write that $f \succ g$ if and only if the ratio $|f(x)/g(x)|$ converges to 0.

With respect to this scale, the *order of growth* of a function can be defined by a subfamily of equivalent functions. More generally, an order of growth is a **cut** in the given scale of functions. That is, it is a partition of this scale into two subfamilies $\mathcal{F}_1$ and $\mathcal{F}_2$ such that every function of $\mathcal{F}_1$ is of a higher order than any function of $\mathcal{F}_2$, and every function of $\mathcal{F}_2$ is of a lower order than any function of $\mathcal{F}_1$ (in algebraic terms we say that the set of the orders of growth is the completion of the quotient set $\mathcal{F}/\simeq$, that is, the set of all cuts defined by the equivalence classes on $\mathcal{F}$ with respect to the relation $\simeq$).

Hardy's scale We consider the functions x^α (α real), $\exp x$, and $\log x$, as a basis of this scale. Let $\mathcal{F}$ be the family of functions constructed from this basis using: finite sums ($(f+g)(x) = f(x) + g(x)$), finite products ($fg(x) = f(x)g(x)$), and finite compositions ($(f \circ g)(x) = f(g(x))$). Apart from the periodic functions, this family contains nearly all the functions encountered in any course of classical calculus. It can be shown that all the functions of this family, which are defined in a neighborhood of 0, are comparable. The proof is not easy (Hardy's theorem). Its main idea is to show that all these functions have limits (eventually infinity) when x tends to 0. Thus, we obtain a scale of functions on the neighborhood of 0. This scale is large, even too large to give an effective notion of quantifying the degrees of *space filling*. However, it is easy to find functions that are not comparable to some functions of $\mathcal{F}$ (such functions do not belong to $\mathcal{F}$). For example, the function $x \sin \frac{1}{x}$ is not in $\mathcal{F}$ and is not comparable to x.

Logarithmic scale A smaller scale of functions in which the comparison of orders of growth is easier is given by the scale of double index:

$$\mathcal{F} = \{f_{\alpha,\beta}(x) = x^\alpha \left(\log_n \frac{1}{x} \right)^\beta , \quad \alpha \text{ real} > 0, \quad n \text{ integer} \geq 0, \quad \beta \text{ real} \} .$$

The notation $\log_n$ indicates the iterated logarithm: $\log_0(x) = 1$, $\log_1(x) = \log(x)$, $\log_2(x) = \log(\log(x))$, ...; $\log_n(x) = \log(\log_{n-1}(x))$ is the n-times iterated logarithm. Here is an example of a *cut* in $\mathcal{F}$: $\mathcal{F}_1$ is the set of functions that grow faster to 0 than a given function of type $|\log x|^{-\beta}$, $\beta > 0$; $\mathcal{F}_2$ is the set of functions that grow slower to 0 than a function of type x^α, $\alpha > 0$. In fact, it is impossible to find in $\mathcal{F}$ a function f whose order of growth is between these two. (While this is possible in Hardy's scale, for example, consider the function $\exp(-(\log_2 \frac{1}{x})^2)$, which is between $\mathcal{F}_1$ and $\mathcal{F}_2$.) This family is useful in some cases of fine analysis of irregular curves, like the trajectories of brownian motion. A simpler and widely used scale is the following.

The power functions scale The power functions scale is defined as

$$\mathcal{F} = \{f_\alpha(x) = x^\alpha, \quad \alpha > 0 \} .$$

The comparison in this scale is straightforward. Given two functions f_α and f_β, they have the same order of growth if and only if $\alpha = \beta$, and f_α grows faster to 0 than f_β if and only if $\alpha > \beta$. Thus each equivalence class contains a unique element. Moreover, the set of real numbers is, by hypothesis, *complete* (that's why the irrationals were defined). In other words, it is equal to the set of its cuts. Therefore, each order of growth in this scale is equal to a real number α.

Given this scale as a reference, we now have a simple method to compute the order of growth of a function, provided this order exists:

The function $f(x)$ has an order of growth α at 0 if:

$$\lim_{x \to 0} \frac{\log f(x)}{\log x} = \alpha .$$

For example, the order of growth of the function $x^{(ax+b)/(cx+d)}$ at 0 is equal to b/d. While the order of growth of the function

$$x^\alpha \left(\log_n \frac{1}{x} \right)^\beta$$

at 0 is α, providing α is not null. It is worth noting that the concept of the order of growth depends to a great extent on the chosen scale.

In a straightforward manner, we can now define a *fractional* dimension, that is, a dimension whose value is not always an integer. The degree of space filling of a linear set E is larger if the order of growth of the function $L(\epsilon)$ at 0, in the scale of power functions, is smaller. This order is the limit of

$$\frac{\log L(E(\epsilon))}{\log \epsilon} ,$$

when ϵ tends to 0. Thus, we may define the fractional dimension as:

$$\text{dimension of } E = 1 - (\text{order of growth of } L(E(\epsilon))) ,$$

if this order exists. The real number, defined by this dimension, is a good measure of the degree of space filling. In the past it was given different names. G. Bouligand, who defined it in 1928, called it the *Cantor-Minkowski order*, because it originated from Cantor's ideas on measure. It was also called *fractional dimension*, *logarithmic density*, and even *entropy* and *capacity*. Now it is known as *fractal dimension*, although there are many different concepts of fractal dimension. It could also be an *index of space filling*. In as much as this book is concerned, it will be called **dimension**, since no confusion is feared. We shall denote it by Δ. Therefore:

$$\Delta(E) = \lim_{\epsilon \to 0} \left(1 - \frac{\log L(E(\epsilon))}{\log \epsilon} \right) \quad \text{is the } \textit{dimension} \text{ of } E,$$

if this limit exists. Using some well-known properties of limits (see Appendix A), we may write $\Delta(E)$ as a critical exponent:

$$\Delta(E) = \inf\left\{\,\alpha \;\; \text{such that} \;\; \epsilon^{\alpha-1}L(E(\epsilon)) \;\; \text{tends to } 0\,\right\}.$$

2.6 Equivalent definitions of the dimension

To make the computation of $\Delta(E)$ easier, it is sometimes interesting to give some equivalent definitions that can be used in different situations.

Assume that for all ϵ, E can be covered by $N(\epsilon)$ intervals of length ϵ with disjoint interiors such that each interval contains at least one element of E. If E is an infinite set, then $N(\epsilon)$ tends to infinity as ϵ tends to 0. No matter how these intervals are chosen:

$$\Delta(E) = \lim_{\epsilon\to 0} \frac{\log N(\epsilon)}{|\log\epsilon|}.$$

▶   As they are disjoint and included in $E(\epsilon)$, these intervals verify:

$$\epsilon\,N(\epsilon) \leq L(E(\epsilon)).$$

By tripling the length of each interval we can cover $E(\epsilon)$ thus:

$$L(E(\epsilon)) \leq 3\,\epsilon\,N(\epsilon).$$

These two inequalities suffice to show that the limit of $\log L(E(\epsilon))/\log\epsilon$ is the same as the limit of $1 + (\log N(\epsilon)/\log\epsilon)$ when ϵ tends to 0. This proves the result.
◀

In other words,

The fractal dimension of E is the order of growth to infinity as ϵ tends to 0 of the number $N(\epsilon)$ of intervals of length ϵ needed to cover E.

It is sometimes more convenient to replace the continuous variable ϵ with a discrete sequence, but this does not come for free, and we need to impose some conditions on the discrete variable. Here is one such condition:

LEMMA   *Let (ϵ_n) be a sequence of positive real numbers that converges to 0. If the sequence*

$$\frac{\log \epsilon_n}{\log \epsilon_{n+1}}$$

converges to 1, then

$$\Delta(E) = \lim_{n \to \infty} \frac{\log N(\epsilon_n)}{|\log \epsilon_n|} .$$

This condition indicates that the sequence ϵ_n has a rather slow rate of convergence toward 0.

▶   To show this, we take any ϵ and an integer n such that:

$$\epsilon_{n+1} < \epsilon \leq \epsilon_n .$$

Each interval of length ϵ is contained in at most two intervals of length ϵ_n. Thus:

$$N(\epsilon_n) \leq 2 N(\epsilon) .$$

Similarly

$$N(\epsilon) \leq 2 N(\epsilon_{n+1}) .$$

These two inequalities imply that:

$$\frac{\log N(\epsilon_n) - \log 2}{|\log \epsilon_{n+1}|} \leq \frac{\log N(\epsilon)}{|\log \epsilon|} \leq \frac{\log N(\epsilon_{n+1}) + \log 2}{|\log \epsilon_n|}$$

and

$$\frac{\log \epsilon_n}{\log \epsilon_{n+1}} \left(\frac{\log N(\epsilon_n)}{|\log \epsilon_n|} - \frac{\log 2}{|\log \epsilon_n|} \right) \leq \frac{\log N(\epsilon)}{|\log \epsilon|}$$
$$\leq \frac{\log \epsilon_{n+1}}{\log \epsilon_n} \left(\frac{\log N(\epsilon_{n+1})}{|\log \epsilon_{n+1}|} + \frac{\log 2}{|\log \epsilon_{n+1}|} \right) .$$

When ϵ tends to 0, the first inequality gives $\limsup_n \log N(\epsilon_n)/|\log \epsilon_n| \leq \Delta$, and the second $\liminf_n \log N(\epsilon_n)/|\log \epsilon_n| \geq \Delta$. ◀

Covers by dyadic intervals   Let F be a perfect set. As in §1, let us call $\omega_n = \omega_n(F)$ the number of dyadic closed sets of length 2^{-n} whose interiors contain at least one element of F. Since the sequence (2^{-n}) satisfies this condition, we obtain:

$$\Delta(F) = \lim_{n \to +\infty} \frac{\log \omega_n(F)}{n \log 2} .$$

A similar formula can be found for any set E, provided we slightly modify the definition of ω_n by considering the semiopen dyadic intervals or the dyadic closed intervals that meet E (possibly on an endpoint). This modification does not change the order of growth of ω_n.

2.7 Examples of computing the dimension

- If E contains only one point, then $\Delta(E) = 0$.

- If E is an interval, or more generally of a nonzero length, then $L(E(\epsilon)) \geq L(E)$ for all ϵ. Hence $\Delta(E) = 1$ is the maximal dimension of a subset of the real line.

- Let E be the set whose points are $(n^{-\beta})$, $n \geq 1$, where β is a fixed positive parameter. If we add 0 to E, the new set is closed and has the same dimension as E. Let $c_n = n^{-\beta} - (n+1)^{-\beta} \simeq n^{-\beta-1}$ be the length of the contiguous intervals. When $c_{n+1} < 2\epsilon \leq c_n$,

$$L(E(\epsilon)) \simeq \text{length of } [0, (n+1)^{-\beta}] + 2n\epsilon$$
$$\simeq n^{-\beta} .$$

Thus
$$\Delta(E) = 1 - \frac{\beta}{\beta+1} = \frac{1}{\beta+1} .$$

For the sequence $(\frac{1}{n})$, $\Delta(E) = \frac{1}{2}$.

- Let E be a symmetrical perfect set of a constant ratio $a < 1/2$. E is covered by 2^n intervals of length a^n, with $N(a^n) = 2^n$. We get:

$$\Delta(E) = \frac{\log 2}{|\log a|} ,$$

which can take all the values between 0 and 1 (not included). For the general symmetrical perfect set defined by the sequence (a_n) of the lengths of its isolating intervals, we have: $N(\epsilon) = 2^n$ for $\epsilon = a_n$. Thus if ϵ is such that $a_n < \epsilon < a_{n-1}$, then $2^{n-1} \leq N(\epsilon) \leq 2^n$ and:

$$\frac{n-1}{n} \frac{n \log 2}{|\log a_n|} \leq \frac{\log N(\epsilon)}{|\log \epsilon|} \leq \frac{n}{n-1} \frac{(n-1) \log 2}{|\log a_{n-1}|} .$$

The limit of $\log N(\epsilon)/|\log \epsilon|$ is the same as the limit of $n \log 2/|\log a_n|$. Therefore

$$\Delta(E) = \lim_{n \to \infty} \frac{n \log 2}{|\log a_n|}$$

if this limit exists.

- Using the formula
$$\Delta(E) = \lim \frac{\log \omega_n}{n \log 2} ,$$

we can compute the dimensions of the perfect sets represented by trees in Chapter 1, §6: Since the set of Figure 1.5 is a symmetrical perfect set of ratio 1/4, its dimension is $\Delta(E) = 1/2$. The set of Figure 1.6 constructed by an initial scheme of rank 2 is such that:

$$\omega_n \simeq x_0^n ,$$

where x_0 is the largest root of $x^2 - x - 1$, that is, $x_0 = (1 + \sqrt{5})/2$. Hence

$$\Delta(E) = \frac{\log \frac{1+\sqrt{5}}{2}}{\log 2} = 0.694\ldots.$$

In general, given an initial scheme of rank k, we construct an infinite tree representing a perfect set E by iterating it infinitely many times. The dimension of E is:

$$\Delta(E) = \frac{\log x_0}{\log 2},$$

where x_0 is the largest root of the polynomial of degree k that corresponds to the initial scheme.

2.8 Some properties of the dimension

Here we discuss some properties of the dimension that do not depend on the structure of the set E. The only assumption we make is the following: E is bounded because $L(E(\epsilon))$ must be finite. Moreover we assume that $\Delta(E)$ exists as a limit.

1. Δ is **monotonous**: if E_1 is contained in E_2, then

$$\Delta(E_1) \leq \Delta(E_2) .$$

2. If $\overline{E}$ is the **closure** of E, then

$$\Delta(\overline{E}) = \Delta(E) .$$

3. Δ is **stable**: given E_1 and E_2,

$$\Delta(E_1 \cup E_2) = \max\{\Delta(E_1), \Delta(E_2)\} .$$

4. For all subsets E of the real line,

$$0 \leq \Delta(E) \leq 1 .$$

5. Δ is **invariant** under a similarity (multiplication by a real number a and addition to a real number b): if $T(E) = a\,E + b$ is such a transformation of the real line, then

$$\Delta(T(E)) = \Delta(E) .$$

6. Δ is **invariant** under a more general type of transformation.

▶   1. Follows from the fact that $L(E_1(\epsilon)) \leq L(E_2(\epsilon))$ for all ϵ.
2. This is true because the Minkowski sausages of E and $\overline{E}$ are the same.
3. Δ monotonous implies that

$$\Delta(E_1 \cup E_2) \geq \max\{\Delta(E_1), \Delta(E_2)\} .$$

On the other hand, if $\alpha > \max\{\Delta(E_1), \Delta(E_2)\}$, the inequality

$$L((E_1 \cup E_2)(\epsilon)) \leq L(E_1(\epsilon)) + L(E_2(\epsilon))$$

implies that

$$\epsilon^{\alpha-1} L((E_1 \cup E_2)(\epsilon)) \longrightarrow 0 .$$

Therefore $\alpha \geq \Delta(E_1 \cup E_2)$, proving that

$$\Delta(E_1 \cup E_2) \leq \max\{\Delta(E_1), \Delta(E_2)\} .$$

4. The set E contains at least one element, thus $L(E(\epsilon)) \geq 2\epsilon$ and $1 - (\log L(E(\epsilon))/\log \epsilon)$ is larger than $\log 2/\log \epsilon$, which converges to 0. E is bounded, as is $E(1)$. For $\epsilon < 1$ we have:

$$L(E(\epsilon)) \leq L(E(1)) .$$

This entails that $\Delta(E) \leq 1$.

5. It is clear that $L(E(\epsilon))$ is invariant under translation. As for the multiplication by a constant a, we note that all lengths will be multiplied by a. Hence $L(T(E)(\epsilon)) = a L(E(\epsilon/a))$, which has the same order of growth to 0 as $L(E(\epsilon))$.

6. It is fairly easy to extend the previous result to any bijective map T that has a derivative (if T^{-1} also has a derivative, we say that T is a *diffeomorphism*). We can do even better, we can show that Δ remains invariant under any map T that verifies the following: for all x, the limit of

$$\frac{\log(T(y) - T(z))}{\log(y - z)}$$

is 1 when y and z both converge to x and $y \neq z$. (We note that the set of such maps together with the composition of functions operation is an algebraic group.)
◄

2.9 Upper and lower dimensions

Up to now we assumed the convergence of

$$\frac{\log L(E(\epsilon))}{\log \epsilon} .$$

But what if this ratio has no limit? In fact, it may have many possible limits (even infinitely many) in $[0, 1]$. We shall be interested in the largest and smallest ones, that is, the upper and lower limits (see Appendix A for more details). Thus we may define two new dimensions; we denote by Δ the larger one:

$$\Delta(E) = \limsup_{\epsilon \to 0} \left(1 - \frac{\log L(E(\epsilon))}{\log \epsilon} \right) \text{ is the } upper \text{ } dimension \text{ of } E$$

$$\delta(E) = \liminf_{\epsilon \to 0} \left(1 - \frac{\log L(E(\epsilon))}{\log \epsilon} \right) \text{ is the } lower \text{ } dimension \text{ of } E \ .$$

If these dimensions are different then $L(E(\epsilon))$ has no order of growth to 0. Nevertheless, each of these dimensions is a cut in the scale of power function x^α because we can write the following equivalent definitions:

$$\Delta(E) = \inf \{ \, \alpha \text{ such that } \lim_{\epsilon \to 0} \epsilon^{\alpha-1} L(E(\epsilon)) = 0 \, \}$$

while

$$\delta(E) = \sup \{ \, \alpha \text{ such that } \lim_{\epsilon \to 0} \epsilon^{\alpha-1} L(E(\epsilon)) = +\infty \, \} \ .$$

Examples Figure 2.1 gives an example of a perfect set where $\Delta(E)$ and $\delta(E)$ are different.

Another example: The symmetrical perfect set E, defined by the sequence (a_n) of the lengths of its isolating intervals where $\log a_n/n$ has no limit. For such sets we have:

$$\Delta(E) = \limsup \frac{n \log 2}{|\log a_n|}$$

$$\delta(E) = \liminf \frac{n \log 2}{|\log a_n|} \ .$$

$\Diamond$ The properties of Δ taken as upper dimension are the same as those of Δ taken as a dimension. Even stability still holds; that is, the following equality holds:

$$\Delta(E_1 \cup E_2) = \max\{\Delta(E_1), \Delta(E_2)\} \ .$$

And this is because given two sequences (a_n) and (b_n), the lim sup of the sequence $(\max\{a_n, b_n\})$ is the maximum lim sup a_n and lim sup b_n (Appendix A).

ATTENTION! The properties of Δ except stability are also satisfied by δ. In fact, it is generally false to write:

$$\liminf \max\{a_n, b_n\} = \max\{\liminf a_n, \liminf b_n\} \ .$$

In Figure 2.2 we give an example of perfect subsets E_1, E_2 of the real line such that: $\delta(E_1 \cup E_2) \neq \max\{\delta(E_1), \delta(E_2)\}$.

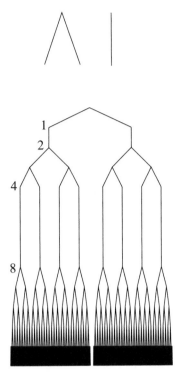

Fig. 2.1. *This tree is constructed using the two schemes alternatively; the alternation is done at rank 2^k. Thus*

$$\omega_{2^{2k}} = \omega_{2^{2k+1}} = 2^{(1+2^{2k+1})/3} \ .$$

We deduce that

$$\Delta(F) = \lim_{k\to\infty} \frac{(1+2^{2k+1})/3}{2^{2k}} = \frac{2}{3} \ ,$$

and

$$\delta(F) = \lim_{k\to\infty} \frac{(1+2^{2k+1})/3}{2^{2k+1}} = \frac{1}{3} \ .$$

2.10 Bibliographical notes

As in Chapter 1 the cited passages of G. Cantor and E. Borel can be found in the works [Cantor], [Borel 1], and [E. Borel 3]. The attempt to define a "qualitative hierarchy" of the sets of null measure is due to [M. Fréchet]. Simultaneously, G. Choquet (in a correspondence with Fréchet) and [Z. Moszner] proved that all perfect sets of null measure have the same order of rarefaction as the one-element set.

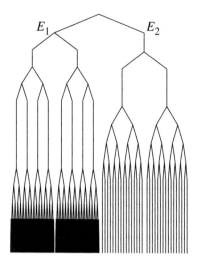

Fig. 2.2. *We reconsider the tree in Figure 2.1, constructed this time on $[0, 1/2]$; this is the set E_1. On $(1/2, 1]$, we place the conjugate E_2 in which we have the same fundamental scheme but applied inversely; that is, on the nodes where E_1 has embranchment E_2 does not, and vice versa. These two sets have the same dimension. In particular*

$$\delta(E_1) = \delta(E_2) = \frac{1}{3}.$$

We find the number of nodes of rank n of $E_1 \cup E_2$. If $2^{2k} + 1 \le n \le 2^{2k+1} + 1$, it equals

$$\omega_n = 2^{(1 + 2^{2k+1})/3} + 2^{n - (4 + 2^{2k+1})/3}.$$

The ratio $\log \omega_n / n \log 2$ is therefore larger than

$$\max\left\{ \frac{(1 + 2^{2k+1})/3}{n}, \frac{n - (4 + 2^{2k+1})/3}{n} \right\}.$$

These two terms become equal when $n = (5 + 2^{2k+2})/3$: we deduce

$$\frac{\log \omega_n}{n \log 2} \ge \frac{(1 + 2^{2k+1})/3}{(5 + 2^{2k+2})/3} \to \frac{1}{2}.$$

We obtain the same evaluation for $2^{2k-1} + 1 \le n \le 2^{2k} + 1$. Thus,

$$\delta(E_1 \cup E_2) = \frac{1}{2}.$$

The theory of the scale of functions constitutes an exciting, but forgotten, part of functional analysis. The pioneer was [P. du Bois–Reymond]; a pleasant and complete survey was written by [G.H. Hardy 1] in 1910. Another interesting reference is Borel's book [Borel 2], where the notation for the order of growth is similar to that used by Cantor to note the transfinite orders. This idea, which does not lack a good meaning, was wrongly criticized by the technicians of Bourbaki's team [N. Bourbaki]. In this last reference we find a correct proof of Hardy's theorem, whose original version is somewhat messy.

In introducing the notion of dimension, we followed its historical evolution. E. Borel, F. Hausdorff, and G. Bouligand—explicitly or not—used the idea of the "scale of functions" to define a dimension. From this point of view, the property that the dimension is a real number is not intrinsic to this notion; it is a consequence of the simplicity of the scale of the power functions. The logarithmic functions can also be used in the fractal analysis of a set.

The first "index of rarefaction," or "fractional dimension," is the dimension of [F. Hausdorff] (1919). This index, though it is a fine tool in a theoretical analysis, presents the practical difficulty of being very hard to compute except when it is equal to the Cantor-Minkowski order of G. Bouligand. This is the reason behind our attachment to the study of the latter index. Its definition seems so simple and natural that we won't be surprised if it appeared previously. Anyway its reincarnations in subsequent works are numerous, and they come in different ways. Consult [C. Tricot 2] for a brief historic note. This dimension appeared as a limit—sometimes upper, sometimes lower—depending on the situation.

The first idea of G. Bouligand appeared in 1925 [G. Bouligand 1]:

In general, we may consider, not a number but a **dimensional order,** *thus giving the word* **order** *the same meaning as that given by M. Emile Borel for the order of growth.*

We recognize here the relation between dimension and order of growth. The most complete articles of G. Bouligand are dated 1928 [G. Bouligand 3 and 4].

G. Bouligand considered only the case where the dimension is a limit. With his literary style he was often criticized for his lack of rigor. We may wonder whether this is a terrible crime after a half-century of the dictatorship of mathematical Logic to the detriment of creativity. In any case, he explained the notion of dimension, gave some equivalent metric definitions, and knew how to compare it to the "similarity exponent" in his not-well-known book [G. Bouligand 5]. This exponent became (with B. Mandelbrot) known as the "similarity dimension." See [G. Bouligand 2] for more details on this subject.

The binary trees of this chapter come from [C. Tricot 1]. The lemma of §6 can be found in [C. Tricot].

3 Contiguous Intervals and Dimension

3.1 Borel's logarithmic rarefaction

A nowhere dense closed subset F (a closed set that contains no interval) of the real line is the complement, in its fundamental segment $[a, b]$, of a family $\mathbf{C}_1$, $\mathbf{C}_2,\ldots,$ $\mathbf{C}_n,\ldots$ of disjoint open intervals. This is the family of the **contiguous** intervals of F. Generally it is infinite, unless F is finite—a case that we discard. If $L(\mathbf{C}_n) = c_n$, the measure (i.e., length) of F is:

$$L(F) = b - a - \sum_{n=1}^{\infty} c_n .$$

Assume that this measure is null and that (c_n) is a decreasing sequence. To evaluate the "rarefaction," E. Borel proceeded in 1948 as follows:

Once the intervals $\mathbf{C}_1,\ldots,\mathbf{C}_{n-1}$, are taken away we are left with n intervals (some of them could be reduced to one-element sets) that cover F. The arithmetical mean of their lengths is equal to:

$$\frac{1}{n}(b - a - \sum_{i=1}^{n-1} c_i) = \frac{1}{n}\sum_{i=n}^{\infty} c_i .$$

This mean converges to 0 when n goes to infinity. The faster the convergence is the rarer the set will be. Borel's logarithmic rarefaction is again an order of growth. We define it as follows:

$$e_B = \lim_{n\to\infty} \frac{\log n}{|\log(\frac{1}{n}\sum_{i=n}^{\infty} c_i)|} ,$$

which can be written as:

$$e_B = \lim_{n\to\infty} \frac{1}{1 + \frac{|\log \sum_{i=n}^{\infty} c_i|}{\log n}} .$$

When this quotient does not converge we consider the lim sup. The index e_B is always equal to:

$$\inf \left\{ \alpha \text{ such that } n^{\frac{1}{\alpha}-1} \sum_{i=n}^{\infty} c_i \text{ tends to } 0 \right\}.$$

We shall see that when F is of null measure this index is equal to Bouligand's, notwithstanding the different methodologies followed by Borel and Bouligand. Similarly for the index in the next section.

3.2 Index of Besicovitch–Taylor

With the same notation as in the preceding section, we start by noting that the rarer F is, the smaller the c_n are. Since the series $\sum c_n$ is equal to b−a, it is convergent. To measure the speed of this convergence, we shall look for the *convergence exponent,*, that is, the critical value α at which the series $\sum c_n^\alpha$ changes from a convergent series to a divergent one. The *index of Besicovitch-Taylor* (1954) is equal to

$$e_{BT} = \inf \left\{ \alpha \text{ such that } \sum_{n=1}^{\infty} c_n^\alpha \text{ converges} \right\}.$$

3.3 Equivalent orders of growth

We can easily find other orders of growth using the sequence (c_n), which could characterize the notion of "rarity" of closed sets of null measure. They may not be of historic importance, like e_B or e_{BT}, but they are of interest by themselves. They are usually found in technical arguments, as in §4. Moreover, the more equivalent definitions we have the better, since we shall be able to choose the most convenient one in order to generate faster numerical algorithms to compute the dimension. They are all theoretically equivalent, as we shall see.

The simplest index that characterizes the convergence of c_n to 0 is:

$$e = \lim_{n \to \infty} \frac{\log n}{|\log c_n|}$$
$$= \inf \left\{ \alpha \text{ such that } n\, c_n^\alpha \text{ tends to } 0 \right\}.$$

We can also define an index in direct relation with the dimension of Bouligand–Minkowski by finding the value of $L(F(\epsilon))$ where $\epsilon \simeq c_n$. Since all the contiguous intervals of rank $\geq n$ are covered by the Minkowski sausage, $\sum_{i=n}^{\infty} c_i$ is an estimate of $L(F(\epsilon))$. Thus, it is reasonable to consider the following index:

$$e_{BM} = \lim_{n \to \infty} \left(1 - \frac{\log \sum_{i=n}^{\infty} c_i}{\log c_n}\right)$$

$$= \inf \left\{ \alpha \text{ such that } c_n^{\alpha-1} \sum_{i=n}^{\infty} c_i \text{ tends to } 0 \right\}.$$

It is important to recall that the sequence (c_n) is assumed to be decreasing, except in the case of e_{BT} where this has no theoretical importance. We shall prove the equality of these indices. On the one hand, this will unify the notion of rarefaction, and on the other hand, it shows the correctness of Borel's idea, advocated some forty years earlier—the tight connection between rarefaction and the convergence of series.

THEOREM   Let (c_n) be a decreasing sequence such that $\sum c_n$ is convergent. Then we always have the following equalities:

$$e_B = e_{BT} = e_{BM} = e.$$

▶   Since $\sum c_n$ is convergent, all these indices are ≤ 1. The index e may be written as

$$e = \inf \left\{ \alpha \text{ such that } n\, c_n^{\alpha} \text{ remains bounded} \right\}$$

(see Appendix 1, §4). Similarly for e_B and e_{BM}. This remark makes the proof much easier.

a) That $e = e_{BT}$ is well known. It can be proved as follows: assume that $e_{BT} < 1$ and let α be a real number such that $e_{BT} < \alpha < 1$. Since $\sum c_n^{\alpha}$ converges, there exists a real number a such that $c_n^{\alpha} \le a/n$ for all n. Therefore $n\, c_n^{\alpha} \le a$, which proves that $e \le \alpha$. Hence $e \le e_{BT}$.

Conversely, if $e < 1$ and $e < \alpha < \beta < 1$, then $n\, c_n^{\alpha} \to 0$, and there is an integer N such that $c_n^{\alpha} \le 1/n$, for all $n > N$. Therefore $\sum c_n^{\beta}$ converges, and $e_{BT} \le \beta$, which proves that $e_{BT} \le e$.

b) We will now show that $e = e_B$. Let $e < \alpha < 1$. There is an integer N such that $c_n \le n^{-1/\alpha}$, for all $n > N$. By comparing a series and an integral we deduce that there exists a constant a such that $\sum_n^{\infty} c_i \le a\, n^{1 - 1/\alpha}$. Hence $n^{(1/\alpha)-1} \sum_n^{\infty} c_i$ is bounded and $e_B \le \alpha$.

Conversely, $\sum_n^{\infty} c_i \le n^{1-1/\alpha}$ implies that $n\, c_{2n} \le n^{1-1/\alpha}$ because (c_n) is decreasing. Thus $c_{2n} \le n^{-1/\alpha}$ and $(2n)\, c_{2n}^{\alpha}$ is bounded. It follows that $e \le \alpha$, proving the equality.

c) Finally we show that all these indexes are equal. The inequality $e \ge e_{BM}$ is simple, because if $e < \alpha < \beta < 1$ and n is sufficiently large, then $c_n \le n^{-1/\alpha}$ and the following holds:

$$c_n^{\beta-1} \sum_{i=n}^{\infty} c_i \le \sum_{i=n}^{\infty} c_i^{\beta} \le \sum_{i=n}^{\infty} i^{-\beta/\alpha}.$$

The series on the right converges to 0, therefore $e_{BM} \le \beta$.

The inequality $e < e_{BM}$ is harder, because of the irregularities that might occur in the convergence of (c_n) to 0. Let $\alpha > e_{BM}$ and $0 < \gamma < 1$. A theorem due to Dini tells us that if the series $\sum_1^\infty c_n$ converges, then so does the series $\sum_1^\infty c_n (\sum_n^\infty c_i)^{-\gamma}$. Therefore for sufficiently large n, we have: $\sum_n^\infty c_i \le c_n^{1-\alpha}$ and we deduce that the series:

$$\sum_1^\infty c_n (c_n^{1-\alpha})^{-\gamma} = \sum_1^\infty c_n^{1-\gamma+\alpha\gamma}$$

converges. This implies that $e = e_{BT} \le 1 - \gamma + \alpha\gamma$. When γ goes to 1 and α to e_{BM} we obtain the desired result.   ◄

3.4 The contiguous intervals and the fractal dimension

We now arrive at the metric interpretation of these orders of growth. They all lead us to the same notion: the dimension.

THEOREM   *Let F be a closed set, $\Delta(F) = \lim_{\epsilon \to 0}(1 - \log L(F(\epsilon))/\log \epsilon)$ be its dimension, and $e = \lim_{n \to \infty}(\log n/|\log c_n|)$ be the order of growth of the lengths of its contiguous intervals. If F is of null measure, then*

$$e = \Delta(F) \,.$$

▶   Assume, as always, that (c_n) is a decreasing sequence. For every ϵ, we can find a sufficiently large integer n such that

$$c_n \le 2\,\epsilon < c_{n-1} \,.$$

Using the additive property of lengths, the computation in Chapter 2, §3 gives us the following:

$$L(F(\epsilon)) = L(F) + 2\,\epsilon\,n + \sum_{i=n}^\infty c_i$$

$$= 2\,\epsilon\,n + \sum_{i=n}^\infty c_i \,,$$

because F is of null measure. Thus we obtain:

$$\epsilon^{\alpha-1} L(F(\epsilon)) = 2\,\epsilon^\alpha\,n + \epsilon^{\alpha-1}\sum_{i=n}^\infty c_i \,.$$

Now if $\alpha > \Delta(F)$, then $n c_n^\alpha$ converges to 0. Therefore $\alpha \ge e$, proving that $e \le \Delta(F)$. Conversely, and without any loss of generality, we may assume that $e < 1$. Let α be such that $e < \alpha < 1$. The previous equality can be transformed into:

$$\epsilon^{\alpha-1} L(F(\epsilon)) \le 2^{1-\alpha}\,c_{n-1}^\alpha\,n + c_n^{\alpha-1}\sum_{i=n}^\infty c_i \,.$$

The term on the right-hand side converges to 0 because $e = e_{BM}$ (§3). Therefore $\alpha \geq \Delta(F)$. We conclude that:

$$e \geq \Delta(F) . \quad \blacktriangleleft$$

◇   This is a fast method to compute the dimension of some sets. For example, if $E = \{n^{-\beta}\}$ where β is a positive parameter, we know the value of c_n: $c_n = n^{-\beta} - (n+1)^{-\beta} \simeq n^{-1-\beta}$. Thus

$$e = \frac{\log n}{|\log n^{-1-\beta}|} = \frac{1}{\beta + 1} .$$

This is the value of the dimension found in Chapter 2, §7.

◇   What can we say about the exponent e when F is of non-null measure? The dimension of F is now 1, but e could be any value between 0 and 1. Given a convergent series $\sum c_n$, we can always arrange the sequence of contiguous intervals of length c_n in the interior of an interval whose length is larger than the sum of their lengths. In this case there is no relation between e and the dimension. Reconsidering the previous computation we realize that e is nothing but the order of growth of $L(F(\epsilon) - F)$, that is, the length of the Minkowski sausage minus the set itself. We call this the *exterior dimension* of F:

$$\Delta^{\text{Ext}}(F) = \lim_{\epsilon \to 0}(1 - \frac{\log L(F(\epsilon) - F)}{\log \epsilon}) .$$

This index is evidently equal to $\Delta(F)$ whenever F is of null measure. The following is a more general result:

$$\boxed{e = \Delta^{\text{Ext}}(F) .}$$

This index measures the degree of proximity of F to its complement; it increases whenever the external space filled by the neighborhood of F gets bigger. This is a real index of space filling. We shall reconsider it later, not in the case of a real line but rather in a plane, when we study the *lateral dimension* of curves (Chap. 17). It is also useful in the study of *porous fractal surfaces* where the *pores* generalize in the plane and the contiguous intervals on the real line.

3.5 Algorithms to compute the dimension

The closed sets of null measure are of practical interest in their own right. For example, the symmetrical perfect sets (like the Cantor set) constitute excellent mathematical models of those sets whose structure remains the same at any scale (sets that are *self-similar*). For more details on different applications, the reader is urged to consult the works of Mandelbrot. These sets are frequently found as *sections* of irregular curves. For example, the closed set with an accumulation point can be seen as the intersection of a spiral with a straight line passing through its center. The perfect sets of null measure can also be the level sets of the graphs of continuous, nowhere differentiable functions (like Weierstrass or brownian functions). If we want to estimate the fractal dimension of such a set F, we always look for a quantity $Q(\epsilon)$ where ϵ takes the value of a decreasing sequence $\epsilon_1, \epsilon_2, \ldots, \epsilon_N$. The quantity $Q(\epsilon)$ will depend on a length (Minkowski sausage), a number of intervals, or a series etc. Theoretically, this quantity should be chosen so that the limit as ϵ goes to 0 and

$$\frac{\log Q(\epsilon)}{|\log \epsilon|}$$

is the fractal dimension $\Delta(F)$. Therefore, in the Cartesian plane the points whose coordinates are

$$(|\log \epsilon_n|, \log Q(\epsilon_n))$$

must form a curve whose asymptotic direction is $\Delta(F)$. The set of these N points is called the *logarithmic diagram*. We estimate its asymptotic direction by its general direction, that is, the least square line.

The value of an algorithm will then depend on the choice and the estimation of the quantity $Q(\epsilon)$. Accordingly, the diagram presents some local irregularities and a global concavity. Surely the ideal will be to have a straight line. However, it is interesting to have different definitions of $\Delta(F)$ and to test the corresponding algorithms on different sets whose dimensions can theoretically be computed. We give an example:

Example Consider the triadic set of Cantor; it is the complement in $[0, 1]$ of a family of contiguous intervals of lengths (c_n) such that:

$$c_1 = 3^{-1}$$
$$c_2 = c_3 = 3^{-2}$$
$$\cdots$$
$$c_{2^k} = \ldots = c_{2^{k+1}-1} = 3^{-k-1} .$$

Before doing any computation, let us note that this set is among those that will have the worst estimation of the dimension because of the repetitions in the sequence (c_n). Its construction contains a form of periodicity that will be reflected in the logarithmic diagram and will hinder the computation. The sets whose intervals are randomly chosen generally give more satisfying results. The advantage of Cantor's set is precisely that: it permits us to distinguish between the good methods and the bad ones.

The exponent e Though it is the simplest to compute

$$\Delta(F) = e = \lim_{n \to \infty} \frac{\log n}{|\log c_n|} ,$$

this method is the worst because of the "jumps" in the sequence (c_n). Here the quantity that converges to 0 is c_n, while the countable quantity is n. We form the diagram $(|\log c_n|, \log n)$: these are the points whose coordinates are

$$((k+1)\log 3, \log n)$$

for every integer $k \geq 0$ and all integers n such that $2^k \leq n < 2^{k+1}$. We see in Figure 3.1 that this diagram cannot be aligned, even though its asymptotical direction is equal to $\log 2 / \log 3$.

The logarithmic rarefaction e_B The method

$$\Delta(F) = e_B = \lim_{n \to \infty} \frac{\log n}{|\log(\frac{1}{n} \sum_{i=n}^{\infty} c_i)|}$$

is a much better method because the values $\epsilon_n = \sum_{i=n}^{\infty} c_i/n$ are more "continuous" than the c_n. Once more $Q(\epsilon) = n$. If $2^k \leq n < 2^{k+1}$, $\sum_{i=n}^{\infty} c_i = (2^{k+1} - n) 3^{-k-1} + \sum_{i=k+1}^{\infty} 2^i 3^{-i-1} = 3^{-k-1}(2^{k+2} - n)$. We form the diagram

$$\left((k+1)\log 3 - \log(\frac{2^{k+2}}{n} - 1) , \log n \right)$$

for every integer k and all integers n such that $2^k \leq n < 2^{k+1}$. This diagram contains all the points $(k \log 3, k \log 2)$, and it is better aligned than the previous one. It presents some periodicity, which is inherited from the set itself.

The Minkowski sausage

$$\Delta(F) = \lim(1 - \frac{\log L(F(\epsilon))}{\log \epsilon}) = \frac{\log \frac{1}{\epsilon} L(F(\epsilon))}{|\log \epsilon|}$$

probably gives the most precise results, because of the continuity of ϵ and of $Q(\epsilon) = L(F(\epsilon))/\epsilon$. This quantity is an estimation of the number of intervals of length ϵ needed to cover F. We recall that

$$L(F(\epsilon)) = 2 n \epsilon + \sum_{i=n}^{\infty} c_i$$

for all ϵ such that $c_n \leq 2\epsilon < c_{n-1}$. We put $x = |\log \epsilon|$. The logarithmic diagram is simply the graph of:

$$y = k \log 2 + \log(2 + 3^{-k} e^x)$$

for every integer $k \geq 0$ and all x such that

$$k \log 3 + \log 2 < x \leq (k+1) \log 3 + \log 2 .$$

This diagram contains the points $(k \log 3 + \log 2, k \log 2 + \log 4)$ (Fig. 3.1).

3.6 Bibliographical notes

When we restrict the development of the real numbers in a given basis, we generally obtain a set of measure 0. It is in the framework of such a study that the logarithmic rarefaction appeared in 1948 [E. Borel 1 and 3]. Some years later, [A.S. Besicovitch and S.J. Taylor] introduced the index, which since then has carried their names, with the aim of finding an upper bound of the Hausdorff dimension of linearly closed sets. It is amusing to observe that, without doubt independently from Borel, the notion of logarithmic rarefaction appeared in the same article, under the form of lim inf. We see how these indices appear and disappear. We must not conclude that there are multiple notions, but rather a unique one under different aspects. A brief historical review can be found in [C. Tricot 2], as well as the first known proof of the equality $e_{BT} = e_B$ (1981!).

Finally we can find a large quantity of different formulations of the exponent of convergence of a series of positive terms; a more or less general treatment of this problem can be found in [C. Tricot 4]. See also the article of [J.B. Wilker].

The first proof of the equality $e_{BT} = \Delta(F)$ was found by [J. Hawkes]. We had to wait until 1974 to show that the two approaches—covers by intervals of equal lengths and formation of the sequence of complementary intervals (which, following A. Denjoy, were called "contiguous intervals")—are characterized by the same dimensional number. The notion of "exterior dimension" is also more interesting in dimensions 2 and 3; see [C. Grebogi et al.], and [C. Tricot 5] for its relation to the contiguous intervals.

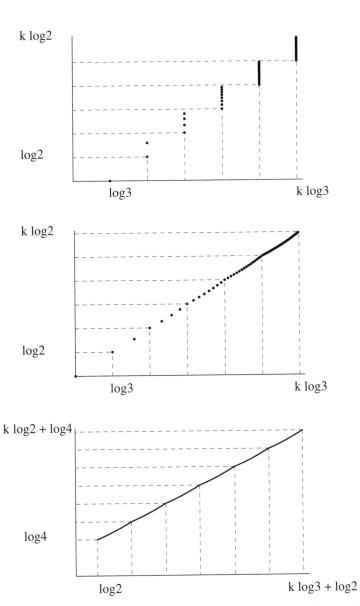

Fig. 3.1. *Three logarithmic diagrams for the dimension of the triadic set of Cantor: 1)* $(|\log c_n|, \log n)$; *2)* $(|\log(\sum_{i=n}^{\infty} c_i/n)|, \log n)$; *and 3)* $(|\log \epsilon|, \log(L(F(\epsilon))/\epsilon))$. *The asymptotic directions are* e, e_B, *and* $\Delta(F)$, *respectively, all three are equal to* $\log 2/\log 3$.

4 What is a Curve?

4.1 Some types of sets in the plane

The topological study of the real line starts, as we have seen, with the intervals. We go on to construct more complicated and sophisticated sets using the fundamental operations of union, intersection, and complement. This is the *constructive* method of building sets.

In the plane we should, in the same manner, start with basic sets that have simple forms such as disks or squares. The interval $[x - \epsilon, x + \epsilon]$, used on the real line, is replaced by the **ball** $B_\epsilon(x)$; this is the set of all points of the plane whose distance to x is at most equal to ϵ. It may have different geometrical forms depending on the notion of distance used. We shall always use the Euclidian distance, where a ball is nothing but a disk. We find the same characterization of closed and open sets as in the case of the real line:

> A set V is **open** if each point of V is the center of a ball (even a very small one) contained in V.
>
> A set F is **closed** if the limit of any convergent sequence of points of F also belongs to F.

In particular, $B_\epsilon(x)$ is a closed set, while the set of points whose distance to x is *strictly less* than ϵ is an open ball; this is the *interior* of $B_\epsilon(x)$.

The closed sets are the complements of the open sets and conversely.

We shall return to the notion of closed sets in Chapter 5, when the Hausdorff distance is discussed. But an important remark should be stated here: If we hope to have a clear idea about the structure of the open and closed sets in the real line, then in the plane the situation is hopeless. In fact, here we have infinitely many different topological types. In what concerns the Borel measure on the plane this is no longer called length, but *area*. Here is how we compute it for closed sets:

If E is a square, its area $\mathcal{A}(E)$ is equal to the square of the length of its sides. Otherwise, we cover the plane with a grid whose squares have sides of length ϵ, and we count the number $\omega_\epsilon(E)$ of the squares that contain a point of E. This set will then be covered by a surface whose total area equals $\epsilon^2 \omega_\epsilon(E)$. By taking finer grids we get a finer cover, and thus the area of E can be written as:

$$\mathcal{A}(E) = \lim_{\epsilon \to 0} \epsilon^2 \, \omega_\epsilon(E) \,.$$

Using this method, we find that the area of a rectangle is the product of the lengths of its sides, and the area of a disk of radius r is πr^2. The area of other sets in the plane can be deduced, as in the case of the real line, by using unions and complements.

Given the diversity of closed sets, we restrict our discussion to a special type of these sets, which is proper to mathematical modeling: the *curves*. More precisely we shall be working with *simple curves*.

4.2 Velocities, trajectories

We may describe a **curve** as the trajectory of an (unidentified) object in the plane or in the space. To each time t a position $\gamma(t)$ corresponds. The set of all these positions constitutes the trajectory. In the observable cases, there is a continuous motion (a motion that contains no *jump*) between the different positions of the object. The resultant curve is thus *continuous*. If the velocity is finite, then so is the **length** of the trajectory; we measure this length by multiplying the velocity of the object by the time needed to run through the whole curve, providing that this velocity is constant. Otherwise, we get this length by integrating the velocity with respect to time. This is the classical approach. In another part of this book, we consider bounded curves that contain no pieces of finite length. By comparison with the previous curves we are led to speak about infinite velocity. This second approach, the *fractal* one, does not come as a simple generalization of ancient ideas; rather, it is inspired by experience, and its worthiness is due to its ability to mirror reality in some types of models (trajectories of brownian motions, boundaries of aggregates, profiles of signals). To every problem, its model: there is no universal geometry. The fractal model appears as natural as any other, but it is recent for the following two reasons: the finer technical tools employed in investigating nature obligate us, as they do in other physical sciences, to review our concepts, concepts that were only based on our visual experience; the arrival of an initiator, Benoît Mandelbrot, whose works describe the enormous potential of applying some ancient, hidden, badly known mathematical theories to the study of nature.

In reality, both types of models are justified by a common notion: the *measure*; which, in the case of trajectory becomes the *measure of the time*. In this part (Chaps. 4 to 9) we shall only consider curves of finite length. We indicate the main methods that have been devised to measure them. This is an inexhaustible historic problem of great interest in its own right. Later, we shall see that these ideas, coupled with the order of growth, introduced in Chapter 2, are very useful in the mathematical analysis of the fractal curves. Thus it is interesting to acquire the habit of working with classical methods before confronting any other type of geometry.

4.3 The definition of a curve

To be precise in our definition of a curve, we need a variable t: the *parameter* (this is the time in the case of the trajectory of a motion) and a function γ that associates a position $\gamma(t)$ to any value t of the parameter. These data constitute a *parameterization* of the curve. The curve itself (or the trajectory) will be denoted by Γ. Since the (unidentified) object can run through the curve with different velocities, there are infinitely many possible parameterizations of the curve Γ.

There are other methods to define a curve; the algebraic definition, for example, goes as follows. A curve is a set of points (x, y) of the plane verifying an equation of the form:

$$F(x, y) = 0.$$

There is also the geometric definition of curves in the space; a curve is the intersection of two surfaces (thus the intersection of two spheres is a circle). These definitions are less general and more ambiguous since the result could be an empty set. In fact, we are not sure that we have defined a curve unless we can establish a map between the set in question and an interval of the real line, i.e., unless we define a parameterization of the curve.

We say that Γ is a **curve** *if it is the image in the plane or in space of an interval $[a, b]$ of real numbers of a continuous function γ.*

The points $A = \gamma(a)$ and $B = \gamma(b)$ are the endpoints of Γ (Fig. 4.1). A **subarc**, or simply an **arc** of Γ, is the image of γ of a subinterval $[c, d]$ of $[a, b]$. If its endpoints are $C = \gamma(c)$ and $D = \gamma(d)$, we denote this arc by $C^{\frown}D$.

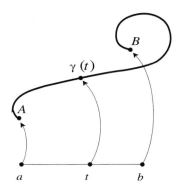

Fig. 4.1. *The curve $A^{\frown}B$ is parameterized by time using the application γ, which relates the position A to the time a, the position $\gamma(t)$ to the time t.*

The point x of Γ is called a *double point* if there are exactly two different values of time t_1 and t_2 such that $\gamma(t_1) = \gamma(t_2) = x$; the object goes back to a previous position in its trajectory. If there are at least two different values of time that correspond to the position x, then we say that x is a *multiple point*.

If there are no multiple points on Γ, then the function γ is a *bijection* from $[a, b]$ onto Γ: to distinct times corresponding to distinct positions. And then the curve Γ is said to be **simple**, or a *Jordan arc*. We may consider closed curves such that $\gamma(a) = \gamma(b)$. In this case, Γ is the image of a circle by a bijection.

Particular case   Given a function $z(t)$ defined on an interval $[a, b]$, we call the set of pairs $(t, z(t))$ contained in a Cartesian plane the *graph* of z. If z is continuous, this graph is a curve that is naturally parameterized by its abscissa t. The function γ relates every value t of $[a, b]$ to the point whose coordinates are:

$$\begin{cases} x_1(t) = t \\ x_2(t) = z(t). \end{cases}$$

Another particular case   The *polygonal curves*, formed by a finite number of line segments whose endpoints are the *vertices* of the curve, are in fact the only curves whose lengths can be computed without any problem. It suffices to add the lengths of its constituent segments. A scaled ruler is enough! It is not a surprise that we always look for polygonal approximations of any given curve.

4.4 Bibliographical notes

The idea of approximating a curve by *inscribed* polygonal curves (whose vertices belong to the initial curve) goes back to Archimedes. It seems that it was not formalized, at least in modern terms, before [G. Peano 1], L. Scheeffer, and [C. Jordan]. Peano proposed defining the length as a lim sup of the inscribed polygonal curves, and Jordan proved that there is always convergence whenever the sides of these polygonal curves tend to 0. Looking for a generalization to compute the area of the surfaces, [H. Lebesgue 1] (doctorate thesis, 1902) preferred defining the length as a lim inf of all the polygonal curves whose "distance" to the initial curve tends to 0.

Surely all these definitions are equivalent, with the Lebesgue method having the inconvenience of being *nonconstructive*. This is, as in the case of the Lebesgue integral, the price of generality.

These ideas, applied to parameterized curves, led Jordan to the notion of *bounded variation* of a function. At that point, the notion of length escaped geometry to go into analysis. We find interesting accounts of this epoch's works in the thesis of Lebesgue, and the works of [W.H. and G. Young] and [E.W. Hobson]. See [J.C. and H. Burkill] or [M.J. Pelling] for an account in a more modern style.

We can generalize the curves in different manners. If we are interested in the length, we can consider the "Y–sets," or the "rectifiable curves" in the large sense (see references in Chap. 7). If we are interested in topological properties, we introduce the notion of "continuum." We note that the definition of a parameterization of a curve as given in this chapter (the image of an interval of a continuous function) is not entirely satisfying; [G. Peano 2] constructed a continuous function defined on an interval whose image is an entire square (Fig. 13.7). Thus it seems that the definition is too loose to have any real significance. That is why we prefer, in this part, to restrict ourselves to the notion of simple curves.

5 Polygonal Curves and Length

5.1 Rectifiability

Is it justified to approximate the length of a curve Γ by the length of a *polygonal approximation* of Γ? Yes, as long as each point of Γ is situated on a small arc that cannot be distinguished from a segment; in other words, at each of its points, (or at least at almost all of its points), Γ has a tangent. The curve is then said to be *rectifiable*. Not all curves are rectifiable; some do not have a tangent at any of their points. These are called *fractal* curves. In the first case, the notion of rectifiability is associated with the idea of *finite length*, at least locally. In the second case, the notion of fractal curves is associated with the idea of *infinite length*. We need to clarify these notions, in particular, the notion of "approximation." But first we will go deeper in the study of the metric of curves: the first important notion is *distance*.

5.2 Hausdorff distance

Consider two populations E_1 and E_2 dispersed on the same territory. The distance between two persons is simply the length to be covered when going from one person to the other. The distance between a person x of E_1 and the whole population E_2, is the distance between x and its nearest neighbor in E_2:

$$\mathrm{dist}(x, E_2) = \inf_{y \in E_2} \mathrm{dist}(x, y) \ .$$

But we can also speak about the distance between the two populations themselves; this is a number that determines the extent to which these populations are amalgamated. It is, for example, less than 1 kilometer if every person of E_1 is a neighbor of a person of E_2 in a radius of 1 kilometer, and vice versa. It is larger than 1 kilometer if there is at least one person belonging to one of the two populations that is isolated from the other by a radius of at least 1 kilometer. Formally, we should take the largest possible distance between an inhabitant of E_1 and the set of the people in E_2; that is:

$$\sup_{x \in E_1} \mathrm{dist}(x, E_2) \ ,$$

similarly for the inhabitants of E_2 and the set of those in E_1:

$$\sup_{x \in E_2} \text{dist}(x, E_1) \, .$$

Finally, the larger of these two numbers is the *Hausdorff distance* between E_1 and E_2:

$$\text{dist}(E_1, E_2) = \max \left\{ \sup_{x \in E_1} \text{dist}(x, E_2) \, , \ \sup_{x \in E_2} \text{dist}(x, E_1) \right\} \, .$$

We apply this definition to sets of points (on a line, in a plane, or in the space). We specify in advance that we only consider *bounded* sets, so that these distances will be finite.

Application to closed sets

• If a point x belongs to a set E, then $\text{dist}(x, E) = 0$.

• The same holds if x does not belong to E but is the limit of a convergent sequence of elements of E. For example, the distance between 1 and the open interval $]0, 1[$ is null; and so is the distance between 0 and the set of all the fractions of the form $\frac{1}{n}$, for every non-null integer n. The point x belongs to the *closure* of E. Therefore we may give the following definition of the closure:

• The **closure** of a subset E (of the line, the plane, etc.) is the set of all points whose distance to E is null. We denote it by $\overline{E}$. A set is **closed** if it is equal to its closure.

• Therefore *a point belongs to a closed set if and only if its distance to this set is null*.

• For every set E:
$$\text{dist}(E, \overline{E}) = 0 \, .$$

That is, Hausdorff distance does not distinguish a set from its closure. In general, for all sets E_1 and E_2,

$$\text{dist}(E_1, E_2) = \text{dist}(\overline{E}_1, \overline{E}_2) \, .$$

• $\text{dist}(E_1, E_2) = 0$ if and only if $\overline{E}_1 = \overline{E}_2$. In particular, if the distance between any two closed sets is null, then these sets are necessarily identical.

Relation with Minkowski sausage To characterize the notion of Hausdorff distance, we may also use the notion of the *Minkowski sausage*, introduced for subsets of the line in Chapter 2, §3. The Minkowski sausage of a set E is the set

$$E(\epsilon) = \bigcup_{x \in E} B_\epsilon(x) \, .$$

Consequently,

$$\sup_{x \in E_1} \mathrm{dist}(x, E_2) \le \epsilon \iff E_1 \subset E_2(\epsilon) \,.$$

Therefore, we can define the Hausdorff distance (Fig. 5.1) as follows:

$$\mathrm{dist}(E_1, E_2) = \inf\{\epsilon \text{ such that } E_1 \subset E_2(\epsilon) \text{ and } E_2 \subset E_1(\epsilon)\}.$$

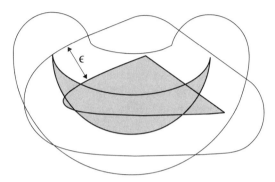

Fig. 5.1. *The Hausdorff distance between two sets E_1 and E_2 is the smallest ϵ such that E_1 is included in the ϵ–sausage of E_2, and E_2 in the ϵ–sausage of E_1.*

Justification of the word "distance"   By definition, a function $f(x,y)$ that associates a positive real number to every pair (x,y) is called *distance* if it verifies the following three properties:

      (i) $f(x,y) = 0 \Leftrightarrow x = y$;

      (ii) $f(x,y) = f(y,x)$;

      (iii) $f(x,z) \le f(x,y) + f(y,z)$ (triangle inequality).

The Hausdorff distance $\mathrm{dist}(E_1, E_2)$ satisfies these three properties, provided the considered sets are all *closed* and *bounded*. These sets are said to be **compact**.

▶   We have seen that if E_1 and E_2 are compact sets, the distance between them is not null if and only if they are distinct. Moreover, the definition of this distance is symmetrical with respect to E_1 and E_2. Therefore (i) and (ii) are verified. We will now show the triangle inequality.

Let E_1, E_2, E_3 be three compact sets, and let ϵ be a real number $\epsilon > \mathrm{dist}(E_1, E_2) + \mathrm{dist}(E_2, E_3)$. It suffices to prove that $\epsilon \ge \mathrm{dist}(E_1, E_3)$. We write ϵ as $\epsilon_1 + \epsilon_2$, where $\epsilon_1 > \mathrm{dist}(E_1, E_2)$, and $\epsilon_2 > \mathrm{dist}(E_2, E_3)$. We deduce that:

$$E_1 \subset E_2(\epsilon_1), \ E_2 \subset E_1(\epsilon_1), \ E_2 \subset E_3(\epsilon_2), \ E_3 \subset E_2(\epsilon_2) \,.$$

The second and third inclusions imply the following:

$$E_2(\epsilon_2) \subset E_1(\epsilon_1 + \epsilon_2), \ E_2(\epsilon_1) \subset E_3(\epsilon_1 + \epsilon_2) \,.$$

Consequently, $E_1 \subset E_3(\epsilon_1 + \epsilon_2)$, and $E_3 \subset E_1(\epsilon_1 + \epsilon_2)$. This shows that $\mathrm{dist}(E_1, E_3) \leq \epsilon_1 + \epsilon_2$. ◀

$\diamond$ This distance allows us to determine the extent to which any given two sets are identical. In particular, it is a measure of the deviation between a curve and its polygonal approximation. It is reasonable to think that if $\mathbf{P}$ is a polygonal curve with the same endpoints as Γ and whose vertices belong to Γ, then the distance between $\mathbf{P}$ and Γ gets smaller whenever the lengths of the segments of $\mathbf{P}$ get smaller. But this needs verification.

5.3 Polygonal approximations

A parameterization of a curve imposes an order on this curve. We can define curves formed by segments whose endpoints in Γ follow the already established order on Γ:

> Given a curve Γ parameterized by the function $\gamma(t)$ where t belongs to the interval $[a, b]$, and $K + 1$ values of time $t_1 = a < t_2 < \ldots < t_{K+1} = b$, a **polygonal approximation** of Γ is a curve $\mathbf{P}$ formed by K line segments whose endpoints are $\gamma(t_i)$, $\gamma(t_{i+1})$, for $i = 1, \ldots, K$.

These are the curves whose lengths can always be computed and used to define the length of Γ. Here is the first important result.

THEOREM   *Let Γ be a simple curve, (ϵ_n) a sequence of positive real numbers that converges to 0, and $\mathbf{P}_n$ a polygonal approximation of Γ whose segments are of length less than ϵ_n. Then $\mathrm{dist}(\Gamma, \mathbf{P}_n)$ converges to 0.*

We make use of the following lemma:

> Let (A_n) be a convergent sequence of points of Γ, and let A^* be its limit. If a_n and a^* are the values of the parameter such that $\gamma(a_n) = A_n$ and $\gamma(a^*) = A^*$, then the sequence (a_n) converges to a^*.

▶   Because otherwise we can extract a convergent subsequence (a_{n_k}) of (a_n) whose limit is $b^* \neq a^*$, the continuity of γ implies that $\mathrm{dist}(A_{n_k}, A^*)$ tends to $\mathrm{dist}(\gamma(b^*), A^*)$, which is not null because γ is a bijection. This contradicts the hypothesis. ◀

This result implies that the inverse function γ^{-1}, which to every position A on Γ associates the corresponding time $\gamma^{-1}(A)$, is continuous. We then say that γ is a *bi-continuous bijection*, or simply a *homeomorphism*.

▶   Now we prove the theorem. Since the sets $[a, b]$ and Γ are closed and bounded, they are *compact*; we may use the fact that both applications γ and γ^{-1} are *uniformly continuous*.
    Let $\epsilon > 0$. We should prove that there exists an integer N such that, for all $n \geq N$, $\mathrm{dist}(\Gamma, \mathbf{P}_n) \leq \epsilon$.

The application γ is uniformly continuous. There exists an η that depends only on ϵ such that:

$$|t' - t''| \leq \eta \Longrightarrow \mathrm{dist}(\gamma(t'), \gamma(t'')) \leq \epsilon \,.$$

We fix η.

The function γ^{-1} is uniformly continuous. There exists a ζ that depends only on η such that, for all points x and y of Γ,

$$\mathrm{dist}(x, y) \leq \zeta \Longrightarrow |\gamma^{-1}(x) - \gamma^{-1}(y)| \leq \eta \,.$$

Choose an integer N such that

$$n \geq N \Longrightarrow \epsilon_n \leq \zeta \,.$$

Let n be such an integer, and let $t_1 = a$, t_2, ..., $t_{K+1} = b$ be values of time whose images by γ are the vertices of $\mathbf{P}_n$. Since ϵ_n is less than ζ, we deduce that $t_{i+1} - t_i \leq \eta$ for all $i = 1$, ..., K (uniform continuity of γ^{-1}). Therefore the distance from any value of t to at least one t_i is $\leq \eta$. It follows (by the uniform continuity of γ) that the distance from any point of Γ to at least one vertex of $\mathbf{P}_n$ is $\leq \epsilon$. Therefore $\mathrm{dist}(\Gamma, \mathbf{P}_n) \leq \epsilon$.   ◄

$\Diamond$ In fact, the above is a proof of a more powerful result, namely, if $\mathbf{S}_n^i$ denote the successive segments of $\mathbf{P}_n$, and Γ_n^i the arc of Γ whose chord is $\mathbf{S}_n^i$, then the number

$$\max \{\, \mathrm{dist}(\mathbf{S}_n^i, \Gamma_n^i) \,\}$$

converges to 0. This means that the shorter the segments of a polygonal approximation are, the more indistinguishable (in the sense of Hausdorff distance) the arcs of Γ and their chords become.

$\Diamond$ We remark that the only condition imposed on the curve Γ is that it be simple. The theorem is true whatever the length of the curve is. We define this length in the next section.

5.4 The length of a curve

We call $L(\mathbf{S})$ the length of the line segment $\mathbf{S}$. Evidently, the length of a polygonal curve $\mathbf{P}$ is the sum of the lengths of its constituent line segments; we denote it by $L(\mathbf{P})$. Given a curve Γ, we shall define its length to be a limit of a sequence:

Let Γ be a simple curve, and let $(\mathbf{P}_n)$ be a sequence of polygonal approximations of Γ such that the maximum length of their line segments converges to 0. The **length** of Γ is by definition the limit of the sequence of the lengths of $\mathbf{P}_n$:

$$L(\Gamma) = \lim_{n \to \infty} L(\mathbf{P}_n) \,.$$

For this definition to be meaningful, one has to show that the limit of $L(\mathbf{P}_n)$ exists and that this limit is independent of the choice of the polygonal approximations. This is what we are going to do.

▶   a) We first prove that $L(\mathbf{P}_n)$ converges to:

$$\sup_n L(\mathbf{P}_n) ,$$

the upper bound of this sequence (whether this number is finite or not). It will be enough to show that for all n and all positive ϵ we can find an integer N_0 such that

$$N \geq N_0 \Longrightarrow L(\mathbf{P}_N) > L(\mathbf{P}_n) - \epsilon .$$

First, we observe that if we replace a polygonal approximation with another containing all the vertices of the first one, then we obtain a longer curve.

Let $\mathbf{S}_n^1, \dots, \mathbf{S}_n^{K_n}$ be the segments of $\mathbf{P}_n$, and $\Gamma_n^1, \dots, \Gamma_n^{K_n}$ the arcs of Γ whose chords are the above segments. Let

$$\rho_n = \max_{i=1,\dots,K_n} \left\{ \operatorname{dist}(\mathbf{S}_n^i, \Gamma_n^i) \right\} .$$

We know from Section 3 that ρ_n converges to 0. We fix the integer n. Let N be an integer larger than n, and let $\mathbf{S}$ be any segment of $\mathbf{P}_N$ whose endpoints are C and D. The arc $C^\frown D$ of Γ may contain a vertex of $\mathbf{P}_n$. If this is the case, we shall call this vertex I. Then

$$\operatorname{dist}(I,\mathbf{S}) \leq \rho_N .$$

If $\mathbf{S}_1$ and $\mathbf{S}_2$ denote the segments CI and ID, we see (Fig. 5.2) that

$$L(\mathbf{S}_1) + L(\mathbf{S}_2) \leq L(\mathbf{S}) + 2\rho_N .$$

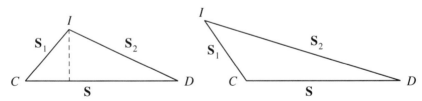

Fig. 5.2. *Two cases of the inequality $L(\mathbf{S}_1) + L(\mathbf{S}_2) \leq L(\mathbf{S}) + 2\operatorname{dist}(I,\mathbf{S})$.*

If there are two vertices I and J of $\mathbf{P}_n$ on the same arc $C^\frown D$, then the segments $\mathbf{S}_1 = CI$, $\mathbf{S}_2 = IJ$, $\mathbf{S}_3 = JD$ verify the following inequality:

$$L(\mathbf{S}_1) + L(\mathbf{S}_2) + L(\mathbf{S}_3) \leq L(\mathbf{S}) + 4\rho_N .$$

In general, if k vertices of $\mathbf{P}_n$ belong to the arc $C^\frown D$ whose cord is the segment $\mathbf{S}$, and if $\mathbf{Q_S}$ is the polygonal approximation whose endpoints are C and D passing through these points, then

$$L(\mathbf{Q_S}) \leq L(\mathbf{S}) + 2k\rho_N .$$

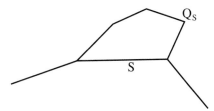

Q_S

S

Fig. 5.3. *If the segment* **S** *of* $\mathbf{P}_N$ *is the chord of an arc of* Γ *containing vertices of* $\mathbf{P}_n$, *we replace* **S** *by the polygonal* $\mathbf{Q_S}$ *of the same endpoints, passing through the vertices of* $\mathbf{P}_n$.

Replacing each segment **S** with such a polygon $\mathbf{Q_S}$ when possible results in a new polygonal curve **Q** approximating Γ passing through all the vertices of $\mathbf{P}_n$ and $\mathbf{P}_N$. In particular,

$$L(\mathbf{P}_n) \leq L(\mathbf{Q}) .$$

Moreover, if we add the lengths $L(\mathbf{Q_S})$, we obtain

$$L(\mathbf{Q}) \leq L(\mathbf{P}_N) + 2K_n \, \rho_N .$$

Choosing a sufficiently large N, we can make the number $2K_n \, \rho_N$ less than ϵ. This proves the desired result.

b) We now prove that $\lim L(\mathbf{P}_n)$ does not depend on the choice of the $(\mathbf{P}_n)$. Let $\mathbf{Q}^*$ be any polygonal approximation of Γ, and choose $\epsilon > 0$. An argument similar to the above one shows that we can always find an integer N such that

$$L(\mathbf{P}_N) > L(\mathbf{Q}^*) - \epsilon .$$

We deduce that $L(\mathbf{Q}^*)$ is always less than $\sup_n L(\mathbf{P}_n)$. Therefore if $(\mathbf{Q}_n)$ is any sequence of polygonal approximations of Γ, then

$$\sup_n L(\mathbf{Q}_n) \leq \sup_n L(\mathbf{P}_n) .$$

By symmetry, we obtain the equality of the two upper bounds. ◀

◇ We note that the length $L(\Gamma)$ as defined earlier could be infinite, even when the curve is a very regular one.

Example Let the spiral be defined by its polar coordinates (ρ, θ) as follows:

$$\begin{cases} \rho(t) = t \\ \theta(t) = \dfrac{2\pi}{t} \end{cases} \qquad 0 < t \leq 1 .$$

We can add to it the origin O (corresponding to the parameter $t = 0$) to obtain a closed set. The curve turns slowly around O (Fig. 5.4). Every spire $\mathbf{S}_k$ corresponds to the values

$$\frac{1}{(k+1)} \leq t \leq \frac{1}{k}$$

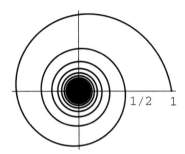

Fig. 5.4. *A curve of locally finite length (except at 0) whose total length is infinite.*

of the parameter and is of finite length, which is larger than $\frac{1}{k}$ (this length is larger than the distance between O and the point $\gamma(\frac{1}{k})$, which is $\frac{1}{k}$). Since the series $\sum \frac{1}{k}$ diverges, the total length of this curve is infinite. We say that the curve is *of locally finite length*, except at O.

In the last example, how can we determine the exact length of each spire? Though our approach using polygonal approximations is satisfying from a geometrical point of view, in practice it is not useful. In the following chapters we shall encounter equivalent definitions of length that are better suited for computations.

5.5 Two distinct notions

We have talked about distance, which is a *metric* notion, and about length, which is a notion of *measure*. These are two distinct ideas, the basis of two distinct mathematical theories. That is why we do not have any general results linking these two notions. As always in such cases, there are too many *counter–examples*. We have used Hausdorff distance to define a measure of the length; we considered the curves $\mathbf{P}_n$ to be polygonal approximations of Γ satisfying $\lim \operatorname{dist}(\mathbf{P}_n, \Gamma) = 0$. But we needed more conditions, namely, $\mathbf{P}_n$ and Γ have the same endpoints, and the vertices of $\mathbf{P}_n$ belong to Γ. If these conditions or any similar ones do not hold, it would be impossible to derive the length of Γ. We can even assert the following:

Let Γ be a curve and Γ_1, Γ_2, ... a sequence of curves such that:

$$\operatorname{dist}(\Gamma_n, \Gamma) \to 0 \,,$$

then it is not always true that

$$L(\Gamma_n) \to L(\Gamma) \,.$$

Counterexample We reconsider the spiral of the previous section. We construct the subspirals Γ_n by taking the parameter t to be between 0 and $\frac{1}{n}$. The distance between Γ_n and the one-element set $\{O\}$ is equal to $\frac{1}{n}$. Therefore the sequence of the curves Γ_n converges to the point O, a limit curve of length 0. However, the length of each Γ_n is infinite.

Another counterexample It is worthwhile to describe a historic counter-example that was fashionable at the end of the last century and that allowed one to claim a "proof" of $2 = 1$! Lebesgue was particularly interested in it, and used it to deduce the above proposition. We start with an equilateral triangle ABC with side length 1. Let $\mathbf{P}_1$ be the curve formed by the segments AC and BC. It is of length 2. The polygonal curve $\mathbf{P}_2$, which has $ADEFB$ as vertices (notation of Fig. 5.5), passes through the midpoints of the sides of ABC. It is also of length 2. Repeating the same operation on each equilateral triangle ADE and EFB, we obtain a new polygonal curve $\mathbf{P}_3$ of length 2, and so forth. The obtained curves $\mathbf{P}_n$ are all of length 2, and they converge (in the sense of Hausdorff distance) to the side AB of the original triangle, whose length is 1.

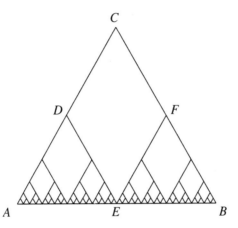

Fig. 5.5. *Construction of a sequence of polygonal curves* $\mathbf{P}_1 = ACB$, $\mathbf{P}_2 = ADEFB$, ...; *all are of length 2. This sequence converges in the sense of Hausdorff distance to the segment* AB, *whose length is 1. Not all the vertices of the curves* $\mathbf{P}_n$ *belong to* AB. *The sides of* $\mathbf{P}_n$ *form the same constant angle* ($\pi/3$) *with* AB. *That is why* $L(\mathbf{P}_n)$ *does not converge to* $L(AB)$.

The only general result that links the *distance* to the *length* is a partial one that holds for any sequence of curves.

THEOREM $\mathrm{dist}(\Gamma_n, \Gamma) \to 0 \Longrightarrow L(\Gamma) \leq \liminf L(\Gamma_n)$.

Here lim inf is the smallest limit of the sequence $(L(\Gamma_n))$ (for more details on lower limit, see Appendix A). Therefore we do not assume that this sequence converges.

▶   It suffices to show that given $\epsilon > 0$ we can find a sufficiently large integer N such that:

$$n \geq N \Longrightarrow L(\Gamma_n) \geq L(\Gamma) - \epsilon .$$

Let $\mathbf{P}$ be a polygonal approximation of Γ, whose vertices $A_1 = A, A_2, \ldots,$ $A_{K+1} = B$ belong to Γ. Let $\rho_n = \mathrm{dist}(\Gamma_n, \Gamma)$. We can always find a large n so that the balls of center A_i and radius ρ_n, are disjoint. Each of these balls contains a point B_i^n of the curve Γ_n. The length of the segment $B_i^n B_{i+1}^n$ is at least equal to $L(A_i A_{i+1}) - 2\rho_n$. Since there are K such segments, we obtain

$$L(\Gamma_n) \geq L(\mathbf{P}) - 2\,K\,\rho_n \ .$$

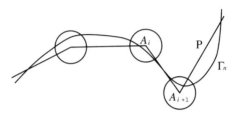

Fig. 5.6. *Approximation of the polygonal* $\mathbf{P}$ *by one of the curves of the convergent sequence* (Γ_n).

We can always choose the polygonal curve $\mathbf{P}$ to be as near Γ as we please, so that for sufficiently large n we obtain the sought inequality.   ◀

5.6 Measuring the length by compass

A particular case of the polygonal approximation method consists of taking equal steps along the curve Γ. We fix ϵ as the step length; we start at the point $A_1 = A$; with a compass we find the next point A_2 on Γ whose distance from A_1 is ϵ; and so forth. To avoid any ambiguity in this construction, we should specify that we follow the same order as the increasing order of the parameter t, that is, the same direction of the motion along the curve. Moreover, we may have more than one point at each step; in this case we systematically choose the one that corresponds to the smallest possible value of the parameter. We halt this construction as soon as we reach a point A_N whose distance from B (the other endpoint of Γ) is less than ϵ.

Fig. 5.7. *We run through the curve by making equal steps. If the length of the step tends to 0 the covered distance is the length of the curve.*

The result is a polygonal path $\mathbf{P}_\epsilon$ whose vertices are $A, A_2, \ldots, A_N, B$. Let $N = N(\epsilon)$. The length of $\mathbf{P}$ is

$$\epsilon\,(N(\epsilon) - 1) \le L(\mathbf{P}) \le \epsilon\,N(\epsilon)\,.$$

We deduce that the length of Γ is:

$$L(\Gamma) = \lim_{\epsilon \to 0} \epsilon\,N(\epsilon) = \sup_\epsilon \epsilon\,N(\epsilon)\,.$$

$\diamond$ Since the birth of geometry, geometers and mathematicians used this method to define the length of a circle as the limit of lengths of a sequence of *regular* (i.e., equal-sided) inscribed polygonal curves whose side length converges to 0.

$\diamond$ This is also an elementary method to draw geographical lines, for example, when these lines are *smooth* enough so that they can be satisfactorily modeled by curves of finite lengths. Nevertheless, when these lines are defined as sets of pixels that are numbered from 1 to K, then it is more natural to fix an integer $k \ll K$, and to consider A_1 as the kth point after the origin, A_2 the $2k$th point and so forth up to the largest integer N such that kN is less than K. The smaller the k is with respect to K, the better the approximation. This method is better than the compass method, because it takes into account the local irregularities of the curve. In terms of parameterization, this is similar to assuming that the motion along Γ is of constant velocity, and that the pixels correspond to the position taken at equal intervals of time. Therefore the polygonal approximation is constructed by means of *temporal* equal steps, and not of *spatial* ones as in the compass method. Finally, more sophisticated methods are used in cartography to emphasize the local geometry of a curve. These methods can use properties of fractals (Chap. 15).

5.7 Bibliographical notes

The notion of *Hausdorff distance* can be found underlying the most ancient reasoning about the limit of sets. It makes precise the idea of the convergence of compact (closed and bounded) sets; in other words, it induces a metric structure on the set of all compact sets. It is used to prove the existence of the attractor of a system of contractions. The reader may find its main properties in most books on topology.

The example in §5.5 of an equilateral triangle in which a sequence of polygonal arcs of length 2 converges to a segment of length 1 was considered a paradox since [H. Lebesgue 2]:

"When I was a high school student, teachers and students were satisfied when they extended their argument to the limit case. Even though this ceased to satisfy me when some of my colleagues, during my fifteenth year, told me of a triangle whose side is equal to the sum of the other two, and that $\pi = 2$."

Historically, the study of the limits of polygonal curves goes hand in hand with the study of rectifiable curves; the reader should consult the references in Chapter 4. For example, the method of polygonal approximation by constant steps (§6) was used by [L. Richardson] and [B. Mandelbrot 1 and 2] to study irregular curves.

6 Parameterized Curves, Support of a Measure

6.1 Parameterization by arc length

The polygonal approximation allowed us to define the length of a simple curve Γ. Length is a global measure of Γ. It is independent from the initial parameterization γ that defines the curve. We assume that it is finite. We also know how to compute the length of any arc of Γ. This allows us to define a special parameterization of the curve, called parameterization by *arc length*, which can be different from γ:

Let A, B be the endpoints of Γ. To each positive value t, which is less than $L(\Gamma)$, we relate a point $\gamma^*(t)$ of the curve such that the length of the arc $A^\frown\gamma^*(t)$ is precisely equal to t. The application $\gamma^* : [0, L(\Gamma)] \longrightarrow \Gamma$ is a bijection. The image of 0 is A, and the image of $L(\Gamma)$ is B. Everything goes as if the curve Γ is the trajectory of an object moving at a constant velocity, and the time unit is adjusted so that this velocity is 1. The length of any subarc of Γ, then is the value of the time spent by the object while it runs through this subarc. Similarly, the *measure* of any part of Γ is the value of the *time* spent while the object runs through this part.

The parameterization by arc length is defined by the equality:

$$L(\gamma^*([0, t])) = t .$$

Here we have a direct method to *transfer measure* from the line to the curve. On the line, the Borel measure is simply called "length." We find an identical notion on the curve by virtue of the definition of an arc "length." For example, a part of Γ is of null measure if for any positive ϵ this part can be covered by arcs whose total sum of lengths is less than ϵ. This will allow us to speak, later in this chapter, of properties that are true **almost everywhere**. Such properties are true on every point of Γ except on a set of null measure.

$\Diamond$ It is possible to choose any of the two endpoints as an origin for the motion, that is, there are two versions of the parameterization by arc length, according to the choice of the direction of the motion. But this has no impact on the measure of Γ. It will always be the same no matter which direction is chosen because the length of an arc is independent of its parameterization.

6.2 Image measure

However, a curve is not always defined by the lengths of its arcs. Usually a parameterization is imposed by the nature of the curve's definition. However, if the curve is not defined by arc length, then computing its length is a problem by itself. In fact, every parameterization γ induces a *measure* on Γ, but this measure is not always related to the length as in the previous case. For any curve Γ of finite or infinite length, here is how we define the image measure induced on Γ by the function γ:

Assume that γ is a function defined on $[a, b]$, the endpoints of Γ being $\gamma(a) = A, \gamma(b) = B$. The total measure of Γ is $b - a$. The measure of the arc $A \frown \gamma(t)$ is equal to $t - a$. In a general manner:

The measure of each part of a trajectory Γ is the **time** *needed to cover this part during the motion.*

Thus, giving a parameterization of Γ is equivalent to giving a measure on Γ. In other words, *defining a curve correctly is the same as defining a measure on this curve.*

Can we always establish a relation between the *image measure* and the *length*, i.e., between the *time* and the *covered distance*? When the velocity is constant, as in §1, the correspondence is straightforward: the length of an arc of Γ is proportional to the time needed to run through this arc. But we may encounter many irregular motions. In particular, the image $\gamma(E)$ of a (temporal) set E of null measure in $[a, b]$ could very well be a (spatial) subset of Γ whose length is not 0; at some moments the motion suddenly receives an infinite acceleration. The next section deals with more regular curves. Later we will give examples of irregular motions.

6.3 Length by instantaneous velocity

Let O be a fixed origin in the space in which the curve is drawn. The position of the point $\gamma(t)$ is totally determined by the vector $\overrightarrow{O\gamma(t)}$. Its *velocity vector*, which gives the direction of the motion, is the derivative:

$$\mathbf{v}(t) = \lim_{h \to 0} \frac{1}{h} \overrightarrow{\gamma(t)\gamma(t + h)}$$

if this limit exists (in particular, if at the point $\gamma(t_0)$ the trajectory is angular, then $\mathbf{v}(t)$ does not exist at t_0). Finally, its *velocity* (the quantity measured by a speedometer) is the length of $\mathbf{v}(t)$. It will be denoted by:

$$v(t) = \lim_{h \to 0} \frac{1}{h} \text{dist}(\gamma(t), \gamma(t + h)) .$$

Therefore if dl indicates the distance $\text{dist}(\gamma(t), \gamma(t + dt))$ covered during the time dt, the velocity $v(t)$ will be the limit of dl/dt when dt tends to 0, and the total length of Γ will be the sum of dl, i.e., the integral of $v(t)$ with respect to time. This argument is valid for the classical types of trajectories and motions; those with continuous velocity $v(t)$ at every t. In particular, those with finite accelerations. We can even generalize a bit more:

If $v(t)$ exists on the interval $[a,b]$, and if the ratio dl/dt is bounded independently of t and dt, then the length of the curve parameterized by γ on $[a,b]$ is given by the integral

$$L(\Gamma) = \int_a^b v(t)\, dt .$$

When γ parameterizes Γ by arc length, this formula is reduced to:

$$L(\Gamma) = \int_0^{L(\Gamma)} 1\, dt .$$

In a Cartesian coordinate system the point $\gamma(t)$ has $x_1(t)$ and $x_2(t)$ as coordinates. The velocity vector, if it exists, has $x_1'(t)$ and $x_2'(t)$ as components. The integral of the instantaneous velocity can be written as:

$$L(\Gamma) = \int_a^b \sqrt{x_1'(t)^2 + x_2'(t)^2}\, dt .$$

We mention here that the above formula could be false if we just assume that $v(t)$ exists *almost everywhere* on the interval $[a,b]$: the integral of $v(t)$ on its domain of definition could be different from the length of the curve. The "devil staircase," in §4, is such a curve. Here are some examples of the above definition:

The circle The parameterization of a circle in the plan can be given as:

$$\overrightarrow{O\gamma(t)} = R\cos t\, \mathbf{i} + R\sin t\, \mathbf{j} , \quad 0 \le t \le 2\pi ,$$

where O is its center, R is its radius, and $\mathbf{i}, \mathbf{j}$ are perpendicular unit vectors. This is an example in which the "velocity" is constant. The total length of this circle is $2\pi R$, and the length of each piece E is the product of R by the time spent in E.

The spiral In Chapter 5, §4, we discussed the spiral defined by polar coordinates:

$$\begin{cases} \rho(t) = t \\ \theta(t) = \dfrac{2\pi}{t} \end{cases} \quad 0 < t \le 1 .$$

Each of its spires $\mathbf{S}_k$ is obtained when the parameter runs through the interval

$$\left[\frac{1}{k+1}, \frac{1}{k}\right] .$$

We have said that $\mathbf{S}_k$ is of finite length, which is larger than $\frac{1}{(k+1)}$. We calculate this length:

▶   In the usual orthonormal basis $\{\mathbf{i}, \mathbf{j}\}$, the parameterization is written as:

$$\overrightarrow{O\gamma(t)} = t \cos \frac{2\pi}{t} \, \mathbf{i} \, + \, t \sin \frac{2\pi}{t} \, \mathbf{j} \, .$$

By differentiating each component we get:

$$\mathbf{v}(t) = \left(\cos \frac{2\pi}{t} + \frac{2\pi}{t} \sin \frac{2\pi}{t} \right) \mathbf{i} \, + \, \left(\sin \frac{2\pi}{t} - \frac{2\pi}{t} \cos \frac{2\pi}{t} \right) \mathbf{j} \, .$$

We deduce that:

$$v(t) = \sqrt{1 + \frac{4\pi^2}{t^2}} \, .$$

Hence the result:

$$L(\mathbf{S}_k) = \int_{1/(k+1)}^{1/k} \sqrt{1 + \frac{4\pi^2}{t^2}} \, dt \, . \quad \blacktriangleleft$$

6.4 The devil staircase

We shall describe a curve of finite length such that the formula computing the length by integrating the velocity will be of no use. This is the graph of an increasing continuous function $z(t)$ defined on the interval $[0, 1]$, which is constant on each contiguous interval of the Cantor set F (see Chapter 1, §§4 and 5).

Recall that in $[0, 1]$, F has one contiguous interval $C(1)$ of length $1/3$; two contiguous intervals $C(1, 0)$ and $C(1, 1)$ of length $1/9$, ...; 2^n contiguous intervals of length 3^{-n-1}, which we denote by $C(1, i_1, \ldots, i_n)$ ($i_j = 0$ or 1); and so on. These are open disjoint intervals. We attribute to z the value $1/2$ on $C(1)$, $1/4$ on $C(1, 0)$, $3/4$ on $C(1, 1)$, etc. ...: the correspondence between the numbering of a contiguous interval and the value of z on it is determined by the correspondence between the two binary trees of Fig. 6.1.

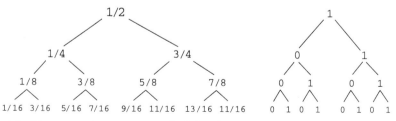

Fig. 6.1. *Every contiguous interval of the Cantor set is symbolized by a vertex in the right-hand tree. On the left-hand tree the vertex of the same situation gives the value taken by z on this contiguous interval. The graph of the function z is the devil staircase.*

The value of z on $C(1, i_1, \ldots, i_n)$ is

$$2^{-n-1} + \sum_{j=1}^{n} 2^{-j} i_j \ .$$

Once defined on the contiguous intervals, i.e., on the complement of F, the function z is completed by "continuity" on F in such a way that the domain of definition of z is the whole interval $[0, 1]$. We can determine the value of $z(t)$ at every point t of F. As we have seen, such a point is determined by an infinite sequence $(i_n)_{n \geq 1}$ of 0 and 1, such that t is the limit of the sequence of contiguous intervals $C(1, i_1)$, $C(1, i_1, i_2)$, ... by continuity we have that:

$$z(t) = \sum_{n=1}^{\infty} 2^{-n} i_n \ .$$

The graph of $z(t)$ This is the set of points of the plane whose coordinates are $(t, z(t))$. It is a curve Γ shown in Fig. 6.2. As we have seen, Cantor set is of null measure. Thus this function is *constant almost everywhere*. However, it is continuous and increases from 0 to 1. Its points of growth are the points of F. At these points it has no derivative. Elsewhere its derivative is 0.

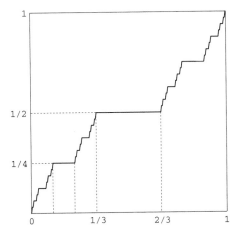

Fig. 6.2. *The devil staircase.*

Parameterization   This curve is naturally parameterized by the function $\gamma(t) = (t, z(t))$. Its velocity vector is not defined on the Cantor set. On the complement of the Cantor set, it is the vector $\mathbf{v}(t) = \mathbf{i} + z'(t)\,\mathbf{j} = \mathbf{i}$, whose length is 1. The velocity is therefore defined *almost everywhere* on $[0, 1]$, but not *everywhere*. The integral

$$\int_{[0,1]-F} v(t)\,dt$$

equals 1. We remark that this result is not equal to the length of Γ. The velocity is not defined everywhere, and the formula relating the velocity to the length of the trajectory does not hold.

Length of Γ   For all n, the set F can be covered by the union F_n of 2^n isolating intervals of length 3^{-n}. We can approximate the function z by a function z_n, whose derivative has the value $(3/2)^n$ on the set F_n, and 0 on the complement $[0,1] - F_n$. The graph of z_n is a polygonal curve $\mathbf{P}_n$, whose vertices belong to Γ (Fig. 6.3).

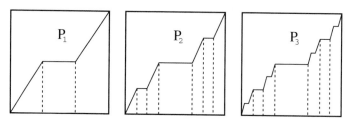

Fig. 6.3. *We can also construct the devil staircase as a sequence of polygonal curves: the nth contains, apart from its horizontal parts, 2^n segments whose slopes are all equal to $(3/2)^n$.*

Among the segments, that form $\mathbf{P}_n$ are those that were not already included in Γ; their lengths are of the same order as 3^{-n}, which tends to 0. We know that in this case the length of $\mathbf{P}_n$ tends to the length of Γ. But $L(\mathbf{P}_n)$ is:

$$\int_{F_n} \sqrt{1 + (3/2)^{2n}}\, dt + \int_{[0,1]-F_n} dt = (2/3)^n \sqrt{1 + (3/2)^{2n}} + (1 - (2/3)^n),$$

which tends to 2 when n goes to infinity. We conclude that:

$$L(\Gamma) = 2 .$$

This is also the length of the path from $(0,0)$ to $(1,1)$ going through the sides of the square.

Irregularity of the measure   The image by γ of the set $[0,1] - F$, the union of the contiguous intervals of the Cantor set, is the union of the horizontal parts of the graph Γ. Its total length is equal to 1. Now the total length of Γ is 2. Consequently, the image by γ of F itself is of length 1. We have here an example of an irregular image measure on the curve Γ, induced by the Borel measure on the interval $[0,1]$. It is irregular in the sense that the image of a set of null measure is not of null measure. This is another proof of the fact that not all parameterizations can be used to compute the length.

Exercise   How can we construct a parameterization of Γ by arc length? When we project this curve on the t axis perpendicular to the diagonal, the set of the projected points is the interval $[0,2]$. The horizontal parts of Γ are projected, while conserving their lengths, to the contiguous intervals of a symmetrical perfect

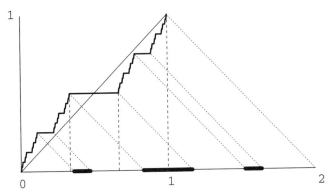

Fig. 6.4. *Parameterization by arc length of the devil staircase. The total length is 2.*

set in $[0, 2]$. This set, which has contiguous intervals the same size as those of Cantor set F, is a perfect nowhere dense set of length 1.

Conversely, the projection of $[0, 2]$ on Γ perpendicular to the diagonal constitutes a parameterization of Γ by arc length. If x_0 is a point of Γ, the length of the arc $O^\frown x_0$ is t_0, where t_0 is the projection of x_0 on $[0, 2]$. The length of any subset of Γ is also the length of its projection on $[0, 2]$.

Variations on the devil staircase   The devil staircase can serve as an example of a parameterization of a curve Γ with no direct relation between the time measure and the length of Γ. In particular, the image of the temporal subset F is of length 1 on Γ, as if the motion reached an infinite velocity peak at each value of time that belongs to Cantor set. We may object that this type of behavior is essentially due to the trajectory whose geometrical complexity affects the regularity of the motion. But that is not it. We can always describe irregular motion on a straight trajectory. Here is an example:

In a plane with Cartesian coordinate system (Ot, Oz), we take the interval of time $[0, 1]$ on the axis Ot, and we consider the Cantor set F_1 in this interval. Then on the axis Oz, we take the interval $[0, 1]$ as the trajectory, and we consider in this interval a symmetrical perfect set F_2, defined by the isolating intervals of length

$$a_n = 2^{-n}b + 3^{-n}(1 - b),$$

where b is a parameter $0 \le b < 1$. The necessary relation for this set to exist, namely, $2\,a_{n+1} < a_n$, is satisfied. The measure (length) of F_2 is equal to $\lim_n 2^n a_n = b$. When $b = 0$, the set F_2 is identical to F_1.

Now we construct a function γ defined on $[0, 1]$, whose values are in $[0, 1]$ such that the image of F_1 is exactly F_2. We may simply start by defining γ as a linear function on the contiguous intervals of F_1: if $]c, d[$ is a contiguous interval of F_1, then it has a unique image $]c', d'[$ that is a contiguous interval of F_2 and that corresponds to the same sequence $1, i_1, \ldots, i_n$ of 0 and 1 (Fig. 6.1). For all t in $]c, d[$, we put

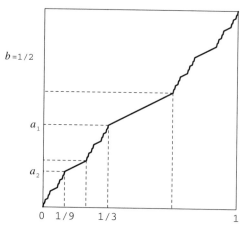

Fig. 6.5. *An intermediate staircase between the devil's and the diagonal of the square. It is obtained by a correspondence between two symmetrical perfect sets, each on one of the axes.*

$$\gamma(t) = \frac{1}{d-c} \left[(d-t)\, c' + (t-c)\, d' \right] .$$

Then by continuity we extend the function γ to all of $[0,1]$. Below we can see the graph of γ: this a "more slippery" devil staircase. The slope of each linear part (the "steps" of the stair) equals $1-b$.

But let us recall that the curve Γ that is under consideration is the segment of length 1 on the axis Oz, and not the graph of γ. Since γ is a strictly increasing function, it is a bijection, and it constitutes a parameterization of Γ. The velocity of the motion along each contiguous interval of F_1 is constant. The measure of the image F_2 by γ of the set F_1, whose length is null, is b. We have here another example of irregular behavior of velocity, but on a straight trajectory. The velocity is infinite at each time that belongs to the Cantor set.

We mention that the length of the graph $\mathbf{G}$ of γ may be computed in the same manner as the length of the devil staircase; we find that

$$L(\mathbf{G}) = b + \sqrt{1 + (1-b)^2} .$$

This is also the length of the path between $(0,0)$ and $(1,1)$, formed by a segment of slope $1-b$ and a vertical segment. When b tends to 1, $\mathbf{G}$ tends to the devil staircase and $L(\mathbf{G})$ tends to 2. When b tends to 0, the function γ becomes the *identity* function because F_2 is then equal to F_1. The graph $\mathbf{G}$ becomes the diagonal of the unit square, whose length is $\sqrt{2}$.

6.5 Length by the average of local velocity

Even if the instantaneous velocity is not defined on the curve Γ, it is possible to obtain a formula for the integral of velocity by using the *average of local velocity*. For this we measure the covered distance:

$$\text{dist}(\gamma(t - \tau), \gamma(t + \tau))$$

during a time 2τ. To define this average at a and b, we define a function $d(t, \tau)$ for all t in $[a, b]$ and all τ in $[0, (b - a)/2]$:

$$d(t, \tau) = \begin{cases} \text{dist}(\gamma(a), \gamma(a + 2\tau)), & \text{if } t - \tau \le a; \\ \text{dist}(\gamma(t - \tau), \gamma(t + \tau)), & \text{if } a \le t - \tau < t + \tau \le b; \\ \text{dist}(\gamma(b - 2\tau), \gamma(b)), & \text{if } b \le t + \tau. \end{cases}$$

Then we estimate the local velocity at the time t by the ratio $d(t, \tau)/2\tau$. The average of the distance covered during the time 2τ is estimated by:

$$\bar{d}_\tau = \frac{1}{b - a} \int_a^b d(t, \tau)\, dt \ .$$

If we divide this by 2τ, we obtain the average of the local velocity, and if we multiply by the total time $b - a$, we obtain an estimation of the length of the curve. In fact, we can prove the following result without any supplementary hypothesis on the simple curve Γ:

$$L(\Gamma) = (b - a) \lim_{\tau \to 0} \frac{\bar{d}_\tau}{2\tau} \ .$$

▶   We shall assume that Γ is of finite length; the infinite case can be proved with similar arguments. To simplify the notation, we take $a = 0$, $b = 1$. Let $\epsilon > 0$, and let $\mathbf{P}$ be a polygonal approximation of Γ whose vertices are images of the times $t_1 = 0$, ..., t_i, ..., $t_{K+1} = 1$. We know that there exists a real $\eta(\epsilon)$ such that if $\text{dist}(\gamma(t_i), \gamma(t_{i+1})) \le \eta(\epsilon)$ for all i, then $L(\Gamma) - L(\mathbf{P}) \le \epsilon$.

On the other hand, the function γ is uniformly continuous on $[0, 1]$; there exists a real $\tau(\epsilon)$ such that

$$\tau \le \tau(\epsilon) \implies d(t, \tau) \le \eta(\epsilon) \ \text{ for all } t \ .$$

Consequently, if $\tau \le \tau(\epsilon)$, then every polygonal curve $\mathbf{P}$ such that $t_{i+1} - t_i \le \tau$ for all i satisfies the inequality

$$L(\Gamma) - L(\mathbf{P}) \le \epsilon \ .$$

The rest of the proof is based on the following idea. We shall construct polygonal approximations of Γ whose lengths are Riemannian sums. This will allow us to compute an approximation of the integral $\int_0^1 d(t, \tau)\, dt$.

We fix a value $\tau \leq \tau(\epsilon)$. We call $\mathbf{P}_0$ the polygonal curve that corresponds to the following partition of $[0, 1]$:

$$t_1 = 0,\ t_2 = 2\,\tau, \ldots,\ t_i = 2\,(i-1)\,\tau, \ldots,\ t_{K_0} = 2\,(K_0-1)\,\tau,\ t_{K_0+1} = 1\ ,$$

where K_0 is the largest integer such that $2\,(K_0 - 1)\,\tau < 1$. From the above argument it follows that:

$$L(\Gamma) - L(\mathbf{P}_0) \leq \epsilon\ .$$

Let N be any integer. Construct the other $N-1$ polygonal curves that correspond to other partitions of $[0, 1]$ into intervals of lengths $2\,\tau$ (except, eventually the first and the last). For all integers n, $1 \leq n \leq N$, the polygonal curve $\mathbf{P}_n$ corresponds to the partition

$$t_1 = 0,\ t_2 = \frac{2\,n\,\tau}{N}, \ldots,\ t_i = \frac{2\,n\,\tau}{N} + 2\,(i-2)\,\tau,$$

$$\ldots,\ t_{K_n} = \frac{2\,n\,\tau}{N} + 2\,(K_n - 2)\,\tau,\ t_{K_n+1} = 1\ ,$$

where K_n is the largest integer such that $(2\,n\,\tau/N) + 2\,(K_n - 2)\,\tau < 1$. Since $\mathbf{P}_N = \mathbf{P}_0$, we obtain a total of N polygonal curves $\mathbf{P}_0, \ldots, \mathbf{P}_{N-1}$, each of which satisfies the inequality:

$$0 < L(\Gamma) - L(\mathbf{P}_n) \leq \epsilon\ .$$

Taking the average of the lengths we get:

$$0 < L(\Gamma) - \frac{1}{N}\sum_{n=0}^{N-1} L(\mathbf{P}_n) \leq \epsilon\ .$$

We shall use another method to evaluate this average, and we shall show that it is related to $\bar{d}_\tau$. In fact the sum of all the lengths of $L(\mathbf{P}_n)$ can be written as

$$\sum_{n=0}^{N-1}\sum_{i=1}^{K_n-2} \operatorname{dist}(\gamma(\frac{2\,n\,\tau}{N} + 2\,(i-1)\,\tau), \gamma(\frac{2\,n\,\tau}{N} + 2\,i\,\tau)) + h(\tau)\ ,$$

where $h(\tau)$ is the sum of lengths of the line segments at the endpoints of $\mathbf{P}_n$. This value is less than $2\,N\,\eta(\epsilon)$. Rearranging the above double sum results in a sum of the $d(s_k, \tau)$, where the sequence (s_K) divides $[0, 1]$ into intervals of length $2\,\tau/N$. Multiplying by $2\,\tau/N$, the latter sum constitutes a Riemannian approximation of the integral of $d(t, \tau)$, with larger N giving better approximations. Finally

$$\lim_{N\to\infty} \frac{2\,\tau}{N}\sum_{k=0}^{N} d(s_k, \tau) = \int_0^1 d(t, \tau)\,dt = \bar{d}_\tau\ .$$

For all real $c > 1$, we can find an integer N^* such that

$$N \geq N^* \implies \frac{1}{c}\left(\frac{1}{N}\sum_{n=0}^{N-1} L(\mathbf{P}_n)\right) \leq \frac{\bar{d}_\tau}{2\,\tau} \leq c\left(\frac{1}{N}\sum_{n=0}^{N-1} L(\mathbf{P}_n) + 2\,\eta(\epsilon)\right)\ .$$

Replacing the average of the lengths of $\mathbf{P}_n$ by the length of Γ gives:

$$\frac{1}{c}\left(L(\Gamma) - \epsilon\right) \leq \frac{\bar{d}_\tau}{2\,\tau} \leq c\left(L(\Gamma) + 2\,\eta(\epsilon)\right).$$

If τ and ϵ tend to 0, then $\eta(\epsilon)$ will tend to 0, and if c tends to 1 then we obtain the desired limit:

$$\lim_{\tau \to 0} \frac{\bar{d}_\tau}{2\,\tau} = L(\Gamma). \quad \blacktriangleleft$$

$\diamondsuit$ If the instantaneous velocity exists at the time t, it can be written as:

$$v(t) = \lim_{\tau \to 0} \frac{d(t,\tau)}{2\,\tau}.$$

The above integral formula, which defines the length, will become:

$$L(\Gamma) = \lim_{\tau \to 0} \frac{\int_a^b d(t,\tau)\,dt}{2\,\tau}.$$

We deduce that it is possible to write the length as an integral of velocity, i.e.,

$$L(\Gamma) = \int_a^b v(t)\,dt\,,$$

provided we can exchange the limit and the integral. For that, the function $v(t)$ must satisfy some hypothesis. First the limit should exist everywhere on $[a, b]$ (at least the set of points where the limit does not exist must be finite). Then we need some conditions of regularity, like:

    The function $v(t)$ is continuous on $[a, b]$,

or

    the ratio $d(t,\tau)/\tau$ is bounded on $[a, b]$, independently of t and τ.

The limit–integral exchange is then possible (by a simple application of a theorem of Lebesgue).

$\diamondsuit$ The formula

$$L(\Gamma) = (b - a) \lim_{\tau \to 0} \frac{\bar{d}_\tau}{2\,\tau}$$

has the advantage of being general. Yet it has another advantage, and that is why we introduced it. It can be associated with a formula that computes the fractal dimension of Γ when Γ is of infinite length (Chap. 15, §7).

6.6 Bibliographical notes

We do not give any references for the relation between the length and the integral of the velocity; no doubt this is one of the most ancient problems of differential geometry as introduced by Newton and Leibniz.

The devil staircase, a graph of the distribution function of a probability measure on a Cantor set, was found a little after Cantor. For this, see [E.W. Hobson] and [E. Hille & J.D. Tamarkin].

In §5, we exchanged the operations of the limit and integration. This is more general than the classical method of integrating the velocity, and it constitutes a preparation for the study of nonrectifiable curves.

7 Local Geometry of Rectifiable Curves

7.1 Tangent, cone, convex hulls

All the curves Γ considered in this chapter are *simple* (with no double point). The parameterization $\gamma : [a, b] \longrightarrow \Gamma$ is a bijective function.

We restate the idea that a curve is *rectifiable at a point* if it is *indistinguishable from a segment of a straight line in a neighborhood of this point*. How can we convert this idea into a rigorous form? In fact, it can be done in many ways. The notion of *tangency* seems to offer a satisfying answer when the curve admits a derivative, but unfortunately this notion totally disappears in the case of fractal curves. We are forced to show more imagination and to rethink our geometrical procedures that analyze curves so that we can find other concepts that are, in a universal sense, easier to generalize.

Choose a point x_0 of a curve Γ. As an example, we shall propose four properties $\mathcal{P}1$ to $\mathcal{P}4$ that could be satisfied by Γ on a neighborhood of this point, and each of them could *locally* characterize the notion of rectifiability. In this chapter, we will see that:

- *Only two of these four properties are equivalent at x_0;*
- *If the curve is of finite length, then all four properties hold almost everywhere.*

Here are the four properties.

Tangent Let x be any point of Γ, and let $T(x_0, x)$ be the straight line passing through x_0 and x. When x tends to x_0, we say that this line tends to a limit $T(x_0)$ if the director vector

$$\frac{\overrightarrow{x_0 x}}{\text{dist}(x_0, x)},$$

whose length is 1, tends to a fixed vector. In this case $T(x_0)$ is the *tangent* to the curve at x_0. The first property can be stated as:

$(\mathcal{P}1)$ *There exists a tangent $T(x_0)$ at x_0.*

Local cone Let $\epsilon > 0$. When the curve is "smooth" in the neighborhood of x_0, we can include a set of points of Γ whose distance to x_0 is less than ϵ in a cone with vertex x_0 and angle θ. The smoother the curve is, the smaller the angle (Fig. 7.2). Let $\theta_\epsilon(x_0)$ be the minimal angle (when this cone does not exist, we can always consider $\theta_\epsilon(x_0) = \pi$ so that the function θ_ϵ will be defined at every

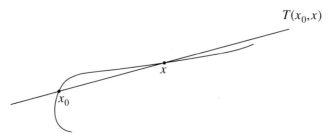

Fig. 7.1. *The limit of the chord $T(x_0, x)$ is the tangent at x_0.*

point of Γ). When ϵ decreases, the value of $\theta_\epsilon(x_0)$ decreases too. Therefore it has a limit when ϵ tends to 0. The second property of rectifiability is to be stated as follows:

($\mathcal{P}2$)   *The limit of $\theta_\epsilon(x_0)$ is null when ϵ tends to 0.*

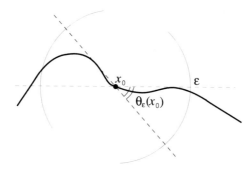

Fig. 7.2. *In the neighborhood of $B_\epsilon(x_0)$, the curve is entirely included in a cone with vertex x_0 and angle $\theta_\epsilon(x_0)$.*

Local length   If Γ is a straight line in a neighborhood of x_0, the arc $x_0{}^\frown x$ is a segment with length equal to the distance between x_0 and x. For any given curve, the fact that these two values are equivalent can be translated at the limit stage by the linear behavior of Γ:

($\mathcal{P}3$)   *The limit of the ratio* $\dfrac{L(x_0{}^\frown x)}{\mathrm{dist}(x_0, x)}$ *is equal to 1 when x tends to x_0.*

Local convex hull   We consider the convex hull $\mathcal{K}(x_0{}^\frown x)$ of the subarc $x_0{}^\frown x$ of Γ (for more details on convex hulls see Appendix C). Its   area $\mathcal{A}(\mathcal{K}(x_0{}^\frown x))$ is not null, unless $x_0{}^\frown x$ is a line segment. The distance between x_0 and x being fixed, this area is smaller whenever the arc is more linear, and the more chaotic the arc is, the bigger the area (Fig. 7.3). In fact, we should compare the area to the square of a length. We state the following condition:

($\mathcal{P}4$)   *The limit of the ratio* $\dfrac{\mathcal{A}(\mathcal{K}(x_0{}^\frown x))}{\mathrm{dist}(x_0, x)^2}$ *is equal to 0 when x tends to x_0.*

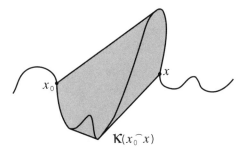

Fig. 7.3. *The convex hull of the arc $x_0 \frown x$ is flatter in the neighborhood of x_0 when the curve is rectifiable at this point.*

Here is all we can say about the relations between these four properties at the point x_0:

$$\mathcal{P}1 \iff \mathcal{P}2 \qquad \text{and} \qquad \mathcal{P}3 \implies \mathcal{P}4 .$$

In general, no other relation is true: $\mathcal{P}1$ implies neither $\mathcal{P}3$ nor $\mathcal{P}4$, $\mathcal{P}4$ implies neither $\mathcal{P}1$ nor $\mathcal{P}3$, and $\mathcal{P}3$ does not imply $\mathcal{P}1$.

We shall first prove the first two results, and then we will give three counterexamples to show the nonimplications. Finally, in §4 and §5 we will prove the following global result.

THEOREM   *If the curve is of finite length, then the properties $\mathcal{P}1$, $\mathcal{P}2$, $\mathcal{P}3$, and $\mathcal{P}4$ are true* **almost everywhere** *on this curve.*

This is an important theorem, and it is the archetype of global results that can be proved in the theory of geometric measure. These are always true *except on a set of null measure.* In our case the *measure* is nothing but the *length.*

7.2 Relations between local properties

▶   We first show that $\mathcal{P}1$ and $\mathcal{P}2$ are equivalent. We fix a point x_0, and we let $\theta_\epsilon(x_0) = \theta_\epsilon$ to shorten the notation.

(i)   Assume that $\mathcal{P}1$ is satisfied: at x_0 there exists a tangent to Γ. That is, for every positive angle ϕ, there exists an ϵ such that:

$$\text{dist}(x_0, x) \leq \epsilon \implies \angle(T(x_0, x), T(x_0)) \leq \phi .$$

Every chord $T(x_0, x)$ is then included in a cone with vertex x_0 and angle 2ϕ, since the angle θ_ϵ is less than 2ϕ. Since ϕ can be as small as we wish, we have shown that $\mathcal{P}2$ holds.

(ii)   Conversely, assume that $\mathcal{P}2$ holds at x_0. Let D_ϵ be the axis of the cone $\mathcal{C}_\epsilon$ with vertex x_0 and angle θ_ϵ. If ϵ' is less than ϵ, then $\mathcal{C}_{\epsilon'}$ is included in $\mathcal{C}_\epsilon$. Therefore,

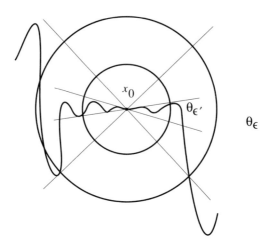

Fig. 7.4. *The cone $\mathcal{C}_\epsilon$, whose minimal angle θ_ϵ contains all the points of Γ whose distance to x_0 is $\leq \epsilon$, it also contains the cone $\mathcal{C}_{\epsilon'}$ for all $\epsilon' < \epsilon$.*

$$\epsilon' < \epsilon \Longrightarrow \angle(D_\epsilon, D_{\epsilon'}) \leq \theta_\epsilon \ .$$

This proves that when ϵ tends to 0 the line D_ϵ tends to a fixed line D_0. But

$$\text{dist}(x_0, x) \leq \epsilon \Longrightarrow \angle(T(x_0, x), D_\epsilon) \leq \theta_\epsilon \ .$$

So when ϵ tends to 0, the chord $T(x_0, x)$ tends to the line D_0, which is therefore a tangent to the curve at x_0. ◄

▶   We show that $\mathcal{P}3$ implies $\mathcal{P}4$. From the following result on convex hulls (justified in Appendix C §7):

$$A(\mathcal{K}(x_0 \frown x)) \leq L(x_0 \frown x)^{3/2} \sqrt{L(x_0 \frown x) - \text{dist}(x_0, x)} \ ,$$

and we immediately deduce that:

$$\frac{A(\mathcal{K}(x_0 \frown x))}{\text{dist}(x_0, x)^2} \leq \left(\frac{L(x_0 \frown x)}{\text{dist}(x_0, x)} \right)^{3/2} \sqrt{\frac{L(x_0 \frown x)}{\text{dist}(x_0, x)} - 1} \ .$$

Since the ratio $L(x_0 \frown x)/\text{dist}(x_0, x)$ tends to 1, the two sides of this inequality tend to 0. ◄

7.3 Counterexamples

The following counterexamples are all of the same type: Γ is the graph of a function $z(t)$ constructed from the cosine function whose amplitude and frequency are to be varied. The point x_0 will be the origin O of the axis. The curve is naturally parameterized by its abscissa t, taken to be between 0 and 1. By the same token, O is an endpoint of Γ. Therefore this is a special point from the topological point of view. But to make O an ordinary point of Γ, it suffices to complete the curve on the left side, i.e., on the negative abscissa.

- $\mathcal{P}1$ **implies neither** $\mathcal{P}3$ **nor** $\mathcal{P}4$ Let $z(t)$ be the function defined as:

$$\begin{cases} z(0) = 0; \\ z(t) = t^\alpha + t^\beta \left(1 + \cos \frac{1}{t}\right) & \text{if } 0 < t \le 1, \end{cases}$$

where α, β are two constants, $0 < \beta < \alpha < 1$.

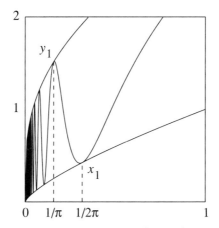

Fig. 7.5. *The graph of the function* $z(t) = t^{3/4} + t^{1/4}\left(1 + \cos \frac{1}{t}\right)$. *This curve is enclosed between the graphs of* $t^{3/4}$ *and* $t^{3/4} + 2\,t^{1/4}$.

Since $\mathcal{P}3$ implies $\mathcal{P}4$, it is enough to show that the graph Γ of z satisfies $\mathcal{P}1$ at 0 but not $\mathcal{P}4$. Then $\mathcal{P}3$ will not hold at O.

▶ Since the graph Γ of z is enclosed between the graph of t^α and the graph of $t^\alpha + 2 t^\beta$, whose slope is infinity at the origin, we deduce that Γ has a tangent at O, which is the axis Oz. Thus $\mathcal{P}1$ holds at O.

Let x_k, y_k be the points of Γ whose abscissas are $1/(2k-1)\pi$ and $1/2k\pi$, respectively; the convex hull of the arc $O\frown x_k$ of the curve contains the triangle Ox_ky_k whose area equals

$$\frac{1}{2}\left| \frac{z(1/2k\pi)}{(2k-1)\pi} - \frac{z(1/(2k-1)\pi)}{2k\pi} \right|.$$

When k goes to infinity, x_k tends to 0, and this area is equivalent to $k^{-1-\beta}$. On the other hand, $\text{dist}(O, x_k)$ is equivalent to $k^{-\alpha}$. We deduce that the ratio:

$$\frac{\mathcal{A}(\mathcal{K}(O^\frown x_k))}{\text{dist}(O, x_k)^2}$$

has the same order of growth as $k^{2\alpha-1-\beta}$, which tends to infinity when

$$2\alpha > 1 + \beta .$$

For this, take $\alpha = 3/4, \beta = 1/4$. In this case the curve does not satisfy the property $\mathcal{P}4$ at O. ◀

• $\mathcal{P}4$ **does not imply** $\mathcal{P}3$ We consider the function z such that $z(0) = 0$, and for all $t > 0$:

$$z(t) = t^2 \cos \frac{1}{t^2} .$$

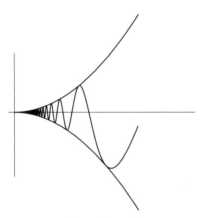

Fig. 7.6. *Graph of the function $z(t) = t^2 \cos \frac{1}{t^2}$.*

▶   We shall show that the graph Γ satisfies $\mathcal{P}4$, but not $\mathcal{P}3$ at the point 0. For every point x of the curve with coordinates $(t, z(t))$, the area of the convex hull of the arc $O^\frown x$ is less than the area t^3 of the triangle whose vertices are O, (t, t^2), $(t, -t^2)$. Since $\text{dist}(O, x) \geq t$, we deduce that:

$$\frac{\mathcal{A}(\mathcal{K}(O^\frown x))}{\text{dist}(O, x)^2} \leq t .$$

Thus, $\mathcal{P}4$ holds at 0.

On the other hand, the length of the part of Γ that corresponds to the abscissas between the values $1/\sqrt{2k\pi}$ and $1/\sqrt{2(k+1)\pi}$ is larger than $z(1/\sqrt{2k\pi})$ $= 1/2k\pi$. The sum of these lengths, for k larger than a given integer n, diverges, as does the harmonic series. Therefore every arc $O^\frown x$ of Γ is of infinite length. Thus Γ does not verify the property $\mathcal{P}3$ at the point O. ◀

• $\mathcal{P}3$ **does not imply** $\mathcal{P}1$ Since $\mathcal{P}3$ implies $\mathcal{P}4$, it is sufficient to show that $\mathcal{P}4$ does not imply $\mathcal{P}1$. Here is our final counterexample.

We consider the function z with $z(0) = 0$, and for all $0 < t \leq 1$:

$$z(t) = t \cos \theta(t) \, ,$$

where the function θ satisfies the following properties:

- $\lim_{t \to 0} \theta(t) = +\infty$.
- θ is differentiable on $]0, 1]$, its derivative is θ'.
- $\lim_{t \to 0} t \, \theta'(t) = 0$.

For example, we can take the function

$$\theta(t) = \log \log \frac{2}{t}$$

whose derivative is

$$\theta' = \frac{1}{t \log(t/2)} \, .$$

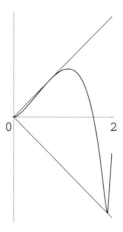

Fig. 7.7. *Graph of the function $z(t) = t \, \cos(\log \log(2/t))$.*

Since θ tends to infinity when t tends to 0, there exists a sequence of values of t that tends to 0 and for which $z(t) = t$. Similarly, there exists another sequence for which $z(t) = -t$. This suffices to show that there is no tangent to the curve at O. We are left with the task of showing property $\mathcal{P}3$; i.e., if x is the point $(t, z(t))$ of the curve Γ, then the ratio

$$\frac{L(O^\frown x)}{\mathrm{dist}(O, x)}$$

tends to 1 when t tends to 0.

▶   We evaluate $L(O^\frown x)$ by using the integral:

$$f(t) = \int_0^t \sqrt{1 + z'(s)^2} \, ds \, ,$$

while $\operatorname{dist}(O, x)$ equals

$$g(t) = t\sqrt{1 + \cos^2 \theta(t)} \ .$$

These two functions are differentiable at all $t > 0$, and their derivatives equal

$$f'(t) = \sqrt{1 + z'(t)^2}$$
$$g'(t) = \sqrt{1 + \cos^2 \theta} t \, \theta' \frac{\cos \theta \sin \theta}{\sqrt{1 + \cos^2 \theta}} \ .$$

Since $t \, \theta'$ tends to 0 with t, these derivatives could be written as:

$$f'(t) = \sqrt{1 + \cos^2 \theta} + \epsilon_1(t)$$
$$g'(t) = \sqrt{1 + \cos^2 \theta} + \epsilon_2(t) \ ,$$

where the functions $\epsilon_1(t)$ and $\epsilon_2(t)$ tend to 0. Since $f(0) = g(0) = 0$, we can apply l'Hôpital's rule and write that:

$$\lim_{t \to 0} \frac{f(t)}{g(t)} = \lim_{t \to 0} \frac{f'(t)}{g'(t)} = \lim_{t \to 0} \frac{\sqrt{1 + \cos^2 \theta} + \epsilon_1(t)}{\sqrt{1 + \cos^2 \theta} + \epsilon_2(t)} = 1 \ . \quad \blacktriangleleft$$

7.4 Tangent almost everywhere

We recall the property $\mathcal{P}2$: at each point x of Γ and for each $\epsilon > 0$ we construct a local cone with vertex x and minimal angle $\theta_\epsilon = \theta_\epsilon(x)$. The value of this angle decreases as ϵ tends to 0, and therefore it has limit $\theta(x)$. There is a tangent to Γ at x if and only if $\theta(x)$ is null. We shall prove the following result.

THEOREM   *If the curve Γ is of finite length, then the property $\mathcal{P}1$ (or $\mathcal{P}2$) is true almost everywhere on Γ.*

Almost everywhere means for all x except a subset of Γ of null length. A Vitali-type argument is necessary for the proof.

▶   A Vitali cover (see Appendix B) of Γ by subarcs is a family $\mathcal{F}$ of arcs such that every point of Γ belongs to a sequence of arcs of $\mathcal{F}$ whose length tends to 0. This notion can be extended to subsets of Γ.

Let E be a subset of Γ, and let $\mathcal{R}$ be a Vitali cover of closed subarcs of E. For all $\epsilon > 0$, we can extract from $\mathcal{R}$ a finite family of disjoint arcs J_1, J_2, ..., J_n, such that:

$$L(E) - \sum_{i=1}^{n} L(J_i) \leq \epsilon \ .$$

Our aim is to show that the set of x for which $\theta(x) \neq 0$ is of length 0. Let $E(\phi)$ be the set of x for which $\theta(x) \geq 2\phi$ for any angle $\phi < \pi/2$. It suffices to show that:

$$L(E(\phi)) = 0.$$

Assume that this set is not empty. Let A and B be the endpoints of Γ, and let D be the straight line passing through A and B.

We first show the following inequality:

$$L(E(\phi)) \leq \frac{L(\Gamma) - \text{dist}(A, B)}{1 - \cos\phi}.$$

Let $\epsilon > 0$. If x is a point of $E(\phi)$, we can always find two points y and z of Γ, such that the lengths of the arcs $x^\frown y$ and $x^\frown z$ are less than ϵ, and

$$\angle(xy, xz) \geq 2\phi.$$

At least one of the segments xy or xz, say xy, form an angle greater than ϕ with D. Let $J_\epsilon(x)$ be the arc $x^\frown y$ of Γ. The orthogonal projection of this arc on D is a segment of length $L(J_\epsilon(x)) \cos\phi$.

Since every $J_\epsilon(x)$ is of length smaller than or equal to ϵ and each contains x, the family of all arcs $J_\epsilon(x)$, for all x of $E(\phi)$ and all non-null values of ϵ constitutes a Vitali cover of $E(\phi)$ on Γ. Thus for all ϵ, we can extract a finite family $J_1, J_2, \ldots, J_n$, such that

$$L(E(\phi)) \leq \sum_{i=1}^{n} L(J_i) + \epsilon.$$

These arcs do not cover all of Γ. Since they are of finite number, the complement of their union is formed by arcs. We denote them by $J_{n+1}, \ldots, J_k$, where $k = 2n - 1$, $2n$ or $2n + 1$. The projections of all these arcs on D cover the segment AB. We deduce that:

$$\text{dist}(A, B) \leq \sum_{i=1}^{k} L(\text{ projection of } J_i)$$

$$\leq \cos\phi \sum_{i=1}^{n} L(J_i) + \sum_{i=n+1}^{k} L(J_i)$$

$$= \cos\phi \sum_{i=1}^{n} L(J_i) + L(\Gamma) \sum_{i=1}^{n} L(J_i)$$

and consequently:

$$L(E(\phi)) \leq \sum_{i=1}^{n} L(J_i) + \epsilon \leq \frac{L(\Gamma) - \text{dist}(A, B)}{1 - \cos\phi} + \epsilon.$$

This is true for all ϵ, and when ϵ tends to 0 we obtain the sought inequality.

Now we consider a polygonal curve $\mathbf{P}$ with endpoints A and B whose vertices belong to Γ. Each of its segments $\mathbf{S}$ is the cord of an arc $\mathbf{C}$ of Γ. Applying this argument to $\mathbf{C}$ we obtain that:

$$L(E(\phi) \cap \mathbf{C}) \le \frac{L(\mathbf{C}) - L(\mathbf{S})}{1 - \cos \phi} \ .$$

Summing up these inequalities over all $\mathbf{C}$, we finally find

$$L(E(\phi)) \le \frac{L(\Gamma) - L(\mathbf{P})}{1 - \cos \phi} \ .$$

But we can always take $\mathbf{P}$ to be as near to Γ as we wish, so that the right-hand side will be as small as we please. This proves that $L(E(\phi)) = 0$. ◀

7.5 Local length, almost everywhere

What is left to show is that every curve Γ of finite length satisfies the property $\mathcal{P}3$ almost everywhere; i.e., the equality:

$$\lim_{x \to x_0} \frac{L(x_0 \frown x)}{\operatorname{dist}(x_0, x)} = 1$$

holds almost everywhere. When $\mathcal{P}3$ is satisfied, the length of any local arc $x_0 \frown x$ is equivalent to the distance between its endpoints. Since this property implies $\mathcal{P}4$ at each point, it follows that $\mathcal{P}4$ is true almost everywhere. This completes the proof of the theorem stated at the beginning of this chapter.

▶   As with the previous proof, this one makes use of Vitali covers. We call $r(x_0)$ the lim sup (eventually infinite) of the ratio $L(x_0 \frown x)/\operatorname{dist}(x_0, x)$, when x tends to x_0. Since the length of the curve is greater than the distance between its endpoints, this limit is at least 1. We shall prove that the set of points of Γ for which this limit is greater than 1 is of length 0. For all $h > 0$, let E_h be the set of all the points x_0 of Γ for which $r(x_0) > 1 + h$. It suffices to show that:

$$L(E_h) = 0 \ ,$$

provided the set is not empty.

Let $\epsilon > 0$. For every point x_0 of E_h we can find a point x of Γ such that:

$$(1 + h) \operatorname{dist}(x_0, x) \le L(x_0 \frown x) \le \epsilon \ .$$

Let $J_\epsilon(x_0)$ be the arc $x_0 \frown x$. The family formed by all the arcs $J_\epsilon(x_0)$, for all x_0 in E_h and for all $\epsilon > 0$, is a Vitali cover of E_h. Therefore, we can extract a finite subfamily of arcs $J_1, J_2, \ldots, J_n$, such that:

$$L(E_h) \le \sum_{i=1}^{n} L(J_i) + \epsilon \ .$$

The complement of these arcs in Γ consists of a union of arcs $J_{n+1}, \ldots, J_k$, provided they are divided into small arcs whose lengths can be considered to be less than ϵ. Let $\mathbf{S}_i$ be the segment with the same endpoints as J_i, and let $\mathbf{P}$ be the polygonal curve formed by the segments $\mathbf{S}_i$, for $i = 1, \ldots, k$. For all i, we have $L(\mathbf{S}_i) \le \epsilon$. Moreover,

$$L(\mathbf{P}) \le \sum_{i=1}^{n} L(\mathbf{S}_i) + \sum_{i=n+1}^{k} L(J_i)$$

$$\le \frac{1}{1+h} \sum_{i=1}^{n} L(J_i) + L(\Gamma) - \sum_{i=1}^{n} L(J_i)$$

$$\le L(\Gamma) - \frac{h}{1+h} \sum_{i=1}^{n} L(J_i) \, .$$

Thus, we can deduce an estimate of the length of E_h:

$$L(E_h) \le \sum_{i=1}^{n} L(J_i) + \epsilon \le \frac{1+h}{h} \left(L(\Gamma) - L(\mathbf{P}) \right) + \epsilon \, .$$

When ϵ tends to 0, the length of $\mathbf{P}$ whose segment lengths are less than ϵ tends to $L(\Gamma)$ (as we have seen in Chap. 5). Thus we deduce the desired equality: $L(E_h) = 0$. ◄

7.6 Rectifiability revisited

If we consult a dictionary, we find that a **rectifiable** curve is a curve of **finite length**. We have just given a summary of the geometrical properties of such a curve, provided it is simple, i.e., a Jordan arc. It would be better if the notion of rectifiable curve was not restricted to such a narrow definition.

For example, at any of its points except the origin, the spiral $\theta = 2\pi/\rho$ has a tangent. Nonetheless the global length of this curve is infinite. It is *locally rectifiable everywhere* except at O.

Also, we can consider more general curves. Those that have a finite number of *double points* can be divided into a finite union of simple curves. Then we say that the curve is *rectifiable* if each of its constituents is of finite length. We may even say that any finite or countable union of simple curves is *rectifiable* if the sum of the lengths of these curves is finite.

What we sometimes mean by a *rectifiable* set in the theory of geometric measure is a set that is almost entirely included in a finite or countable union of simple curves whose total lengths is finite. Besicovitch called them the *Y–sets*, no doubt in reference to the geometrical structure of Fig. 7.8. "Almost entirely" means that the part of the set that is not included in this union is of null measure. The difficulty here is due to the fact that this part is not always included in a simple curve. What is, in this case, the meaning of "length"? This motivates the generalization of the notion of *length* to that of *measure in dimension 1*, used on sets of any topological type. This explains the theoretical interest of the measure invented by Caratheodory and later developed by Hausdorff.

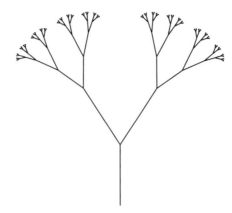

Fig. 7.8. *This tree Y is formed by a countable set of segments; the total sum of their lengths is finite. If Y is to be considered a closed set, we must add to it the "leaves," the limit points of the sequences of the segments; this is a totally disconnected set of null measure in the sense of the Hausdorff 1–measure. Thus the set Y is of finite 1–measure and therefore is rectifiable.*

Finally, we do not only generalize the support, but the measure itself. Thus, rectifiability becomes a notion relative to a given measure. If this is an *area* of a surface, we speak of 2–rectifiable surfaces, etc. Quite often the word *rectifiable* is replaced by *regular*.

7.7 Bibliographical notes

The theorem on the existence of tangency almost everywhere is due to Lebesgue [H. Lebesgue 1] (1903). Lebesgue's proof consists of finding the characteristic properties of functions with bounded variation. In this chapter, we have given a more geometrical proof based on local cones due to [A.S. Besicovitch].

Rectifiability constitutes a chapter of the Geometric Theory of Measure. A.S. Besicovitch had started to write a book on the "regular" and "irregular" sets, which, due to his sudden death, was left unfinished. The articles of A.S. Besicovitch, which are full of interesting ideas, are unfortunately presented in a hermetic style. The classification of Besicovitch papers was undertaken by R.O. Davies. The main ideas of Besicovitch's book have since been put in a condensed form in the book of [K. Falconer].

Equally, we should state the rival American school and its chief [H. Federer]. For a long time, his book was considered to be the infallible and complete Bible of the subject. This may have been far beyond his intentions. However, we note the dangerous influence on sciences of a book whose authority may discourage the reader from wandering out of its trite paths.

In this theory, which takes the points of view of both the measure and the geometry (and tries to conceal them) some results have a surprisingly complicated proof. Constructing them is, without a doubt, an incontestable honor to human intelligence. In the interest of the progress of science, are they worth the energy put into them? The technique finally overtakes the search for new ideas.

8 Length, by Intersections with Straight Lines

8.1 Intersections, projections

The method of computing lengths that we shall discuss in this chapter is essentially of a different nature than the previous ones. As was proved by Steinhaus (1930), this method is of double interest: theoretical and practical. We consider the intersection of the curve with *random* straight lines. This intersection, if it is not empty, generally contains a finite number of points. Counting these points will give us an estimate of the length of this curve. We shall see that it is possible to formulate this method in terms of *projections*. This idea gave rise to the definition of *integral–geometric measure*, whose development constitutes an important branch of the geometric theory of measure.

We have spoken of random straight lines; to understand the employed methodology, let us recall one of the most ancient problems of integral geometry.

Buffon needle On a plane, we draw a set of equidistant parallel lines. We call ϵ the distance between any two successive lines. On this plane, we "randomly" throw a "needle," whose length l is less than ϵ. We know how to calculate the probability that the needle will touch a line of the set:

$$p = \frac{2l}{\pi\epsilon}.$$

An estimate of the frequency of hitting a line is obtained after a large number of manual experiments. This formula allowed Buffon to find an approximate value of the number π. We look at this formula from a different perspective; assuming we know the value of π, we should be able to deduce the length l. We shall prove this formula, but first let us extend this problem. Instead of a segment that is "thrown" on the lines, a "random" line shall be "thrown" on a segment, or for that matter on any curve. We should first define a measure on the set of lines so that it will make sense to speak of *random lines*.

8.2 The measure of families of straight lines

Duality   Choose an axis in the plane, with O as its origin. Let **D** be a line that does not pass through the origin. Let H be the orthogonal projection of O on **D**. The polar coordinates of H are (ρ, θ). To make things easier we take the angle θ to be between 0 and π. Therefore ρ could be any real number; ρ is considered positive if H is situated either above the axis or on the right side of O, and it is negative elsewhere. This point H uniquely determines the line D, and conversely. Thus we can define a **duality** between lines and points. To stress this relation the above line is often denoted by $\mathbf{D}(\rho, \theta)$.

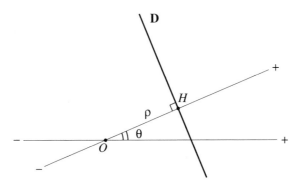

Fig. 8.1. *The polar coordinates (ρ, θ) of the point H uniquely determine the position of the line* **D***. In a Cartesian coordinate system of the same origin, the equation of this line is: $x \cos \theta + y \sin \theta - \rho = 0$. Here $0 \leq \theta < \pi$, and ρ can be any real number.*

Probability   There are many ways to introduce the notion of a random line. Assume that this line meets the unit disk $B_1(O)$. We may decide that the two polar coordinates ρ and θ of the point H are independent random variables with uniform distribution, the first is on $[-1, 1]$, the other on $[0, \pi]$. The density function is then constant. For any set E included in $B_1(O)$, the probability of H being in E is equal to $(1/2\pi) \int_E d\rho\, d\theta$. But for a general study, we prefer not to use the basic hypothesis that the lines will intersect the unit disk $B_1(0)$ or any fixed set. In this case it is much better to speak in terms of *measure* than *probability*.

Measure   Inspired by the above, here is how we define a measure μ on the families of straight lines:

> For every pair ρ_0, θ_0, we attribute the measure: $d\rho\, d\theta$ to the family of straight lines $\mathbf{D}(\rho, \theta)$ such that: $\rho_0 \leq \rho \leq \rho_0 + d\rho$, $\theta_0 \leq \theta \leq \theta_0 + d\theta$.

This measure, which will be called μ, corresponds to the *product measure* of the Borel measure on the real line to which the values of ρ belong, and the Borel measure on the interval $[0, \pi]$ to which the values of θ belong. We use the same notation μ for measures of lines or of the subsets of their dual points.

Measure $\mu(E)$ If E is any set of points, then its μ–measure is

$$\mu(E) = \int_E d\rho \, d\theta \ .$$

This measure is different from the *area*, i.e., *the Borel measure* on a plane. The latter is given by the integral:

$$A(E) = \int_E \rho \, d\rho \, d\theta \ .$$

For example, the shaded surfaces of Fig. 8.2 are all of the same μ–measure but of different areas.

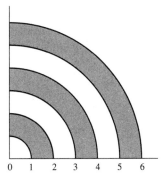

0   1   2   3   4   5   6

Fig. 8.2. *The shaded surfaces are all of the same measure $\mu(E) = \int_E d\rho \, d\theta = \pi/2$ but of different areas $\int_E \rho \, d\rho \, d\theta$.*

Measure $\mu(\mathcal{D})$ The value of μ at any family of lines is the same as its value at its dual set:

Let $\mathcal{D}$ be a set of straight lines. Its measure $\mu(\mathcal{D})$ is equal to the integral

$$\int_E d\rho \, d\theta \ ,$$

where E denotes the set of the orthogonal projections of O on the lines of $\mathcal{D}$.

The most important property of μ is the following: it is independent of the choice of axis and origin. Therefore $\mu(\mathcal{D})$ is invariant under translation, rotation, or symmetry on $\mathcal{D}$. This can be verified by a simple change of variables.

◇   In a Cartesian coordinate system, the area is given by the integral of the area element: $dx_1 \, dx_2$. The measure μ must then be written (in this system) as:

$$\mu(E) = \int_E \frac{1}{\sqrt{x_1^2 + x_2^2}} \, dx_1 \, dx_2 \ .$$

◇   We retain the following

- *When a translation, a rotation, or a symmetry is applied to a set of lines, its μ–measure does not change.*
- *Also, the μ–measure of the dual set E will not change, but its geometric structure may change completely.*

We can observe this change in the following example.

Example Let $\mathbf{C}$ be the circle with center O and radius r. Take $\mathcal{D}$ to be the set of lines that intersect $\mathbf{C}$. A line belongs to $\mathcal{D}$ if and only if the orthogonal projection H of O on this line belongs to the disk $B_r(O) = E$. The measure of E is $\pi \int_{-r}^{r} d\rho = 2\pi r$, which is the value of the perimeter of $\mathbf{C}$. Now let $\mathbf{C}'$ be a circle with radius r and center different from O. Denote the set of lines intersecting $\mathbf{C}'$ by $\mathcal{D}'$. Since $\mathcal{D}'$ is obtained by a simple translation of $\mathcal{D}$, it has the same measure $2\pi r$, as does its dual set E', which we represent in Fig. 8.3. It is not a disk.

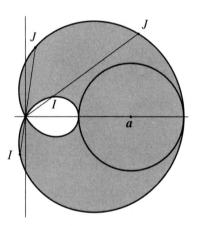

Fig. 8.3. *Representation of the dual set of the family of lines intersecting the circle of radius r and center $(a, 0)$, $a > r$. The intersection of this set with any line directed from the origin is a segment IJ of length $2r$. Its boundary is a cardioid.*

▶   We can directly verify that the measure of E' is $2\pi r$; we know that for every θ, the intersection of E' with a line passing through O and making an angle θ with the axis is a segment IJ of length $2r$. In fact, if we assume that the center of $\mathbf{C}'$ has $\rho = a$ and $\theta = 0$ as polar coordinates, then I runs through a curve that is parameterized by $\rho = a \cos \theta - r$, and J runs through the curve $\rho = a \cos \theta + r$, where $0 \le \theta \le \pi$. ◀

◇   Yet there exists a relation between the measure μ and the area; they are both null on the same sets.

   Let E be a bounded set:

$$\mu(E) = 0 \Longleftrightarrow \mathcal{A}(E) = 0 \, .$$

▶   If E is bounded, then it is included in a ball $B_K(O)$. Let $0 < \epsilon < K$. Every point with polar coordinates (ρ, θ) in $E - B_\epsilon(O)$ satisfies:

$$\epsilon \leq \rho \leq K .$$

Therefore

$$\int_E d\rho\, d\theta - 2\,\pi\,\epsilon \leq \int_{E-B_\epsilon(O)} d\rho\, d\theta \leq \frac{1}{\epsilon}\int_E \rho\, d\rho\, d\theta \leq \frac{K}{\epsilon}\int_E d\rho\, d\theta .$$

Since ϵ is arbitrarily small, the two integrals $\int_E d\rho\, d\theta$ and $\int_E \rho\, d\rho\, \theta$ will be null on the same sets E.   ◀

8.3 Family of lines intersecting a set

We are particularly interested in families of straight lines that intersect a given set of points. We have already seen the example of straight lines that intersect a circle. Let $\mathbf{T}$ be a set of points, $\mathcal{D}(\mathbf{T})$ be the family of all lines intersecting $\mathbf{T}$, and $E_{\mathbf{T}}$ be the dual set. We shall give an easy method to compute the measure $\mu(\mathcal{D}(\mathbf{T}))$.

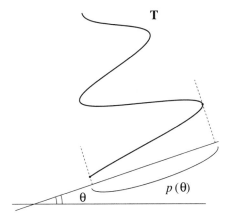

Fig. 8.4. *Projection of a set* $\mathbf{T}$ *(here, a curve) on a line of angle* θ.

Let $\mathbf{D}$ be a straight line passing through the origin. A point H of $E_{\mathbf{T}}$ belongs to $\mathbf{D}$ if and only if there exists a line perpendicular to $\mathbf{D}$ passing through H that meets $\mathbf{T}$, i.e., if and only if H is the orthogonal projection on $\mathbf{D}$ of a point of $\mathbf{T}$. Thus, the intersection $E_{\mathbf{T}} \cap \mathbf{D}$ is nothing but the orthogonal projection of $\mathbf{T}$ on $\mathbf{D}$, and it depends on only one variable: the angle θ made by the line $\mathbf{D}$ and the reference axis. We denote by $p(\theta)$ the measure (length) of this projection:

$$p(\theta) = \int_{E_{\mathbf{T}} \cap \mathbf{D}} d\rho .$$

Therefore, we obtain the measure of $E_{\mathbf{T}}$ by integrating $p(\theta)$ with respect to θ. Thus we have the following result:

The measure of the family $\mathcal{D}(\mathbf{T})$ of the lines intersecting a given set $\mathbf{T}$ is

$$\mu(\mathcal{D}(\mathbf{T})) = \int_0^\pi p(\theta)\, d\theta \, ,$$

where $p(\theta)$ is the measure of the orthogonal projection of $\mathbf{T}$ on the line of angle θ.

Example If $\mathbf{T}$ is a circle or a disk of radius r, its projection on any line is of length $2r$, and $\mu(\mathcal{D}(\mathbf{T}))$ is then $2\pi r$.

Example If $\mathbf{T}$ is a segment of length l that is contained in an axis, its projection on a line making an angle θ with the axis has a length $l|\cos\theta|$. Therefore $\mu(\mathcal{D}(\mathbf{T}))$ is $l \int_0^\pi |\cos\theta|\, d\theta = 2l$. This result should not depend on the position of $\mathbf{T}$. Figure 8.5 illustrates the case where the support line of the segment $\mathbf{T} = AB$ does not pass through 0. The dual set $E_{\mathbf{T}}$ is then equal to

$$E_{\mathbf{T}} = (\mathbf{B}_1 \cup \mathbf{B}_2) - (\mathbf{B}_1 \cap \mathbf{B}_2)$$

where $\mathbf{B}_1$ and $\mathbf{B}_2$ are the disks of diameters OA and OB.

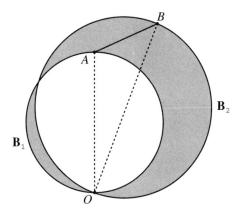

Fig. 8.5. *The dual set of the family of lines intersecting a segment AB. Its boundary is formed by two circles. Its μ–measure is $2\operatorname{dist}(A,B)$.*

8.4 The case of convex sets

More details on convex sets can be found in Appendix C. We simply recall the following property:

The intersection of a straight line **D** *with a convex set* K *is a segment.*

This implies a very interesting result on the measure of a family $\mathcal{D}(K)$ of straight lines intersecting K; it is precisely equal to the perimeter, or the length of the boundary of K.

$$\mu(\mathcal{D}(K)) = L(\partial K) .$$

$\Diamond$ This formula was effectively verified in the case of a disk of radius r, where $\mu(\mathcal{D}(K)) = 2\pi r$, and in the case of a segment of length l, considered as a limit case of convex set whose perimeter is $2\,l = \mu(\mathcal{D}(K))$.

▶   a) We first prove that the formula holds in the case where the boundary of $\partial K = \mathbf{P}$ is a polygonal curve formed by N segments $\mathbf{S}_1$, ..., $\mathbf{S}_N$. The family of lines whose intersection with $\mathbf{P}$ contains a unique point is μ–negligible; in fact, these lines are "support lines" of K, and their points of contact are among the N vertices of $\mathbf{P}$. The measure of the family of lines passing through a given point is null. The measure of the family of lines passing through one of the N vertices is also null.

On the other hand, because of the convexity there is no line that intersects $\mathbf{P}$ of a finite number of points ≥ 3. Finally, the set of lines intersecting $\mathbf{P}$ at an infinite number of points is reduced to the N lines supporting the segments of $\mathbf{P}$. It is μ–negligible.

It follows that $\mathcal{D}(K)$ has the same measure as $\mathcal{D}_2$, the family of lines intersecting $\mathbf{P}$ at exactly two points. Such a line meets two different segments of $\mathbf{P}$. Therefore, if we call $\mathcal{D}(\mathbf{S}_i)$ the family of lines intersecting $\mathbf{S}_i$, we obtain

$$\sum_{i=1}^{N} \mu(\mathcal{D}(\mathbf{S}_i)) = 2\,\mu(\mathcal{D}(K)) .$$

But we have already calculated $\mu(\mathcal{D}(\mathbf{S}_i))$ to be $2\,L(\mathbf{S}_i)$. Thus we conclude

$$\sum_{i=1}^{N} L(\mathbf{S}_i) = \mu(\mathcal{D}(K)) .$$

It follows that if $p(\theta)$ is the length of the orthogonal projection of $\mathbf{P}$ (or K) on a line of angle θ,

$$L(\mathbf{P}) = \int_0^{\pi} p(\theta)\,d\theta .$$

b) Now we consider any convex set K. By taking vertices on ∂K, we can construct a sequence $\mathbf{P}_n$ of convex polygonal curves such that:

$$\text{dist}(\mathbf{P}_n, \partial K) \to 0 \text{ and } L(\mathbf{P}_n) \to L(\partial K) .$$

Take a line of angle θ. Let $p_n(\theta)$ be the length of the orthogonal projection of $\mathbf{P}_n$, and let $p(\theta)$ be the orthogonal projection of ∂K on this line. Then $p_n(\theta) \to p(\theta)$. Since the functions $p_n(\theta)$ are bounded (the upper bound is the diameter of K), using Lebesgue's theorem we can deduce that

$$\int_0^\pi p_n(\theta)\, d\theta \to \int_0^\pi p(\theta)\, d\theta .$$

But the left-hand-side term equals $L(\mathbf{P}_n)$, which tends $L(\partial K)$. Thus we obtain

$$L(\partial K) = \int_0^\pi p(\theta)\, d\theta ,$$

which is the value of $\mu(\mathcal{D}(K))$. ◀

◇ From this result, we can deduce the following, which holds for all curves Γ with convex hull $\mathcal{K}(\Gamma)$:

$$\boxed{\mu(\mathcal{D}(\Gamma)) = L(\partial \mathcal{K}(\Gamma)) .}$$

The measure of the family of lines intersecting Γ is equal to the perimeter of its convex hull.

▶   In fact, a line intersects Γ if and only if it intersects $\mathcal{K}(\Gamma)$. And the measure of $\mathcal{D}(\mathcal{K}(\Gamma))$ is equal to the perimeter. ◀

◇ Here is another corollary that will be used in the next section:

Consider a curve Γ with endpoints A and B. Let $\mathcal{H}$ be the family of lines that intersect Γ without meeting the segment AB. We always have:

$$\mu(\mathcal{H}) \leq L(\Gamma) - \text{dist}(A, B) .$$

▶   If $\mathcal{D}(\Gamma)$ denotes the family of lines that intersect Γ, then $\mu(\mathcal{D}(\Gamma)) = \mu(\mathcal{D}(\mathcal{K}(\Gamma)))$ is equal to the perimeter of $\mathcal{K}(\Gamma)$, because this is a convex set. By a result of Appendix C (Chap. 3, §6), this perimeter is at most equal to $\text{dist}(A, B) + L(\Gamma)$. Thus

$$\mu(\mathcal{D}(\Gamma)) \leq \text{dist}(A, B) + L(\Gamma) .$$

On the other hand, $\mathcal{D}(\Gamma)$ contains the family $\mathcal{D}(AB)$ of straight lines intersecting the segment AB. This family has a measure $2\,\text{dist}(A, B)$. Since $\mathcal{D}(\Gamma)$ is the union of two disjoint families $\mathcal{D}(AB)$ and $\mathcal{H}$, we obtain:

$$\mu(\mathcal{H}) + 2\,\text{dist}(A, B) \leq \text{dist}(A, B) + L(\Gamma) ,$$

which proves the desired inequality. ◀

8.5 Length by secant lines

Now we can compute the length of any simple curve.

THEOREM   *Let Γ be a curve. We divide the family $\mathcal{D}(\Gamma)$ of lines intersecting Γ into a union $\mathcal{D}_1 \cup \mathcal{D}_2 \cup \ldots \cup \mathcal{D}_k \cup \ldots$, where $\mathcal{D}_k$ is the family of lines intersecting Γ on exactly k points. Then*

$$L(\Gamma) = \frac{1}{2} \sum_{k=0}^{\infty} k\,\mu(\mathcal{D}_k) \ .$$

$\diamond$ We remark that this result is a generalization of the case of convex curves. In fact if Γ is the boundary of a convex set, then the family of lines intersecting this curve in one or at most three points is μ–negligible. Thus, we obtain $\mu(\mathcal{D}(\Gamma)) = \mu(\mathcal{D}_2)$. By the results in §4, the length is

$$L(\Gamma) = \mu(\mathcal{D}_2) \ ,$$

which is a special case of the formula.

▶   a) As in the convex case, we start by proving the formula for polygonal curves. Let $\mathbf{P}$ be a polygonal curve formed by N line segments. A line cannot meet $\mathbf{P}$ at more than N points, unless it is the support of one of its segments, a negligible case with respect to the measure μ. Therefore $\mu(\mathcal{D}_k) = 0$ whenever k is larger than N.

Let $\mathbf{S}_i$, $i = 1, \ldots, N$, be the segments of $\mathbf{P}$, and let $\mathcal{E}_i = \mathcal{D}(\mathbf{S}_i)$ be the family of lines that intersects $\mathbf{S}_i$. We know that $L(\mathbf{S}_i) = \mu(\mathcal{E}_i)/2$. Therefore

$$L(\mathbf{P}) = \frac{1}{2} \sum_{i=1}^{N} \mu(\mathcal{E}_i) \ .$$

In this sum, a line that cuts $\mathbf{P}$ at only one point is counted once. A line that cuts it at two points is counted twice because these points belong to an $\mathbf{S}_i$ and an $\mathbf{S}_j$ so that it is counted in $\mu(\mathcal{E}_i)$ and in $\mu(\mathcal{E}_j)$. Similarly, a line that intersects $\mathbf{P}$ at k points is counted k times. We conclude that

$$\sum_{i=1}^{N} \mu(\mathcal{E}_i) = \sum_{k=1}^{N} k\,\mu(\mathcal{D}_k) \ .$$

A more algebraically–directed proof can be given, using *characteristic functions*. If E is a given set, then the characteristic function of E, denoted by 1_E, is defined as follows:

$$1_E(x) = \begin{cases} 1, & \text{if } x \text{ belongs to } E \\ 0, & \text{otherwise.} \end{cases}$$

For any measure μ, we can write $\mu(E)$ as an integral of this function:

$$\mu(E) = \int 1_E \, d\mu \; .$$

Applying this notion to the sets of lines and using the additivity of integrals, we get that:

$$\sum_{i=1}^{N} \mu(\mathcal{E}_i) = \int \sum_{i=1}^{N} 1_{\mathcal{E}_i} \, d\mu \; .$$

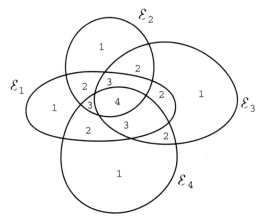

Fig. 8.6. *Values taken by the characteristic function $\sum_{i=1}^{4} 1_{\mathcal{E}_i}$.*

But $\sum_{i=1}^{N} 1_{\mathcal{E}_i}$ is the function that associates to each line $\mathbf{D}$ of the plane; the value 0 if $\mathbf{D}$ does not intersect $\mathbf{P}$; the value 1 if $\mathbf{D}$ intersects $\mathbf{P}$ at one point, i.e., if $\mathbf{D}$ belongs to $\mathcal{D}_1$; and the value k if $\mathbf{D}$ intersects $\mathbf{P}$ at k points, i.e., if $\mathbf{D}$ belongs to $\mathcal{D}_k$. Its integral becomes

$$\sum_{i=1}^{N} \mu(\mathcal{E}_i) = \sum_{k=1}^{N} k \, \mu(\mathcal{D}_k) \; .$$

Therefore the theorem holds for polygonal curves.

b) When Γ is any given curve, it is the limit of a sequence of polygonal approximations $\mathbf{P}_n$. We will show that the sought result holds for τ by using a limit argument.

We are going to use the following notation:

- $\mathcal{D}_k$ *is the set of lines that intersect Γ at exactly k points;*
- $\mathcal{F}_k$ *is the set of lines that intersect Γ at a finite number at least equal to k points;*
- $\mathcal{D}_{n,k}$ *is the set of lines that intersect $\mathbf{P}_n$ at exactly k points;*
- $\mathcal{F}_{n,k}$ *is the set of lines that intersect $\mathbf{P}_n$ at a finite number at least equal to k points.*

b1) Since $\mathcal{F}_k = \mathcal{D}_k \cup \mathcal{D}_{k+1} \cup \cdots$ is a union of disjoint sets, we have:

$$\sum k\,\mu(\mathcal{D}_k) = \sum \mu(\mathcal{F}_k)\,,$$

and similarly

$$\sum k\,\mu(\mathcal{D}_{n,k}) = \sum \mu(\mathcal{F}_{n,k})\,.$$

When n tends to infinity, $2L(\mathbf{P}_n)$ tends to $2L(\Gamma)$. On the other hand, since the vertices of $\mathbf{P}_n$ belong to Γ, every straight line intersecting $\mathbf{P}_n$ on k points intersects Γ on k or more points. Therefore $\mathcal{F}_{n,k} \subset \mathcal{F}_k$, and

$$\mu(\mathcal{F}_{n,k}) \le \mu(\mathcal{F}_k)\,.$$

Since the sum over k on the left-hand side is equal to $2L(\mathbf{P}_n)$, we deduce the following inequality:

$$2L(\mathbf{P}_n) \le \sum \mu(\mathcal{F}_k)\,.$$

Making n go to infinity:

$$2L(\Gamma) \le \sum \mu(\mathcal{F}_k)\,.$$

b2) For an inequality in the other direction, assume that $L(\Gamma)$ is finite. We use a result from the previous section:

> If AB is a segment of $\mathbf{P}_n$ that is the chord of the arc $A^\frown B$ of Γ, the measure of the set of lines intersecting the arc $A^\frown B$ without intersecting AB is less than $L(A^\frown B) - \mathrm{dist}(A,B)$.

Adding all the segments of $\mathbf{P}_n$:

> The measure of the set $\mathcal{H}_n$ of lines intersecting at least one arc of Γ but not its chord (which is a segment of $\mathbf{P}_n$) is less than $L(\Gamma) - L(\mathbf{P}_n)$.

Let ϵ_n be the maximal length of the segments of $\mathbf{P}_n$. We can assume without any loss of generality that ϵ_n tends to 0. Consider any line $\mathbf{D}$ that intersects Γ at exactly k points. To each point x of the intersection there corresponds a segment $\mathbf{S}_n(x)$ of $\mathbf{P}_n$ that is the chord of the arc of Γ containing x (if x is a vertex of $\mathbf{P}_n$, we should choose one of the two possible segments). Once $2\epsilon_n$ is less than the smallest distance between any two points of the intersection, the segments $\mathbf{S}_n(x)$ are distinct. Therefore, for all n larger than an integer $N(\mathbf{D})$ (depending on $\mathbf{D}$), the above holds.

For all integers p, we call $\mathcal{G}_p$ the family of lines intersecting Γ for which $N(\mathbf{D}) \le p$. Then

$$\mathcal{G}_p \subset \mathcal{G}_{p+1}\,,$$

and the union of all these families $\mathcal{G}_p$ is exactly the set of lines intersecting Γ at a finite number of points.

We fix a sufficiently large integer p so that $\mathcal{G}_p$ is not empty. Let n be larger than p. Let $\mathbf{D}$ be a line of $\mathcal{G}_p$ that intersects Γ at k points. If x is one of these points, then x is the unique point that belongs to the intersection with the arc $\Gamma_n(x)$. We deduce that if $\mathbf{D}$ does not belong to $\mathcal{H}_n$, i.e., if for every x this line meets the segment $\mathbf{S}_n(x)$, then it intersects $\mathbf{P}_n$ at exactly k points. We can write that

$$n \ge p \implies \mathcal{D}_k \cap \mathcal{G}_p \subset \mathcal{D}_{n,k} \cup \mathcal{H}_n$$

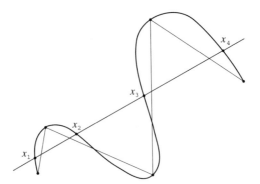

Fig. 8.7. *If the polygonal curve* $\mathbf{P}_n$ *with vertices on* Γ *is formed with small segments, then the number of points of the intersection* $\mathbf{D} \cap \Gamma$ *is equal to that of the intersection* $\mathbf{D} \cap \mathbf{P}_n$.

for all k; and we deduce

$$\mu(\mathcal{D}_k \cap \mathcal{G}_p) \leq \mu(\mathcal{D}_{n,k}) + L(\Gamma) - L(\mathbf{P}_n) \ .$$

Now we fix an integer N. It follows that

$$\sum_{k=1}^{N} k \, \mu(\mathcal{D}_k \cap \mathcal{G}_p) \leq \sum_{k=1}^{N} k \, \mu(\mathcal{D}_{n,k}) + \frac{N(N+1)}{2} \left(L(\Gamma) - L(\mathbf{P}_n) \right) \ ,$$

where $\sum_{k=1}^{N} k \, \mu(\mathcal{D}_{n,k})$ is less than $2L(\mathbf{P}_n)$, and hence it is less than $2L(\Gamma)$. Finally we have

$$\sum_{k=1}^{N} k \, \mu(\mathcal{D}_k \cap \mathcal{G}_p) \leq 2L(\Gamma) + \frac{N(N+1)}{2} \left(L(\Gamma) - L(\mathbf{P}_n) \right) \ .$$

When n tends to infinity, the term on the right tends to $2L(\Gamma)$. When p tends to infinity, the sequence of sets $\mathcal{D}_k \cap \mathcal{G}_p$, which are of finite measure, increases to $\mathcal{D}_k$, which is also of finite measure. The sequence of measures $\mu(\mathcal{D}_k \cap \mathcal{G}_p)$ converges to $\mu(\mathcal{D}_k)$. The left-hand-side term of the inequality tends to $\sum_{k=1}^{N} k \, \mu(\mathcal{D}_k)$, which is less than $2L(\Gamma)$. Finally, when N goes to infinity we get the desired inequality:

$$\sum_{k=1}^{\infty} k \, \mu(\mathcal{D}_k) \leq 2L(\Gamma) \ ,$$

which ends the proof. ◀

◇   The theorem is also true in the case of closed curves, considered as a limit of simple curves.

8.6 The length by projections

We may present the same result in a different way by generalizing to any curve Γ the method in Section 4 that was used for boundaries of convex sets. Let $\mathbf{D}(\rho, \theta)$ be the line whose point H has polar coordinates ρ and θ. H is the orthogonal projection of O on this line. We shall define an integer-valued function $N_\Gamma(\rho, \theta)$ in the following manner:

$$N_\Gamma(\rho, \theta) = \text{number of points in the intersection } \Gamma \cap \mathbf{D}(\rho, \theta) \ ,$$

$N_\Gamma(\rho, \theta)$ *is the number of points of Γ that are orthogonally projected to the same point (ρ, θ) in the direction θ.*

We can call this number the **multiplicity** of the projection.

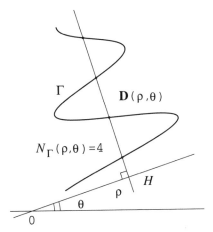

Fig. 8.8. *The line defined by the polar coordinates (ρ, θ) of the point H intersects Γ at four points.*

Consider the set of all points (ρ, θ) such that $N_\Gamma(\rho, \theta)$ is equal to a given integer k. This set is precisely the dual set of $\mathcal{D}_k$, that is, the set of all the orthogonal projections of O on the lines of $\mathcal{D}_k$. We call it E_k. The integral

$$\int_{E_k} d\rho \, d\theta$$

is equal to the measure $\mu(\mathcal{D}_k)$. If we integrate the function $N_\Gamma(\rho, \theta)$, we get

$$\int_{E_k} N_\Gamma(\rho, \theta) \, d\rho \, d\theta = k \, \mu(\mathcal{D}_k) \ .$$

Now we shall assume that Γ is of finite length, in which case the set of straight lines that intersect Γ at infinitely many points is of null measure. This implies that the set of points (ρ, θ) for which $N_\Gamma(\rho, \theta)$ is not null has a measure equal to the sum of the measures $\mu(\mathcal{D}_k)$. Computing this sum, the previous equality becomes:

$$\int_{-\infty}^{+\infty} \int_0^\pi N_\Gamma(\rho,\theta)\, d\rho\, d\theta = \sum_0^\infty k\, \mu(\mathcal{D}_k) \,,$$

and finally we have:

$$L(\Gamma) = \frac{1}{2} \int_{-\infty}^{+\infty} \int_0^\pi N_\Gamma(\rho,\theta)\, d\rho\, d\theta \,.$$

We interpret these integrals as follows:

- $\int_{-\infty}^{+\infty} N_\Gamma(\rho,\theta)\, d\rho$ is the measure of the projection of Γ on the line of angle θ, each point of the projection is counted with its multiplicity;
- $\frac{1}{\pi} \int_{-\infty}^{+\infty} \int_0^\pi N_\Gamma(\rho,\theta)\, d\rho\, d\theta$ is the average (mean) of this projection with respect to θ.

Therefore, we obtain the length by multiplying this mean by $\pi/2$.

The length of the curve is proportional to the mean value over θ of the measure of the projection of Γ on a line with direction θ. This measure is computed by counting each point of the projection with its multiplicity.

Particular case In the case where Γ is a convex set, the function N_Γ takes the value 2 for almost all the lines intersecting Γ. If $p(\theta)$ is the length of the projection of Γ on the line of direction θ, we find that:

$$\int_{-\infty}^{+\infty} N_\Gamma(\rho,\theta)\, d\rho = 2\, p(\theta) \,.$$

Therefore with §4:

$$L(\Gamma) = \int_0^\pi p(\theta)\, d\theta \,.$$

8.7 Application: practical computation of length

We take a transparent piece of paper on which we draw a set of parallel equidistant lines. Let ϵ be the distance between lines. We place this paper on top of the curve Γ, whose length is to be calculated, in such a manner that the direction of these lines forms an angle $\theta + \frac{\pi}{2}$ with a reference axis. We count the number $M(\epsilon,\theta)$ of points of intersection of Γ with these parallel lines. Then the quantity $\epsilon\, M(\epsilon,\theta)$ is an approximation of the measure of the projection of Γ where each point is counted with its multiplicity so that

$$\epsilon\, M(\epsilon,\theta) \simeq \int_{-\infty}^{+\infty} N_\Gamma(\rho,\theta)\, d\rho \,.$$

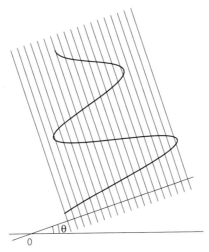

Fig. 8.9. *When we vary the angle θ, the average of the total number of points in the intersection of the set of lines with the curve gives an estimation of its length.*

Now we choose an integer K and a corresponding sequence of angles defined by $\theta = 0$, $\frac{\pi}{K}$, $2\frac{\pi}{K}$, $\cdots$, $(K-1)\frac{\pi}{K}$. The quantity

$$\frac{\epsilon}{K} \sum_{i=0}^{K-1} M(\epsilon, i\frac{\pi}{K})$$

constitutes a mean of the previous one. Following the results of Section 6, we can say that the number

$$\frac{\pi\epsilon}{2K} \sum_{i=0}^{K-1} M(\epsilon, i\frac{\pi}{K})$$

is an approximation of the length of Γ. The larger K is and the smaller ϵ is, the better the approximation is. Steinhaus was the first to apply this method successfully to geographical lines and curves seen under a microscope supplied with a micrometer. However, we can get better performance by exactly calculating the integral

$$\int_{-\infty}^{+\infty} N_\Gamma(\rho, \theta)\, d\rho \ .$$

For this, it suffices to consider, for each θ, a partition of the curve as a union of subarcs Γ_i, such that each perpendicular line to the direction θ intersects Γ_i at at most one point (Fig. 8.10). This integral is then the sum of the lengths of the projections of these subarcs. We call it $L(\theta)$. If $\hat{L}$ is the new approximation of the length of the curve, that is,

$$\hat{L} = \frac{\pi}{2K} \sum_{i=0}^{K-1} L(\theta) \ ,$$

then it is possible to estimate the error resulting from using $L(\Gamma)$ in place of $\hat{L}$. When K is an even integer, this error is given by the following inequality

$$\frac{\pi}{K}\cos(\frac{\pi}{K})\left(\sin(\frac{\pi}{K})\right)^{-1} \le \frac{\hat{L}}{L(\Gamma)} \le \frac{\pi}{K}\left(\sin(\frac{\pi}{K})\right)^{-1} .$$

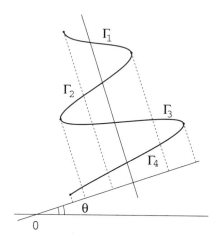

Fig. 8.10. *The curve Γ is divided into four arcs, such that every line perpendicular to the direction θ intersects each arc at one point at most.*

8.8 The length by random intersections

From this measure of families of lines in the plane, we can go directly to probabilities by taking a bounded field of "experiments." If $\mathcal{E}$ and $\mathcal{F}$ are two families of lines such that $\mu(\mathcal{F})$ is finite and not null, we can compute the probability of a line of $\mathcal{F}$ to be in $\mathcal{E}$. This is:

$$\frac{\mu(\mathcal{E}\cap\mathcal{F})}{\mu(\mathcal{F})} .$$

To apply this to the computation of lengths, we assume that the curve Γ is inside a circle of radius 1. The family $\mathcal{F}$ of all lines intersecting this circle has a measure equal to 2π. If $\mathcal{D}_k$ is the family of lines intersecting Γ at exactly k points, we obtain that: *the probability that a random line whose distance to the origin is less than 1 intersects Γ at exactly k points is equal to*

$$p_k = \frac{\mu(\mathcal{D}_k)}{2\pi} .$$

The results in Section 5 give us the following:

$$L(\Gamma) = \pi\sum k\,p_k .$$

We interpret this as:

> *The length of a curve is proportional to the mathematical mean of the number of points of its intersection with a random line.*

In general, the constant ratio is equal to $\frac{1}{2}\mu(\mathcal{F})$; therefore it varies according to the chosen domain. For example, if $\mathcal{F}$ is the set of the straight lines that intersect a square of side a that includes the curve Γ, then this square is a convex set whose perimeter is $4a$ and equal to the measure of $\mathcal{F}$. In this case, the constant is $2a$.

◇ This result on the mathematical mean can be found without using the theorem of the previous section.

• Consider a segment $\mathbf{S}$ of length l included in a circle of radius 1. The measure of the family of lines that intersect $\mathbf{S}$ is $2l$. The measure of the family of lines that intersect the circle is 2π. The probability that a line that intersects the circle will meet $\mathbf{S}$ is $\frac{l}{\pi}$. Ignoring the negligible case in which the line is the support of $\mathbf{S}$ we have that D meets $\mathbf{S}$ at either 0 or 1 point. Then the number X of points of intersection is a Bernoulli variable whose mean is $\mathrm{E}(X) = \frac{l}{\pi}$. We obtain

$$L(\mathbf{S}) = \pi \, \mathrm{E}(X) \, .$$

• If $\mathbf{P}$ is a polygonal curve with N segments $\mathbf{S}_1, \ldots, \mathbf{S}_N$, the number X of points of intersection is the sum $X_1 + \cdots + X_N$ of N Bernoulli variables. Thus

$$L(\mathbf{P}) = \sum_{i=1}^{N} L(\mathbf{S}_i) = \pi \sum_{i=1}^{N} \mathrm{E}(X_i) = \pi \, \mathrm{E}(X) \, .$$

• If Γ is any curve that is a limit of a sequence $\mathbf{P}_n$ of polygonal approximations, we must show that the number X of points of the intersection of a random line with Γ is the limit (in the probability sense) of the number $X^{(n)}$ of the points of intersection with $\mathbf{P}_n$. We then find

$$L(\Gamma) = \pi \, \mathrm{E}(X) \, .$$

8.9 Buffon needle

We reconsider a needle or segment $\mathbf{S}$ of length l that is thrown randomly on a set of parallel equidistant lines separated by a distance $\epsilon \geq l$. The event: $\mathbf{S}$ *meets one of the lines* is the same as the following: *given a segment $\mathbf{S}$ included in a circle of radius $\epsilon/2$, a random line that intersects the circle will intersect $\mathbf{S}$.*

In fact, in the first case the observer is placed in a fixed position with respect to the lines, and the segment $\mathbf{S}$ has a random position. In the second case the observer is on a fixed position with respect to $\mathbf{S}$, while the parallel lines have a random position; one of them is necessarily intersecting the circle of radius $\epsilon/2$. In whatever way we put the problem, the probability of intersection is the same. Since the measure of the family of the lines intersecting $\mathbf{S}$ is $2l$, and the measure of the family intersecting the circle is $\pi\epsilon$, this probability is equal to

$$\frac{2l}{\pi \epsilon} \, .$$

This is the result stated at the beginning of this chapter.

8.10 Bibliographical notes

The domain of geometric probabilities is extremely instructive. One of the best references on this subject is the book of [L. Santaló]. For a brief history on the length, we reproduce the following paragraph from [H. Steinhaus 3]:

> "P.S. Laplace suggested as early as 1812 that probabilistic methods should be applied in measuring lengths. This challenge was answered in 1868 by M. Crofton who defined the measures of sets of straight lines in a plane. Crofton's discoveries met only in part with the appreciation they deserved; his methods were applied to questions belonging to geometrical probabilities; E. Ozuber devotes a book to them. R. Deltheil published in 1926 a monograph in Borel's collection; his "probabilités géométriques" are based on the work of Elie Cartan on the principle of duality and the point of view of differential geometry is maintained throughout the whole work—in this chain no author is t horoughly conscious of his debt towards his immediate predecessor. A great part of these investigations belongs to the integral calculus as is shown by the title of Crofton's first paper ' ... the methods used being also extended to the proof of certain new theorems in the integral calculus.' W. Blashke and his collaborators have derived many new results from Crofton's basic idea. In Blaschke's "integral geometrie" we find the names of H. Lebesgue and J. Favard (1932) as the first to propose the definition of the length of an arc on Crofton's principle—a short pamphlet to the same effect published by the author of this note in 1930 [H. Steinhaus 1] seems to have escaped general notice."

Applied to dual sets, the probabilty associated to the families of straight lines, as was defined by Crofton, becomes the *Deltheil measure*.

This paragraph of Steinhaus needs to be completed. In fact, we have seen, using the same notation in this chapter (§6), that both of the formulas

$$\frac{1}{2} \sum_{1}^{\infty} k \, \mu(\mathcal{D}_k)$$

$$\frac{1}{2} \int_{-\infty}^{+\infty} \int_{0}^{\pi} N_\Gamma(\rho, \theta) \, d\rho \, d\theta$$

equivalently define the length of a curve. To go from one to the other is simple. In 1912 [H. Lebesgue 1] noted it and cited [A. Cauchy], who in fact discovered the second formula:

> "In a lithographed memoir, in 1832, I have given the following proposition:

THEOREM 1. —*Let p be the polar angle made by any line OO' drawn at random in a given plane; let S be a system of one or more lengths measured on one or more straight lines or curves, closed or not; A the sum of the absolute projections of the different elements of S on OO'; and π the ratio of the circumference to t he diameter, we shall have:*

$$S = \frac{1}{4} \int_{-\pi}^{\pi} A\,dp \ \ldots .\text{''}$$

This proposition is fairly clear, once we know that Cauchy uses the same symbol to denote the set and its measure, that he uses "the sum of the projections" to mean "the measure of the projection, where each projected point is counted with its multiplicity," and that he integrates between $-\pi$ and π instead of between 0 and π so that one must divide by 4 instead of 2. We note that Cauchy has forseen an application of this result to sets other than simple arcs. Applied to any set the above formula is called, by [L. Santaló], "the Favard length," and by [S. Sherman] " the Steinhaus length." Later, this will be "the integral–geometric measure," which was studied in [S. Sherman] and [H.Federer] in comparison to the Hausdorff measure.

The interest in Steinhaus's works stems from their wide range of applications and the computation of the relative error $\hat{L}/L$ [H. Steinhaus 2]. Other methods to compute the error, using unbaised estimators, can be found in [P.A.P. Moran].

It is rare to find a complete proof relating the length to the above formula; another one that uses analytical arguments may be found in [S. Sherman].

9 The Length by the Area of Centered Balls

9.1 Minkowski sausage

In Chapter 1, we considered the "Minkowski sausages" of a set P on the real line. In the plane, its geometric form is more apparent. The "balls" $B_\epsilon(x)$ are disks of radius ϵ that are centered at x. For every curve Γ, the union of all these balls centered at Γ:

$$\Gamma(\epsilon) = \bigcup_{x \in \Gamma} B_\epsilon(x)$$

is the ϵ–*Minkowski sausage* of Γ. This is also the set of all the points of the plane whose distance to Γ is $\leq \epsilon$. Relative to its diameter, the larger the area of this set is, the more chaotic the curve.

$\diamondsuit$ The **diameter** of a set E is the largest distance between any two of its points:

$$\text{diam}(E) = \sup_{x,y \in E} \text{dist}(x, y) \ .$$

The diameter of E is always equal to the diameter of its convex hull $\mathcal{K}(E)$ (Appendix C). For a curve Γ, the diameter is the length of the longest chord of Γ. If Γ is a segment of length l, then $\mathcal{A}(\Gamma(\epsilon))$ is $2l\epsilon + \pi\epsilon^2$. This is the smallest area of a Minkowski sausage of a curve with diameter l. This can be proved as follows.

▶ Let Γ be any curve. Let A and B be two points of Γ such that $\text{dist}(A, B) = \text{diam}(\Gamma)$. Every line perpendicular to AB that intersects Γ intersects $\Gamma(\epsilon)$, and the intersection contains a segment whose length is at least equal to 2ϵ. We deduce that the area of the part of $\Gamma(\epsilon)$ that can be projected orthogonally on AB is at least equal to $2\epsilon \, \text{diam}(\Gamma)$. On the other hand, there is at least a half-disk at each endpoint, centered at A and B and disjoint from the previous part (Fig. 9.1). We conclude that:

$$\mathcal{A}(\Gamma(\epsilon)) \geq 2\epsilon \, \text{diam}(\Gamma) + \pi\epsilon^2 \ . \quad ◀$$

For a segment of length l, the ratio

$$\frac{\mathcal{A}(\Gamma(\epsilon))}{2\epsilon}$$

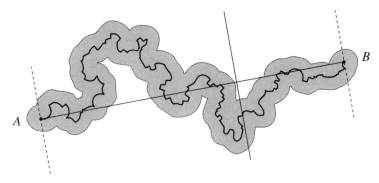

Fig. 9.1. *The area of the ϵ-sausage of Minkowski of a curve Γ passing through the points A and B is larger than $2\,\epsilon\,\mathrm{dist}(A,B) + \pi\,\epsilon^2$.*

tends to l when ϵ tends to 0. It is interesting to observe that this result can be generalized to curves.

THEOREM   *Let Γ be a simple curve of finite length. Then this length is given by*

$$L(\Gamma) = \lim_{\epsilon \to 0} \frac{\mathcal{A}\left(\Gamma(\epsilon)\right)}{2\,\epsilon} \ .$$

The next section is devoted to the proof of this result.

9.2 Length by the area of sausages

To prove the above formula, our main argument consists of using the local properties discussed in Chapter 7, which hold *almost everywhere* on any curve of finite length. We approximately calculate the value of the area $\mathcal{A}(\Gamma(\epsilon))$ by covering $\Gamma(\epsilon)$ with rectangles. We take advantage of the fact that they are aligned in a neighborhood of the points where the curve has a tangent.

Fix a real $a > 1$ and an angle θ, $0 < \theta \le \pi/4$. We use the notation of Chapter 7. Then almost everywhere on Γ, $\lim_{r \to 0} \theta_r(x) = 0$ and $\lim_{y \to x} L(x \frown y)/\mathrm{dist}(x,y) = 1$. Thus if E_r is the set

$$E_r = \Big\{ x \in \Gamma \ : (i)\ \theta_{2r}(x) \le \theta$$

$$(ii)\ \mathrm{dist}(x,y) \le r \Rightarrow L(x \frown y) \le a\,\mathrm{dist}(x,y) \Big\} ,$$

then the length $L(E_r)$ tends to $L(\Gamma)$ when r tends to 0.

▶   a) We first show that

$$L(\Gamma) \le \liminf_{\epsilon \to 0} \frac{\mathcal{A}\left(\Gamma(\epsilon)\right)}{2\,\epsilon} \ .$$

We fix a value of r and divide the curve Γ into subarcs of length r (if the division is not exact, a negligible correction at the endpoints could be performed). Let $\Gamma_1, \Gamma_2, \ldots, \Gamma_N$ be these subarcs whose intersections with E_r are not empty. Let A_i, B_i be the endpoints of Γ_i, and $\mathbf{S}_i = A_i B_i$ be the corresponding chord of Γ. Let x_i be a point of $\Gamma_i \cap E_r$. We have

$$L(E_r) \leq N\, r \leq L(\Gamma)$$

and

$$r = L(\Gamma_i) \leq \frac{a}{\cos\theta} L(\mathbf{S}_i) .$$

The last equality can be easily verified by observing that by (i), the angle $\angle(A_i x_i, \mathbf{S}_i)$ is less than θ, and so is the angle $\angle(B_i x_i, \mathbf{S}_i)$ (Fig. 9.2). Therefore, $L(\mathbf{S}_i) \geq \cos\theta\,(\mathrm{dist}(A_i, x_i) + \mathrm{dist}(B_i, x_i))$. Now we use (ii) to get: $L(A_i \frown x_i) \leq a\,\mathrm{dist}(A_i, x_i)$, and $L(B_i \frown x_i) \leq a\,\mathrm{dist}(B_i, x_i)$, thus proving the above inequality.

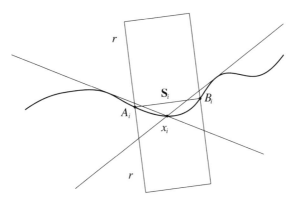

Fig. 9.2. *At the neighborhood of the point x_i that belongs to the arc $A_i \frown B_i$ of length r, we construct a rectangle C_i of length $2r$.*

We construct a rectangle on each side of $\mathbf{S}_i$ with basis $L(\mathbf{S}_i)$ and width r so that their union forms a unique rectangle C_i of dimensions $L(\mathbf{S}_i) \times 2\,r$. Condition (i) implies that the distance between each point of $\mathbf{S}_i$ and Γ_i is less than $\theta\, L(\mathbf{S}_i)$. Therefore it is less than $\theta\, r$. Thus the distance between any point of C_i and Γ is less than $r(1 + \theta)$. We deduce that for all i,

$$C_i \subset \Gamma(r(1 + \theta)) .$$

Finally, the fact that all the points of $\Gamma \cap B_{2r}(x_i)$ are included in a cone with vertex x_i and angle less than θ implies that any two rectangles C_i and C_k cannot intersect unless the segments $\mathbf{S}_i$ and $\mathbf{S}_k$ are consecutive cords of Γ. The fact that the angle $\angle(\mathbf{S}_i, \mathbf{S}_k)$ is less than 2θ implies that the common part of these two rectangles has an area less than $r^2 \sin\theta$ (Fig. 9.3).

To end, let $\epsilon = r(1 + \theta)$. The surface $\Gamma(\epsilon)$ contains N rectangles, whose areas of common parts were already evaluated. We deduce that

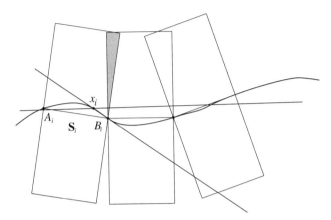

Fig. 9.3. *The curve is paved by the rectangles C_i whose common parts are negligible, provided Γ is rectifiable.*

$$\mathcal{A}(\Gamma(\epsilon)) \geq \sum_{i=1}^{N} \mathcal{A}(C_i) - N r^2 \sin \theta$$

$$= 2r \sum_{i=1}^{N} L(\mathbf{S}_i) - N r^2 \sin \theta$$

$$\geq \frac{2}{a} N r^2 \cos \theta - N r^2 \sin \theta .$$

The previous estimates of the quantity $N r$ allow us to write:

$$\frac{\mathcal{A}(\Gamma(\epsilon))}{2 \epsilon} \geq \frac{\cos \theta}{a(1 + \theta)} L(E_r) - \frac{\sin \theta}{2(1 + \theta)} L(\Gamma) .$$

When ϵ tends to 0, r will also tend to 0 and $L(E_r)$ tends to $L(\Gamma)$. Thus:

$$\liminf_{\epsilon \to 0} \frac{\mathcal{A}(\Gamma(\epsilon))}{2 \epsilon} \geq \frac{1}{1 + \theta} \left(\frac{1}{a} \cos \theta \frac{1}{2} \sin \theta \right) L(\Gamma) .$$

We get the desired result by making θ tend to 0 and a to 1.

b) For an inequality in the other direction, we find the largest integer N, for which there exists a polygonal $\mathbf{P}$ whose vertices are on Γ and whose N segments are of length r. To make the computation easier we shall assume that $\mathbf{P}$ and Γ have the same endpoints (otherwise, a negligible correction can be performed). Let $\mathbf{S}_1, \mathbf{S}_2, \ldots, \mathbf{S}_N$ be these segments, and let $\Gamma_1, \Gamma_2, \ldots, \Gamma_N$ be their corresponding arcs on Γ. The endpoints of the segment $\mathbf{S}_i$ (and of Γ_i) are x_i and x_{i+1}. We assume that these points are numbered in the same order as the parameterization. The distance between any point of Γ_i and $\mathbf{S}_i$ is less than r. The Γ_i are divided into two classes: the Γ_i^*, which contain a point of E_r, and the Γ_i^{**}, which do not. Let M be the number of the arcs of the first class. By (ii), their lengths are less than $a r$. Since they cover E_r, the number M is at least equal to $L(E_r)/a r$. We also note that by (i), we have for this type of arc that:

$$\text{dist}(\Gamma_i^*, \mathbf{S}_i^*) \leq r\,\theta\ .$$

We construct on both sides of the $\mathbf{S}_i$ a square of side r. These form a rectangle C_i of dimensions $r \times 2\,r$.

Let $\epsilon = r(1 - \theta)$, and let y be a point of $\Gamma(\epsilon)$. We consider two cases:

1.   There exists an arc Γ_i^{**} such that $\text{dist}(y, \Gamma_i^{**}) \leq \epsilon$. Then y belongs to the ball $B_{r+\epsilon}(x_i)$. There are $N - M$ such arcs.

2.   There is no such arc. Then the point y is at a distance less than ϵ from an arc Γ_i^*. Let z be the orthogonal projection of y on the support line of the segment $\mathbf{S}_i^*$. There are several possibilities:

• Either z belongs to $\mathbf{S}_i^*$; then y belongs to C_i.

• Or z is outside $\mathbf{S}_i^*$, where $i \neq 1$, $i \neq N$. We assume, for example, that z is on the same side as x_{i+1}. Hypothesis 2 implies that the arc Γ_{i+1} is necessarily of the first class: $\Gamma_{i+1} = \Gamma_{i+1}^*$. The angle $\angle(\mathbf{S}_i^*, \mathbf{S}_{i+1}^*)$ is of measure $\alpha \leq 2\theta$. The point y belongs either to C_{i+1} or to the circular sector with radius r and angle α situated between the rectangles C_i and C_{i+1} (see Fig. 9.4). The area $\mathcal{A}_i$ of this sector is less than that of the intersection $C_i \cap C_{i+1}$. We deduce that:

$$\mathcal{A}(C_i \cup C_{i+1}) + \mathcal{A}_i \leq \mathcal{A}(C_i) + \mathcal{A}(C_{i+1})\ .$$

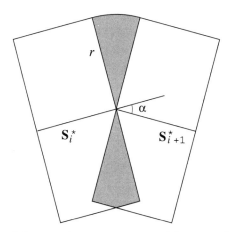

Fig. 9.4. *The area of the circular sector situated between the upper two squares is less than the area of the intersection of the lower ones.*

• We are left with only one case: where $i = 1$ or $i = N$. But in this case the distance between y and one of the endpoints of Γ is less than $r + \epsilon$.

Putting all these cases together, we finally deduce that:

$$\mathcal{A}(\Gamma(\epsilon)) \leq \sum_{\mathbf{S}_i^*} \mathcal{A}(C_i) + \pi(N - M + 2)(r + \epsilon)^2$$

$$= 2\,M\,r^2 + \pi(N - M + 2)(r + \epsilon)^2\ .$$

Since $L(E_r)/a \leq M\,r \leq N\,r \leq L(\Gamma)$, we find that

$$\frac{\mathcal{A}(\Gamma(\epsilon))}{2\,\epsilon} \leq \frac{1}{1-\theta} L(\Gamma) + \pi \frac{(2-\theta)^2}{1-\theta} \left(L(\Gamma) - \frac{1}{a} L(E_r) + 2\,r \right) .$$

If ϵ tends to 0, then so does r, and we get:

$$\limsup \frac{\mathcal{A}(\Gamma(\epsilon))}{2\,\epsilon} \leq \left(\frac{1}{1-\theta} + \pi \frac{(2-\theta)^2}{1-\theta} \left(1 - \frac{1}{a} \right) \right) L(\Gamma) .$$

Since θ is as small as we wish and a is as near to 1 as we wish, we obtain:

$$\limsup_{\epsilon \to 0} \frac{\mathcal{A}(\Gamma(\epsilon))}{2\,\epsilon} \leq L(\Gamma) ,$$

which ends the proof. ◀

◇ We can even prove that if the ratio $\mathcal{A}(\Gamma(\epsilon))/2\,\epsilon$ has no limit when ϵ tends to 0, then it tends to infinity. In this case we say that Γ has an infinite length (Chap. 10).

9.3 Convergence of the algorithm of the sausages

Sometimes it takes a long time to compute the area of the sausage $\Gamma(\epsilon)$. The simplest method is to obtain an approximation by constructing this sausage on a screen and then to count the number of lightened pixels. In principle, the formula

$$L(\Gamma) = \lim_{\epsilon \to 0} \frac{\mathcal{A}\left(\Gamma(\epsilon)\right)}{2\,\epsilon}$$

will serve us to compute $L(\Gamma)$ as a limit of a sequence. However, it is necessary to take some numerical precautions. The convergence of the sequence in question could be painfully slow. Also we often observe a concavity in the sequence of values $\mathcal{A}(\Gamma(\epsilon))$ that hinders the evaluation of the limit.

◇ When Γ is a segment:

$$\mathcal{A}(\Gamma(\epsilon)) = 2\epsilon\, L(\Gamma) + \pi\epsilon^2 .$$

The error committed when we replace $L(\Gamma)$ by the ratio $\mathcal{A}\left(\Gamma(\epsilon)\right)/2\,\epsilon$ is linear (proportional to ϵ).

◇ When Γ is a polygonal curve, the same will hold. If, for example, it is formed by two segments whose obtuse angle is $\pi - 2\theta$, as in Fig. 9.5, we obtain

$$\mathcal{A}(\Gamma(\epsilon)) = 2\epsilon\, L(\Gamma) + (\pi + 2\theta - \tan\theta)\epsilon^2 .$$

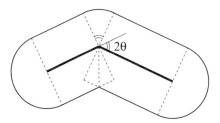

Fig. 9.5. *Computing the area of $\Gamma(\epsilon)$, when Γ is formed by two segments.*

For sufficiently small ϵ and a large number of segments, the error made on the evaluation of $L(\Gamma)$ remains linear.

We can systematically correct this error by observing that the limit of the ratio $\mathcal{A}(\Gamma(\epsilon))/2\,\epsilon$ is equal to the slope of the function $\mathcal{A}(\Gamma(\epsilon))$ at 0. Thus, we can take two successive values ϵ_1 and ϵ_2; construct the parabola passing through the points of coordinates $(0,0)$, $(2\epsilon_1, \mathcal{A}(\Gamma(\epsilon_1)))$; and $(2\epsilon_2, \mathcal{A}(\Gamma(\epsilon_2)))$, and estimate $L(\Gamma)$ by the slope at the origin of this parabola. This leads to the estimate

$$L(\Gamma) \simeq \frac{1}{\epsilon_1 - \epsilon_2} \left[\epsilon_1 \frac{\mathcal{A}(\Gamma(\epsilon_2))}{2\,\epsilon_2} - \epsilon_2 \frac{\mathcal{A}(\Gamma(\epsilon_1))}{2\,\epsilon_1} \right] .$$

This formula gives the exact result for all curves Γ satisfying

$$\mathcal{A}(\Gamma(\epsilon)) = 2\epsilon\, L(\Gamma) + C\,\epsilon^2 ,$$

in particular for all polygonal curves, or for all closed convex curves.

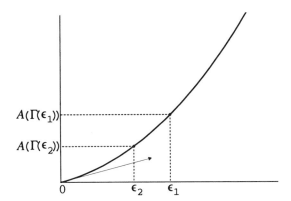

Fig. 9.6. *The length of Γ is the slope of the function $\mathcal{A}(\Gamma(\epsilon))$ at the origin.*

9.4 Reduction of balls to parallel segments

What is the direct relation between the method of this chapter (the Minkowski sausage method) and the Steinhaus method seen in the previous chapter (the intersection by lines)?

We recall (Chap. 8, §6) that the length can be written as:

$$L(\Gamma) = \frac{1}{2} \int_{-\infty}^{+\infty} \int_{0}^{\pi} N_\Gamma(\rho, \theta) \, d\rho \, d\theta \ ,$$

where $N_\Gamma(\rho, \theta)$ is the number of points that belong to both the curve Γ and the line $\mathbf{D}(\rho, \theta)$ (Fig. 8.8).

In this formula, we do not take the positions of these $N_\Gamma(\rho, \theta)$ points on $\mathbf{D}(\rho, \theta)$ into account. It seems reasonable to confuse any two points that are *very near* each other, and to distinguish the *isolated* points in a sense to be determined. To establish a threshold between *near* and *isolated* we may proceed as follows: we fix an $\epsilon > 0$, and say that a point is isolated if the distance between this point and any other point is larger than ϵ.

New limit formula for the length Thus we consider the union of all segments of $\mathbf{D}(\rho, \theta)$ of length ϵ that are centered at one of the $N_\Gamma(\rho, \theta)$ points of the intersection. This is a unidimensional sausage (Chap. 2, §3) constructed on the line $\mathbf{D}(\rho, \theta)$ around $\Gamma \cap \mathbf{D}(\rho, \theta)$. We call $L_\Gamma(\rho, \theta, \epsilon)$ its length. In this measure of length it is not the points that count but their neighborhoods. If the distance between any two of them is larger than 2ϵ, then $L_\Gamma(\rho, \theta, \epsilon)$ reaches its maximum, that is, $2\epsilon \, N_\Gamma(\rho, \theta)$. The function $L_\Gamma(\rho, \theta, \epsilon)/2\epsilon$ increases when ϵ tends to 0, and

$$\lim_{\epsilon \to 0} \frac{L_\Gamma(\rho, \theta, \epsilon)}{2\epsilon} = N_\Gamma(\rho, \theta) \ .$$

By exchanging limit and integral, we obtain

$$L(\Gamma) = \lim_{\epsilon \to 0} \frac{1}{4\epsilon} \int_{0}^{\pi} \int_{-\infty}^{+\infty} L_\Gamma(\rho, \theta, \epsilon) \, d\rho \, d\theta \ .$$

Geometrical interpretation If the variables θ and ϵ are fixed, the integral $\int_{-\infty}^{+\infty} L_\Gamma(\rho, \theta, \epsilon) \, d\rho$ is nothing but the area of the set $\Gamma(\theta, \epsilon)$ formed by the union of all the segments of length 2ϵ centered on Γ and making an angle $\theta + \pi/2$ with the reference axis. This is the area scanned by Γ when it is displaced by a length 2ϵ in the direction $\theta + \pi/2$.

The set $\Gamma(\theta, \epsilon)$ is always included in the Minkowski sausage $\Gamma(\epsilon)$ made by centered balls. Moreover,

$$\bigcup_{0 \le \theta \le \pi} \Gamma(\theta, \epsilon) = \Gamma(\epsilon) \ .$$

In fact, by varying the orientation of the segments we cover all the balls. The average of the areas $\mathcal{A}(\Gamma(\theta, \epsilon))$ over θ is

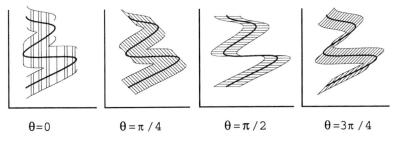

Fig. 9.7. *Construction of the ϵ–sausages by flat covering elements, with different orientations.*

$$\frac{1}{\pi} \int_0^\pi \int_{-\infty}^{+\infty} L_\Gamma(\rho, \theta, \epsilon)\, d\rho\, d\theta \ .$$

The formula of the length by the Minkowski sausage

$$L(\Gamma) = \lim_{\epsilon \to 0} \frac{\mathcal{A}(\Gamma(\epsilon))}{2\,\epsilon}$$

allows us to estimate this mean by:

$$\frac{2}{\pi} \mathcal{A}(\Gamma(\epsilon)) \ .$$

To obtain an approximation of the area $\mathcal{A}(\Gamma(\epsilon))$, we multiply $\pi/2$ by the mean over θ of the areas of sausages $\Gamma(\theta, \epsilon)$ formed by segments of orientation $\theta + \pi/2$.

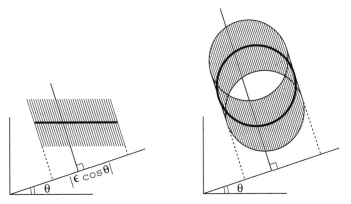

Fig. 9.8. *The surface $\Gamma(\theta, \epsilon)$ in the case of a segment of length l and in the case of a circle.*

Thus, with flat structural elements, we can pass from the method of intersections to the method of the Minkowski sausage.

Case of a segment If Γ is a segment of length l, the area $\mathcal{A}(\Gamma(\theta, \epsilon))$ is $2\,l\,\epsilon|\cos\theta|$. Its mean over θ is equal to $4\,l\,\epsilon/\pi$. By multiplying by $\pi/2$, we obtain $2\,l\,\epsilon$, which is equivalent to $\mathcal{A}(\Gamma(\epsilon))$ when ϵ tends to 0.

Case of a circle The area scanned by a circle of diameter D when it is translated by a length 2ϵ in any given direction can be written as $4\,D\,\epsilon + g(\epsilon)$, where $g(\epsilon)$ is of order ϵ^2 and therefore negligible. By multiplying $4\,D\,\epsilon$ by $\pi/2$, we obtain $2\,\pi\,D\,\epsilon$, which is exactly $\mathcal{A}(\Gamma(\epsilon))$.

9.5 Bibliographical notes

We have already seen, in a more general context, the sausage $P(\epsilon)$ of G. Cantor (1884) (Chap. 2, §4). In what concerns curves, this notion appeared earlier. H. Lebesgue attributes it to C.W. Borchardt (1854). Lebesgue called the length of Γ computed by the limit of the ratio $\mathcal{A}(\Gamma(\epsilon))/2\,\epsilon$ the "Borchardt–Minkowski length." But it was impossible for us to find Borchardt's article. Minkowski's article was published in 1901, and contained no formal proofs. However, it contains a generalization to three-dimensional space. For example, the "sausage" becomes a volume, the curve's length is computed by $\mathcal{V}(S(\epsilon))/\pi\,\epsilon^2$, and the area of a surface S by $\mathcal{V}(S(\epsilon))/2\,\epsilon$.

In the case of a convex curve, we can exactly compute the area of the Minkowski sausage; see, for example, [L. Santaló, Chap. 1].

10 Curves of Infinite Length

10.1 What is infinite length?

Now we will consider bounded simple curves to which it is impossible to attribute a length. These curves could be *locally rectifiable*, like spirals, or nowhere rectifiable like fractals. Examples of these two categories will be given in §2. The tools used to analyze such curves are not similar to those used in the case of rectifiable curves. We cannot speak of *tangent* or *length*, but rather of *local size* and *dimension*.

One essential and common point to these curves is the following: they are all defined by a parameterization. A plane curve Γ is the image of an interval $[a, b]$ (the *temporal* interval) by a continuous map $\gamma(t)$ that is bijective (for Γ to be a simple curve) with values in the plane. When Γ, which is entirely covered during the finite time $b - a$, is of infinite length, the velocity of the motion along Γ is necessarily unbounded. It is better not to speak about *velocity*. This notion is replaced by the notion of *measure* on Γ; the measure induced by the parameterization. We recall (Chap. 6, §2) that

> The measure of any part of the trajectory Γ is the **time** needed to cover this part during the motion.

This is an important remark to clearly set the problem, and it is more important when applied in the computation of the characteristic coefficients of the curve. Every trajectory has infinitely many possible parameterizations. The "best" are those that are well-suited to characterize the curve.

There are many ways to determine whether a curve is of infinite length:

• Given $K + 1$ real numbers $t_1 = a < t_2 < \ldots < t_{K+1} = b$, we use the *polygonal approximation* $\mathbf{P}$ of Γ formed by K segments whose endpoints are $\gamma(t_i), \gamma(t_{i+1})$, $i = 1, \ldots, K$ (Chap. 5, §3). Let (ϵ_n) be a sequence of positive real numbers converging to 0. We can always construct a sequence $(\mathbf{P}_n)$ of such curves so that the maximal length of the segments of $\mathbf{P}_n$ is less than ϵ_n. We then know that the Hausdorff distance $\mathrm{dist}(\Gamma, \mathbf{P}_n)$ tends to 0. We shall say that:

> Γ is of infinite length if the length $L(\mathbf{P}_n)$ tends to infinity,

this is equivalent to

$$\sup_n L(\mathbf{P}_n) = +\infty .$$

This definition does not depend on a particular choice of the sequence $(\mathbf{P}_n)$.

• We can also use the parameterization of Γ. We have defined (Chap. 6, §5) a distance function $d(t, \tau)$ as follows

$$d(t, \tau) = \begin{cases} \text{dist}(\gamma(a), \gamma(a + 2\tau)), & \text{if } t - \tau \leq a; \\ \text{dist}(\gamma(t - \tau), \gamma(t + \tau)), & \text{if } a \leq t - \tau < t + \tau \leq b; \\ \text{dist}(\gamma(b - 2\tau), \gamma(b)), & \text{if } b \leq t + \tau. \end{cases}$$

The average with respect to the time of this function is:

$$\bar{d}_\tau = \frac{1}{b - a} \int_a^b d(t, \tau) \, dt .$$

This is always a finite integral. The average velocity $\bar{d}_\tau / 2\tau$ tends to $L(\Gamma)/(b - a)$ in the case of finite length. Otherwise:

Γ *is of infinite length if* $\bar{d}_\tau / \tau$ *tends to infinity when* τ *tends to 0.*

• A third characterization is possible, thanks to the Minkowski sausage,

$$\Gamma(\epsilon) = \bigcup_{x \in \Gamma} B_\epsilon(x) .$$

This is the union of balls of radius ϵ centered on Γ. Its area is $\mathcal{A}(\Gamma(\epsilon))$, and the ratio $\mathcal{A}(\Gamma(\epsilon))/2\epsilon$ tends to $L(\Gamma)$ in the case of finite length. Otherwise

Γ *is of infinite length if* $\mathcal{A}(\Gamma(\epsilon))/\epsilon$ *tends to infinity when* ϵ *tends to 0.*

◇   The first of these characterizations has a logical utility, but the other two are in fact more interesting because they allow a classification of the curves of infinite length. We shall use the *order of growth* to 0 of functions $\bar{d}_\tau$ or $\mathcal{A}(\Gamma(\epsilon))$ for this classification. The function $\mathcal{A}(\Gamma(\epsilon))$ has the advantage of being independent from the parameterization of the curve; we shall encounter it in a most natural way when we give the general definition of the dimension (§3). The function $\bar{d}_\tau$, which depends on the parameterization, will prove to be more useful in a genuine analysis of the curve. However, in the computation of the dimension, we prefer to consider an average over the diameter rather than over a distance (Chap. 11, §4).

10.2 Two examples

1.   We have already encountered (Chap. 5, §4) the spiral

$$\begin{cases} \rho(t) = t \\ \theta(t) = \dfrac{2\pi}{t} \end{cases} \quad 0 < t \leq 1 .$$

Each of its spires $\mathbf{S}_k$, which corresponds to the values of the parameter

$$\frac{1}{k + 1} \leq t \leq \frac{1}{k} ,$$

is of length greater than $1/k$, and $\Sigma\frac{1}{k}$ is a divergent series. Thus, its total length is infinite. From the parameterization we see that the velocity is finite at each point, but it is unbounded on any neighborhood of 0. The function $d(t,\tau)$ has the same order of growth as $\rho(t)[\frac{2\pi}{t-\tau} - \frac{2\pi}{t+\tau}] \simeq \tau/t$ when $t \geq \tau$ (Fig. 10.1). It follows that: $\bar{d}_\tau \simeq \tau|\log\tau|$. The ratio $d(t,\tau)/\tau$ tends to infinity with logarithmic growth. If we estimate $\mathcal{A}(\Gamma(\epsilon))$, we will find

$$\mathcal{A}(\Gamma(\epsilon)) \simeq \epsilon|\log\epsilon|,$$

and the ratio $\mathcal{A}(\Gamma(\epsilon))/\epsilon$ has a logarithmic order of growth.

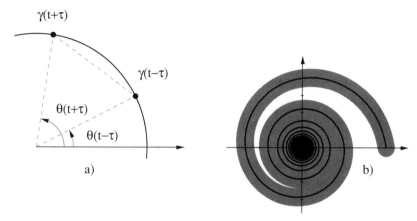

Fig. 10.1. *In a), computation of the function $d(t,\tau) = \mathrm{dist}(\gamma(t-\tau),\gamma(t+\tau))$ on a spiral; in b), the Minkowski sausage of a spiral looks like a comet, with a central nucleus and a tail.*

2. The curve Γ that we are going to describe is constructed with the help of polygonal approximations.

Given an oriented segment **S** and an angle ϕ, $0 \leq \phi \leq \pi/2$, let $\mathcal{T}_\phi$ denote the operation that consists of replacing **S** by a polygonal curve $\mathcal{T}_\phi(\mathbf{S})$ with the same endpoints as **S**. The polygonal curve $\mathcal{T}_\phi$ is formed by four segments of equal length, making with **S** the angles 0, ϕ, $-\phi$ and 0, respectively (Fig. 10.2). The length of each of these segments is $L(\mathbf{S})/2(1+\cos\phi)$.

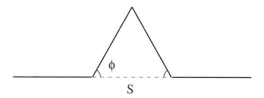

Fig. 10.2. *The segment **S** is replaced by the polygonal curve $\mathcal{T}_\phi(\mathbf{S})$ with the same endpoints and formed by four equal segments.*

Given a sequence of angles (ϕ_k), we start with an initial segment $\mathbf{P}_0$ of length 1. We construct the polygonal curve $\mathbf{P}_1 = \mathcal{T}_{\phi_1}(\mathbf{P}_0)$. We apply on each segment of $\mathbf{P}_1$ the operation $\mathcal{T}_{\phi_2}$. This gives a polygonal curve $\mathbf{P}_2$ containing 16 segments, and so forth. $\mathbf{P}_k$ is obtained from $\mathbf{P}_{k-1}$ by replacing each of its segments by its image under the operation $\mathcal{T}_{\phi_k}$. This curve $\mathbf{P}_k$ contains 4^k segments of length l_k, where

$$l_k = 2^{-k} \prod_{i=1}^{k} \frac{1}{1 + \cos \phi_i} \, .$$

When k goes to infinity, the curve $\mathbf{P}_k$, whose total length is $L(\mathbf{P}_k) = 4^k l_k$, converges to Γ in the sense of Hausdorff distance. The length $L(\Gamma) = \lim L(\mathbf{P}_k)$ could be finite or infinite, depending on the choice of the sequence (ϕ_k). If, for example, $\phi_k = 0$ for all k, then $\mathbf{P}_k = \mathbf{P}_0$, and Γ is equal to the segment $\mathbf{P}_0$. If ϕ_k converges very quickly to 0, then the local irregularities of Γ are small, and the curve is rectifiable. If ϕ_k does not converge to 0, then Γ is of infinite length. It is then a fractal curve (Chap. 11). When ϕ_k is constant, Γ has a *self-similar structure* (Chap. 13). Some examples are described in Fig. 10.3.

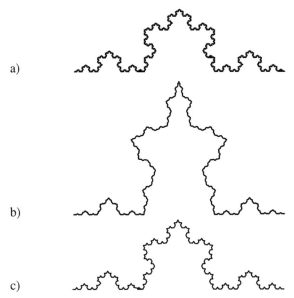

Fig. 10.3. *Pictures of a curve Γ obtained by a sequence of operations $\mathcal{T}_{\phi_k}$ in the following three cases: a) $\phi_k = \pi/3 : L(\mathbf{P}_k) = (4/3)^k$; b) $\phi_k = \arccos(2 \log(k + 1)/\log(k + 2) - 1) : L(\mathbf{P}_k) = \log(k + 2)/\log 2$; c) $\phi_k = \arccos(2 \exp(-1/(k + 1)^2) - 1) : L(\mathbf{P}_k) \simeq 1$. The curve is of infinite length in the cases a) and b) and of finite length in the case c), where the tangent exists almost everywhere.*

10.3 Dimension

The dimension on the real line gives a classification of the sets of null measure (Chap. 2, §5). We extend this concept to the plane in a manner that allows the classification of curves or even of bounded sets in the plane. We start by evaluating the ϵ–Minkowski sausage of E; that is, the set of all the points whose distance to E is less than ϵ:

$$E(\epsilon) = \bigcup_{x \in E} B_\epsilon(x) ,$$

which is a domain of positive area in the plane. This area $\mathcal{A}(E(\epsilon))$ is a function of ϵ, and it is larger when it occupies a larger part of the plane at the scale ϵ. The order of growth to 0 of the function $\mathcal{A}(E(\epsilon))$ is the limit, if it exists, of the ratio

$$\frac{\log \mathcal{A}(E(\epsilon))}{\log \epsilon}$$

when ϵ tends to 0. This limit is null if the area of the set E is not null; it is 2 whenever E is reduced to a point, because in this case, $\mathcal{A}(E(\epsilon)) = \pi\epsilon^2$. We define the **dimension** of E as

$$\text{dimension of } E \;=\; 2 - (\text{order of growth of } \mathcal{A}(E(\epsilon)))$$

when this order exists. Otherwise, we must define two dimensions:

$$\Delta(E) = \limsup_{\epsilon \to 0} \left(2 - \frac{\log \mathcal{A}(E(\epsilon))}{\log \epsilon} \right) \text{ is the } \textit{upper dimension} \text{ of } E$$

$$\delta(E) = \liminf_{\epsilon \to 0} \left(2 - \frac{\log \mathcal{A}(E(\epsilon))}{\log \epsilon} \right) \text{ is the } \textit{lower dimension} \text{ of } E .$$

We find that these indexes have properties similar to those defined on the real line:

1. Δ *is* **monotonous***: if E_1 is included in E_2, then*

$$\Delta(E_1) \leq \Delta(E_2) .$$

2.   *If $\overline{E}$ is the* **closure** *of the set E, then*

$$\Delta(\overline{E}) = \Delta(E) .$$

3. Δ *is* **stable***: given two sets E_1 and E_2,*

$$\Delta(E_1 \cup E_2) = \max\{\Delta(E_1), \Delta(E_2)\} .$$

4.   *For all E in the plane,*

$$0 \le \Delta(E) \le 2 \ .$$

5. Δ is **invariant** *under some transformations* T *of the plane:*

$$\Delta(T(E)) = \Delta(E) \ .$$

Among these transformations, we mention the following:
  — *the similarities formed by a translation, rotation, symmetry, and homothety of non-null ratio (Chap. 13);*
  — *the affine applications, which in a Cartesian coordinate system can be written in their matrix form*

$$T(x) = Ax + B \ ,$$

where B *is a vector and* A *is a matrix of non-null determinant;*
  — *more generally, all the bijections from the plane onto itself such that for all* x *the ratio*

$$\frac{\log(\mathrm{dist}(T(y), T(z)))}{\log(\mathrm{dist}(y, z))}$$

converges to 1 when y *and* z *both tend to* x, *with* $y \ne z$.

The lower dimension δ also satisfies these properties, except the stability. In §4 we shall see an example of the union of two curves for which δ is unstable.

Finally, we note that Δ can be written as a critical exponent:

$$\Delta(E) = \inf \{ \alpha \text{ such that } \epsilon^{\alpha - 2} \mathcal{A}(E(\epsilon)) \text{ tends to } 0 \} \ .$$

10.4 Some examples of dimensions of curves

• If Γ is a curve of finite length, the area $\mathcal{A}(\Gamma(\epsilon))$ is equivalent to ϵ, whose order of growth is 1. Thus $\Delta(\Gamma) = 1$.

• For any given curve Γ, we know that (Chap. 9, §1)

$$\mathcal{A}(\Gamma(\epsilon)) \ge 2\,\epsilon\,\mathrm{diam}\,\Gamma \ .$$

This implies that the order of growth is at most equal to 1. Therefore

$$1 \le \Delta(\Gamma) \le 2 \ .$$

• For the spiral

$$\begin{cases} \rho(t) = t \\ \theta(t) = \dfrac{2\pi}{t} \end{cases} \qquad 0 < t \le 1 \ ,$$

which is of infinite length (§2), the area $\mathcal{A}(\Gamma(\epsilon))$ is equivalent to $\epsilon\,|\log\epsilon|$ whose order of growth is the limit of $\log(\epsilon\,|\log\epsilon|)/\log\epsilon$ when ϵ tends to 0. This limit is 1, and therefore $\Delta(\Gamma) = 1$. Thus the dimension Δ does not permit us to distinguish these curves from any curve of finite length. The scale of the power functions that we used to define Δ somehow lacks the needed precision. With a finer scale, like the double-indexed logarithmic scale (Chap. 2, §5)

$$\mathcal{F} = \{f_{\alpha,\beta}(x) = x^{\alpha}\left(\log_n \frac{1}{x}\right)^{\beta}, \quad \alpha > 0, \quad n \text{ integer} \geq 0, \quad \beta \text{ real}\},$$

we can make the distinction. The spiral will be characterized by the three values: $\alpha = 1$, $\beta = 1$, $n = 1$, while the curves of finite length correspond to $\alpha = 1$, $\beta = 0$, $n = 0$.

- The spiral

$$\begin{cases} \rho(t) = t^{\alpha} \\ \theta(t) = \dfrac{2\pi}{t} \end{cases} \quad 0 < t \leq 1$$

where α is a real number, $0 < \alpha < 1$, winds around the origin with a smaller velocity than the $\alpha = 1$ case, and thus, it occupies more space. We may expect it to be of larger dimension. In fact:

$$\Delta(\Gamma) = \frac{2}{\alpha + 1}.$$

▶   This curve intersects the x-axis at the points of polar coordinates $(\rho, \theta) = (k^{-\alpha}, 2k\pi)$. The kth spire $\mathbf{S}_k$ that corresponds to the values of the parameter $1/(k+1) \leq t \leq 1/k$ therefore has a length of order $k^{-\alpha}$, and $\text{dist}(\mathbf{S}_k, \mathbf{S}_{k+1}) \simeq k^{-\alpha-1}$. Thus if $\epsilon = k^{-\alpha-1}$, the sausage $\Gamma(\epsilon)$ can be divided into:

(i) a "nucleus" of a diameter equivalent to $k^{-\alpha} = \epsilon^{\alpha/(\alpha+1)}$ and area equivalent to $\epsilon^{2\alpha/(\alpha+1)}$; and

(ii) a "tail" of area equivalent to $\epsilon\,k^{1-\alpha} = \epsilon^{2\alpha/(\alpha+1)}$.

Thus, nucleus and tail are of the same order and

$$\Delta(\Gamma) = \delta(\Gamma) = 2 - \frac{2\alpha}{\alpha + 1} = \frac{2}{\alpha + 1}. \quad ◀$$

When $\alpha \to 1$, the dimension tends to 1. When $\alpha \to 0$, the dimension tends to 2 without ever reaching it in this particular case. To obtain $\Delta(\Gamma) = 2$, we need a function $\rho(t)$ whose rate of convergence to 0 is slower than any power function; the function $\rho(t) = 1/|\log t|$ will do.

- Similarly, we can analyze the graph Γ of the function

$$z(t) = \begin{cases} t^{\alpha}\,\cos t^{-\beta} & \text{if } 0 < t \leq 1; \\ 0 & \text{if } t = 0, \end{cases}$$

where $0 < \alpha < \beta$. We find

Fig. 10.4. *Minkowski sausage of the spiral* $\rho(t) = \sqrt{2\pi/t}$, *for different values of* ϵ. *The nucleus and tail have equivalent areas. The dimension is* $4/3$.

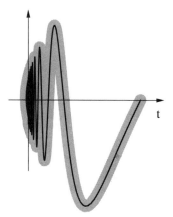

Fig. 10.5. *Minkowski sausage of the graph of the function* $z(t) = \sqrt{t} \cos \frac{1}{t}$. *Its dimension is* $5/4$.

$$\Delta(\Gamma) = \delta(\Gamma) = 2 - \frac{\alpha + 1}{\beta + 1} \ .$$

- We can also construct spirals whose upper and lower dimensions are different.

First, we take a strictly decreasing sequence (t_n) such that $t_0 = 1$, $1/t_n$ is an integer and (t_n) converges sufficiently quickly to 0 so that $\lim_n (\log t_n / \log t_{n+1}) = 0$. For example, $t_n = 2^{1-(n+1)^n}$ will do.

Given the spiral Γ, defined by

$$\begin{cases} \rho(t) = t^\alpha \\ \theta(t) = \dfrac{2\pi}{t} \end{cases} \quad 0 < t \le 1 \, ,$$

we call Γ_n the part of Γ that corresponds to the values of the parameter $t_{n+1} \le t < t_n$, and let $\mathbf{S}_n$ be the segment joining the endpoints of Γ_n; the polar coordinates of these endpoints are $(t_n^\alpha, 0)$ and $(t_{n+1}^\alpha, 0)$. We can obtain new spirals by replacing Γ_n by $\mathbf{S}_n$ for some values of n. We define the following two:

$$\Gamma^* = \Gamma_0 \cup \mathbf{S}_1 \cup \Gamma_2 \cup \mathbf{S}_3 \cup \ldots \cup \Gamma_{2n} \cup \mathbf{S}_{2n+1} \cup \ldots ,$$
$$\Gamma^{**} = \mathbf{S}_0 \cup \Gamma_1 \cup \mathbf{S}_2 \cup \Gamma_3 \cup \ldots \cup \mathbf{S}_{2n} \cup \Gamma_{2n+1} \cup \ldots .$$

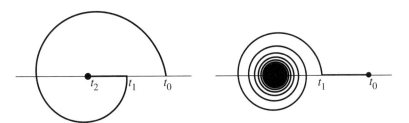

Fig. 10.6. *The two spirals Γ^* and Γ^{**}; in the possible limit of precision, the two spiral-segment schemes alternate along the construction of the two curves.*

When estimating the area of the sausage $\Gamma^*(\epsilon)$, it is either the segment, or the spire, that takes over in the computation according to the size of ϵ. Similarly for Γ^{**}, with alternating values of ϵ. We finally get:

$$\delta(\Gamma^*) = \delta(\Gamma^{**}) = 1 \ , \ \Delta(\Gamma^*) = \Delta(\Gamma^{**}) = \frac{2}{\alpha + 1} \ .$$

• The union $\Gamma^* \cup \Gamma^{**}$ of the two previous spirals is precisely equal to the union of the spiral Γ and the segment whose endpoints are $(0,0)$ and $(1,0)$. We conclude that:

$$\delta(\Gamma^* \cup \Gamma^{**}) = \Delta(\Gamma^* \cup \Gamma^{**}) = \frac{2}{\alpha + 1} \ .$$

This is an example of the instability of the index δ; the δ-measure of this union of curves is strictly larger than that of any of its two components.

10.5 Classical covers: balls and boxes

Many formulations of the dimension of bounded subsets of the real line were presented in Chap. 2, §6. We shall reconsider them in this section in the context of the plane.

Let E be a bounded subset of the plane. It is not always easy to evaluate the area of a set of the type $E(\epsilon)$, which is a union of disks. Other geometrical figures, which can also be used to cover E, might be better suited for this computation. To simplify the discussion, we shall always assume that the order of growth of the quantity $\mathcal{A}(\Gamma(\epsilon))$ exists when ϵ tends to 0; that is, we assume that $\delta(E) = \Delta(E)$. In the other cases, the same considerations remain true when we replace lim by lim sup or lim inf in the corresponding formula.

Centered boxes Assume that the plane is supplied with two orthogonal axes, which permits us to associate to each point x its coordinates (x_1, x_2). We can replace the usual (Euclidian) distance with the following one:

$$d_\infty(x, y) = \max\{\, |x_1 - y_1|, |x_2 - y_2| \,\} \,,$$

which is a distance in the mathematical sense. The set

$$C_\epsilon(x) = \{\, y \text{ such that } d_\infty(x, y) \leq \epsilon \,\}$$

is no longer a disk, but rather a square whose sides are parallel to the axis. We shall call it a *box*.

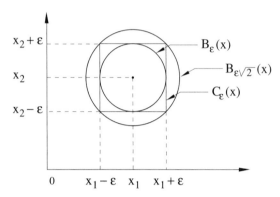

Fig. 10.7. *The inclusions, $B_\epsilon(x) \subset C_\epsilon(x) \subset B_{\epsilon\sqrt{2}}(x)$, prove the equivalence of the Euclidian distance and the distance d_∞.*

We can cover E with a *Minkowski sausage of square elements* by forming the union $\cup_{x \in E} C_\epsilon(x)$. The inequalities,

$$\mathcal{A}(E(\epsilon)) \leq \mathcal{A}(\cup_{x \in E} C_\epsilon(x)) \leq \mathcal{A}(E(\sqrt{2}\epsilon))$$

(Fig. 10.7), prove that the two functions $\mathcal{A}(E(\epsilon))$ and $\mathcal{A}(\cup_{x \in E} C_\epsilon(x))$ are of the same order of growth; thus the dimension of E can also be written as:

$$\Delta(E) = \lim_{\epsilon \to 0}\left(2 - \frac{\log \mathcal{A}(\cup_{x \in E} C_\epsilon(x))}{\log \epsilon}\right).$$

Disjoint boxes   Effecting a grid of squares of side ϵ, we can, to fix the ideas, assume that all boxes are closed (i.e., they contain their boundaries). This implies that any two distinct boxes are either disjoint or **adjacent**; that is, their boundaries touch (while the interiors are disjoint). Let us call (Chap. 4, §1) $\omega_\epsilon(E)$ the number of the squares whose intersections with the set E are not empty. This number tends to infinity when ϵ tends to 0, provided E contains an infinite number of points. Its order of growth to infinity is the dimension of the set E

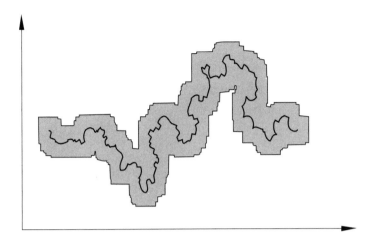

Fig. 10.8. *The union of all boxes of side 2ϵ centered on Γ constitute an approximation of the Minkowski sausage of E.*

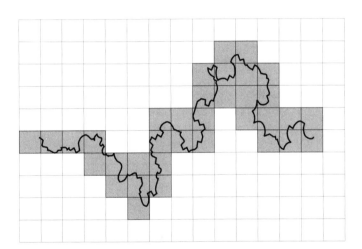

Fig. 10.9. *The union of all boxes intersecting E in a grid of side ϵ constitute an approximation of the Minkowski sausage of E.*

$$\Delta(E) = \lim_{\epsilon \to 0} \frac{\log \omega_\epsilon(E)}{|\log \epsilon|} .$$

▶   It suffices to observe that every square of side ϵ, containing a point x of E, is included in the ball $B_{\epsilon\sqrt{2}}(x)$, which, in turn, is included in at most 16 boxes of the grid. We deduce:

$$\epsilon^2 \, \omega_\epsilon(E) \leq \mathcal{A}(E(\epsilon\sqrt{2})) \leq 16 \, \epsilon^2 \, \omega_\epsilon(E) .$$

Thus the functions $\mathcal{A}(E(\epsilon))$ and $\epsilon^2 \, \omega_\epsilon(E)$ are equivalent, and

$$\Delta(E) = \lim_{\epsilon \to 0}(2 - \frac{\log(\epsilon^2 \, \omega_\epsilon(E))}{\log \epsilon}) = \lim_{\epsilon \to 0} \frac{\log \omega_\epsilon(E)}{|\log \epsilon|} \, . \quad \blacktriangleleft$$

Some variants of this formula use discrete sequences of values of ϵ (same proof as in Chap. 2, §6):

For all sequences (ϵ_n) of real numbers that converge to 0 and satisfy

$$\frac{\log \epsilon_n}{\log \epsilon_{n+1}} \to 1 \, ,$$

we have

$$\Delta(E) = \lim_{n \to \infty} \frac{\log \omega_{\epsilon_n}}{|\log \epsilon_n|} \, .$$

In particular, the **dyadic boxes** of order n are the squares of type $[j\,2^{-n}, (j+1)2^{-n}] \times [k\,2^{-n}, (k+1)2^{-n}]$ where j and k are integers. Each box of order n contains exactly four boxes of order $n+1$. The dimension

$$\Delta(E) = \lim_{n \to \infty} \frac{\log \omega_{2^{-n}}}{n \log 2}$$

is sometimes called *the box dimension*, but it is still the same dimension.

Maximum of disjoint balls We can replace the Minkowski sausage by disjoint balls. Let $M_\epsilon(E)$ be the maximum number of points situated in E such that the distance between any two of them is larger than $2\,\epsilon$. The balls $B_\epsilon(x)$ centered at these particular points, then, are disjoint or adjacent. We can write:

$$\boxed{\Delta(E) = \lim_{\epsilon \to 0} \frac{\log M_\epsilon(E)}{|\log \epsilon|} \, .}$$

$\blacktriangleright$ Since the disks in question are disjoint and included in $E(\epsilon)$, we find

$$\pi \, \epsilon^2 \, M_\epsilon(E) \leq \mathcal{A}(E(\epsilon)) \, .$$

On the other hand, since $M_\epsilon(E)$ is a maximum, each point of E is a distance $\leq \epsilon$ from a disk and, therefore each point of $E(\epsilon)$ is a distance $\leq 2\epsilon$ from a disk. If we multiply the radius of these disks by three we obtain a cover of $E(\epsilon)$. This gives:

$$\mathcal{A}(E(\epsilon)) \leq 9 \, \pi \, \epsilon^2 \, M_\epsilon(E) \, .$$

Thus the functions $\mathcal{A}(E(\epsilon))$ and $\epsilon^2 \, M_\epsilon(E)$ are equivalent when ϵ tends to 0. $\blacktriangleleft$

Minimum of covering balls Let $N_\epsilon(E)$ be the minimum number of disks $B_\epsilon(x)$ centered on E that cover E. We can write:

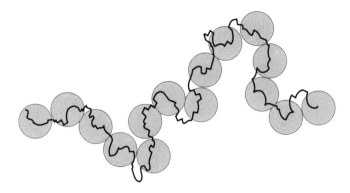

Fig. 10.10. *The union of a maximum number of disjoint or adjacent disks centered on E of radius ε constitute an approximation of the Minkowski sausage of E.*

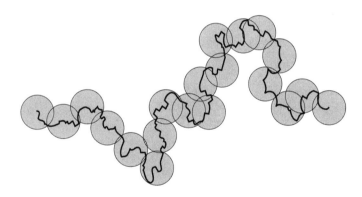

Fig. 10.11. *The union of the minimum number of disks of radius ε centered on E and covering E constitute an approximation of the Minkowski sausage of E.*

$$\Delta(E) = \lim_{\epsilon \to 0} \frac{\log N_\epsilon(E)}{|\log \epsilon|} .$$

▶   Since E is included in the union of these disks, we cover $E(\epsilon)$ by multiplying their radius by 2:

$$\mathcal{A}(E(\epsilon)) \leq 4\,\pi\,\epsilon^2\,N_\epsilon(E) .$$

On the other hand, E can be covered by $M_\epsilon(E)$ balls of radius 2ϵ or by $M_{\epsilon/2}(E)$ balls of radius ϵ. This proves that $N_\epsilon(E) \leq M_{\epsilon/2}(E)$, and therefore

$$\frac{\pi}{4} \epsilon^2 N_\epsilon(E) \leq \frac{\pi}{4} \epsilon^2 M_{\epsilon/2}(E) \leq \mathcal{A}(E(\frac{\epsilon}{2})) \leq \mathcal{A}(E(\epsilon)) .$$

The functions $\mathcal{A}(E(\epsilon))$ and $\epsilon^2 N_\epsilon(E)$ are equivalent when ϵ tends to 0. ◀

10.6 Covers by figures of any kind

We may want to cover E with figures other than disks and squares. In general, let us call any closed set $\mathbf{D}$ whose boundary $\partial\mathbf{D}$ is a simple closed curve (a homeomorphic image of a circle) a domain. When we cover E with domains of any form, we create a *generalized sausage*. The form of these domains does not have much importance; what really counts is their size and their flatness, which shall be interpreted by a relation between the **diameter** and the **interior diameter**. The notion of diameter was already discussed in Chap. 9, §1; this is the greatest distance between any two points of the set.

The **interior diameter** *of* $\mathbf{D}$ *is the diameter of the greatest disk included in* $\mathbf{D}$:

$$\mathrm{diam\,int}(\mathbf{D}) = \sup\{\, r \text{ such that for some } x,\ B_{r/2}(x) \subset \mathbf{D} \,\} .$$

We shall encounter this notion with the *breadth* of a convex set (Appendix C, §3).

For all $\epsilon > 0$, we cover each point x of E with a domain $\mathbf{D}_\epsilon(x)$, whose diameter is less than ϵ. If $x \neq y$, the domains $\mathbf{D}_\epsilon(x)$ and $\mathbf{D}_\epsilon(y)$ could be distinct (in the case of balls $B_\epsilon(x)$ centered on E or squares $C_\epsilon(x)$ centered on E), or identical (in the case of boxes of a grid, when x and y are both in the same box). We do not assume that the domains have the same form. The union $\cup_{x \in E}\mathbf{D}_\epsilon(x)$ forms a sausage covering E. Can we use its area to compute the dimension? Here is a sufficient condition.

THEOREM *For every* ϵ, *let*

$$d_\epsilon = \inf_{x \in E} \mathrm{diam\,int}(\mathbf{D}_\epsilon(x)) .$$

If $\lim_{\epsilon \to 0} \log d_\epsilon / \log \epsilon = 1$, *then*

$$\Delta(E) = \lim_{\epsilon \to 0}(2 - \frac{\log \mathcal{A}(\cup_{x \in E}\mathbf{D}_\epsilon(x))}{\log \epsilon}) .$$

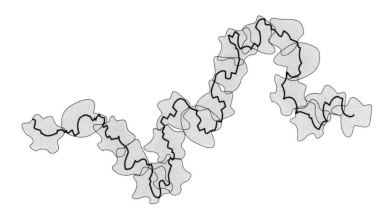

Fig. 10.12. *A cover of E by domains of arbitrary shapes and comparable sizes.*

◇   Since we always have:

$$d_\epsilon \leq \operatorname{diam} \operatorname{int}(\mathbf{D}_\epsilon(x)) \leq \operatorname{diam}(\mathbf{D}_\epsilon(x)) \leq \epsilon \;,$$

the condition of the theorem implies that for all x in E,

$$\frac{\log \operatorname{diam}(\mathbf{D}_\epsilon(x))}{\log \operatorname{diam} \operatorname{int}(\mathbf{D}_\epsilon(x))} \longrightarrow 1 \;.$$

That is, the domains $\mathbf{D}_\epsilon(x)$ are not too flat. Also, for all x, y in E,

$$\frac{\log \operatorname{diam}(\mathbf{D}_\epsilon(x))}{\log \operatorname{diam}(\mathbf{D}_\epsilon(y))} \longrightarrow 1 \;.$$

That is, the domains $\mathbf{D}_\epsilon(x)$ are of comparable size.

◇   The condition of the theorem is satisfied if for every ϵ all the domains $\mathbf{D}_\epsilon(x)$, $x \in E$, are of non-null areas and the same form (one can be generated from another by translation, rotation, or symmetry).

▶   We shall use a technical lemma whose proof can be found in Appendix B.

LEMMA   *Let (B_i) be a family (finite or not) of disks of radius ϵ, let c be a constant larger than 1, and for all i, let B_i^* be the ball with the same center as B_i and radius $c\epsilon$. If the set $\cup_i B_i$ is bounded, then*

$$\mathcal{A}(\cup_i B_i^*) \leq c^2 \mathcal{A}(\cup_i B_i) \;.$$

This inequality becomes an equality if the B_i^* are disjoint. We deduce the following immediate corollary.

For every E and every $\eta \geq \epsilon > 0$,

$$\mathcal{A}(E(\eta)) \leq \left(\frac{\eta}{\epsilon}\right)^2 \mathcal{A}(E(\epsilon)) \;.$$

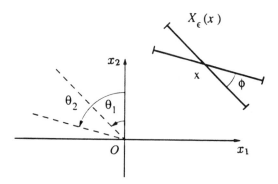

Fig. 10.13. *The cross $X_\epsilon(x)$ is formed by the two segments of length $2\,\epsilon$ centered at x with directions θ_1 and θ_2.*

Now, every domain $\mathbf{D}_\epsilon(x)$ contains a disk of radius $d_\epsilon/2$. Let $y(x)$ be its center: the distance between any point of the domain and $y(x)$ is $\leq \epsilon$. The ball $B_\epsilon(y(x))$ then contains $\mathbf{D}_\epsilon(x)$. Using the corollary, we deduce that:

$$\mathcal{A}(E(\epsilon)) = \mathcal{A}(\bigcup_{x \in E} B_\epsilon(x)) \leq (\frac{2\,\epsilon}{d_\epsilon})^2 \mathcal{A}(\bigcup_{x \in E} B_{d_\epsilon/2}(y(x))) \leq (\frac{2\,\epsilon}{d_\epsilon})^2 \mathcal{A}(\bigcup_{x \in E} \mathbf{D}_\epsilon(x)) \ .$$

On the other hand, $\mathcal{A}(\cup_{x \in E} \mathbf{D}_\epsilon(x)) \leq \mathcal{A}(E(\epsilon))$. Thus, for $\epsilon < 1$ we have the following inequalities:

$$\frac{\log \mathcal{A}(E(\epsilon))}{\log \epsilon} \leq \frac{\log \mathcal{A}(\cup_{x \in E} \mathbf{D}_\epsilon(x))}{\log \epsilon} \leq \frac{\log \mathcal{A}(E(\epsilon))}{\log \epsilon} + \frac{\log(d_\epsilon/2\epsilon)^2}{\log \epsilon} \ .$$

Since

$$\frac{\log(d_\epsilon/2\epsilon)^2}{\log \epsilon} = 2\,(\frac{\log d_\epsilon}{\log \epsilon} - 1 - \frac{\log 2}{\log \epsilon})$$

tends to 0, the functions $\mathcal{A}(E(\epsilon))$ and $\mathcal{A}(\cup_{x \in E} \mathbf{D}_\epsilon(x))$ have the same order of growth when ϵ tends to 0. ◀

10.7 Covering curves by crosses

In the previous sections, the covering domains of a set E were of non-null area. In the particular case of a curve we can take more liberty. We shall create the elements of the cover to be of null area: they will be formed with two segments, but their union along the curve will create (by continuity of the curve) a "sausage" of non-null area. We shall use this type of cover in Chapters 12 and 16. Here is how we will proceed.

In a plane with Cartesian coordinate axes Ox_1 and Ox_2, we consider two angles $\theta_1 \neq \theta_2$, assumed to be in the interval $[0, \pi]$. For every $\epsilon > 0$ and every x in the plane, let $X_\epsilon(x)$ be the cross formed by the two segments of length 2ϵ, centered at x, and carried by the two straight lines making the angles θ_1 and θ_2 with the axis Ox_2.

Let Γ be a curve and

$$X_\epsilon = \bigcup_{x \in \Gamma} X_\epsilon(x)$$

be the union of all the crosses centered on Γ. We shall prove that the area of X_ϵ is not zero. Moreover, it is equivalent to the area of the Minkowski sausage. This will prove the following formula:

$$\Delta(E) = \lim_{\epsilon \to 0} \left(2 - \frac{\log \mathcal{A}(X_\epsilon)}{\log \epsilon} \right).$$

a)

b)

c)

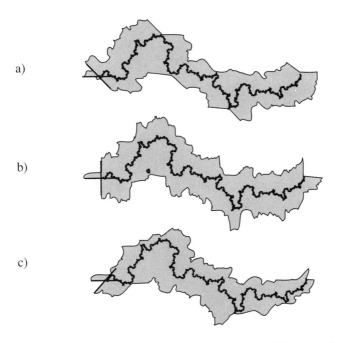

Fig. 10.14. *The surface X_ϵ for different values of θ_1 and θ_2: a) $\theta_1 = \pi/2$, $\theta_2 = \pi/4$; b) $\theta_1 = \pi/2$, $\theta_2 = 0$; and c) $\theta_1 = \pi/2$, $\theta_2 = 3\pi/4$.*

▶ Since $X_\epsilon(x) \subset B_\epsilon(x)$ for all x, we immediately obtain:

$$\mathcal{A}(X_\epsilon) \leq \mathcal{A}(\Gamma(\epsilon)).$$

We look for an inequality in the other direction. Let ϕ be the acute angle formed by the two branches of the cross $X_\epsilon(x)$. Let x_0 be a fixed point of Γ. We construct

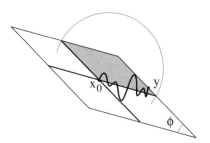

Fig. 10.15. *When x runs through the arc $x_0 \frown y$, the crosses $X_\epsilon(x)$ cover a parallelogram of side ϵ. Its intersection with the disk $B_\epsilon(x_0)$ is always of area $\geq \frac{1}{2}\phi\epsilon^2$.*

a parallelogram $L_\epsilon(x_0)$ centered on x_0 whose sides are of length equal to 2ϵ and are parallel to the branches of the cross.

If ϵ is sufficiently small, we can always find a point y that belongs to Γ on the boundary of this parallelogram such that the arc $x_0 \frown y$ of Γ is entirely included in $L_\epsilon(x_0)$ (see Fig. 10.15). When x runs through the arc $x_0 \frown y$, the crosses $X_\epsilon(x)$ cover at least one of the four equal parallelograms whose union is $L_\epsilon(x_0)$. This covered parallelogram has an angle ϕ at its vertex, and its intersection with the disk $B_\epsilon(x_0)$ is always a domain of area $\geq \frac{1}{2}\phi\epsilon^2$.

We know that there are $M_\epsilon(\Gamma)$ balls of disjoint interiors centered on Γ and of radius ϵ (§5). The previous construction applied to the centers x_0 of these balls allows the construction of $M_\epsilon(\Gamma)$ disjoint domains. Therefore,

$$A(X_\epsilon) \geq \frac{1}{2}\phi\,\epsilon^2 M_\epsilon(\Gamma) \;.$$

On the other hand, we have already used the inequality (§5):

$$A(\Gamma(\epsilon)) \leq 9\,\pi\,\epsilon^2\,M_\epsilon(\Gamma) \;.$$

This gives:

$$A(X_\epsilon) \geq \frac{\phi}{18\pi}A(\Gamma(\epsilon)) \;.$$

We conclude that:

$$\frac{\phi}{18\pi}A(\Gamma(\epsilon)) \leq A(X_\epsilon) \leq A(\Gamma(\epsilon)) \;.$$

The two areas are equivalent. The constant $\phi/18\pi$ is not necessarily the best possible; it can be rendered nearer to 1 by refining the proof. Its maximal value is obtained when $\phi = \pi/2$, the case where the crosses $X_\epsilon(x)$ have perpendicular branches. In this case, the sausage X_ϵ seems to best "fill" the Minkowski sausage $\Gamma(\epsilon)$. ◄

◇   This result is not just true for simple curves; it is generally true for every set E that is *arcwise connected* (that is, any couple of points of E can be joined by a curve entirely included in E). The proof is the same as the earlier one.

10.8 Bibliographical notes

The dimension defined with the help of covers by Minkowski sausages (or by boxes) in the plane or space appears for the first time, it seems, in the articles of G. Bouligand (1928). The main references are cited at the end of Chapter 2. Since then, this dimension was rediscovered many times: in particular, in [L. Pontrjagin & L. Schnirelmann] it appeared under the form $\liminf \log N(\epsilon)/|\log \epsilon|$ and was named "metric order"; and in [A.N. Kolmogorov & V.M. Tihomirov], it appeared under the form of $\limsup \log M(\epsilon)/|\log \epsilon|$ and the name "metric dimension."

We find a more general study of the dimensions of spirals in [M. Mendès–France, Y. Dupain, & C. Tricot].

This notion of *fractional dimension*, or *fractal dimension*, is introduced in this chapter independently from the notion of *fractal set*. This is, in fact, a tool of analysis that can be used to analyze the whole class of compact nowhere dense sets (those not containing any open set). However, its most interesting applications are in the domain of fractals (Chap. 11).

11 Fractal Curves

11.1 What is a fractal curve?

When investigating an experimental curve given by a finite number of points like coastlines, profiles of a surface, boundary of an aggregate, etc., we have to choose in advance the type of mathematical model that can be associated with this curve. Thus we will be facing the general problem of passing from a **discrete** set of numerical data to a **continuous** model. The continuity, if assumed, is mathematically more significant and allows the use of limit theorems. Such a passage can be done by using an **interpolation** procedure, more or less suggested by the data set itself. For example, if we have good reason to believe that our best model is a curve of finite length, then we should look for a rectifiable type of interpolation between the data; we may seek a straight line or a spline function to join successive points so that the theoretical curve has a tangent almost everywhere. In reality, we know that such models do not cover all cases, and we may be better off with curves of infinite length. There are two main families of such curves: the curves that are *locally rectifiable*, like the spirals of Chapter 10 (§2, Example 1); and those that are *nowhere rectifiable*, whose subarcs are of infinite length (Chapter 10, §2, Example 2), which can be called *fractals*. In this case, another interpolation procedure is used; we join the numeric data by an arc (assumed invisible at the scale of the data acquisition) whose geometric structure, in the absence of any other information, is similar to the whole curve or to the curve in the neighborhood of the point in question.

The choice of these models is usually subjective; it depends on the taste or science of the experimenter. In all cases, it is important to know all the characteristics of the chosen model. This is why, before taking any measurement, we feel that a *qualitative* characterization of a fractal curve is most useful. The nearest to our immediate vision is the following:

> *Fractal curves are characterized by the following two properties:*
> *i) Nonrectifiable*
> *ii) Homogeneous.*

Nonrectifiable curves Different interpretations of the word "rectifiable" were given in Chapter 7, §1. We can take it in a global sense to mean "of *finite length*"; or, in the local sense of one of the properties $\mathcal{P}1$ to $\mathcal{P}4$. Thus we can say that globally a curve is nonrectifiable if it has infinite length (Chap. 10), or locally if it satisfies the negation of the properties $\mathcal{P}i$. Moreover, we shall say that it is nowhere rectifiable if it is locally nonrectifiable at any point (this will be discussed in §2).

Homogeneous curves "Homogeneous" means, according to the dictionary, "of the same structure at each point." That is why we say that "any part of a *fractal curve* is similar to the whole." What lies behind the diversity of definitions of fractal curves is the interpretation of "same structure" and "similarity."

To give an exact meaning to these words, one must use *quantitative* methods. We can define one or more parameters, or a family of "test–curves," that allow us to define equivalence relations between the structures. In general, this type of definition will be restrictive, in the sense that a family of parameters, or a class of reference curves, are unable to give a complete description of the object under consideration.

Thus a first method, of statistical type, consists of defining geometric parameters and saying that two structure are "similar" if the parameters have equal values on these structures (or values of the same order).

• For example, take the **fractal dimension** as a parameter. We can say that a curve is homogeneous if each of its subarcs has the same dimension as the whole curve. In other words, the local dimension remains constant at any point (Chap. 18). Moreover, it is fractal if this dimension is larger than 1, since the curve has infinite length and is not rectifiable.

• In fact, it is better to choose parameters that are directly measurable, because this is better suited to diversified applications. We often use the **diameter** as a characteristic parameter of a curve (Chaps. 11 to 15), or more generally, any parameter associated with the notion of **size** (described in §3). At first, two curves are said to be similar if they are of the same size. This is the criteria used to define fractal curves in §4. Then, this notion is completed by that of **deviation** (Chap. 14, §2). Two curves are similar if they have the same size and deviation. Using this, we deduce some interesting consequences about the dimension (Chap. 15).

A second method, favored by mathematicians, consists of translating "similar structures" as follows: structures that can be induced from another by a simple application of some well-defined transformations of the plane. Depending on the chosen transformations, we find the family of *self–similar* curves (Chap. 13) or of *self-affine* curves, or other more general curves (see some examples of less classical types of curves in Figs. 15.2, 15.3, and 15.4). Depending on this method is also more restrictive than the previous one, but it allows us to construct good models. Their structure is, in general, easy to characterize, and they are useful as pedagogical tools to describe fractals.

11.2 A fractal curve is nowhere rectifiable

Since a fractal curve Γ is nowhere rectifiable and homogeneous, then the property of nonrectifiability holds everywhere, in particular on every subarc of Γ. Such a curve is said to be "nowhere rectifiable." We reconsider the properties $\mathcal{P}1$ to $\mathcal{P}4$ of Chapter 7, §1 (local rectifiability). We write their negations (local non-rectifiability), and express it everywhere on Γ (nowhere rectifiablity). This leads to the following characterizations:

We say that Γ is **nowhere rectifiable** *if it satisfies any of the following three properties:*

$\mathcal{Q}1$. Every subarc of Γ that is not reduced to one point is of infinite length.

$\mathcal{Q}2$. At every point x_0 of Γ, the angle $\theta_\epsilon(x_0)$ does not converge to 0, that is,

$$\limsup_{\epsilon \to 0} \theta_\epsilon(x_0) > 0 .$$

$\mathcal{Q}3$. At every point x_0 of Γ, the ratio $\mathcal{A}(\mathcal{K}(x_0{}^\frown x))/(\mathrm{diam}\,(x_0{}^\frown x))^2$ does not converge to 0, that is,

$$\limsup_{\epsilon \to 0} \frac{\mathcal{A}(\mathcal{K}(x_0{}^\frown x))}{\mathrm{diam}\,(x_0{}^\frown x)^2} > 0 .$$

Commentary Though $\mathcal{Q}1$ is induced by $\mathcal{P}3$, it is not its direct negation because it is impossible to use $L(x_0{}^\frown x)$, which is eventually infinite. Thus we have simply replaced (not $\mathcal{P}3$) by $L(x_0{}^\frown x) = +\infty$ for all x_0, x.

$\mathcal{Q}2$ is a direct consequence of (not $\mathcal{P}2$).

$\mathcal{Q}3$ follows from (not $\mathcal{P}4$), but we have replaced $\mathrm{dist}(x_0, x)$ by the diameter of the arc $x_0{}^\frown x$. Choosing the distance or the diameter does not have any real importance when the curve is rectifiable, since the ratio $\mathrm{diam}\,(x_0{}^\frown x)/\mathrm{dist}(x_0, x)$ tends to 1 almost everywhere. But in the case of fractal curves, the function $\mathrm{dist}(x_0, x)$ may have a very irregular behavior when x tends to x_0. Thus it seems better to use the diameter, which has the advantage of being an increasing function of arcs. That is:

For every point y of $x_0{}^\frown x$, $\mathrm{diam}\,(x_0{}^\frown y) \leq \mathrm{diam}\,(x_0{}^\frown x)$.

This property allows the diameter to characterize the size of an arc well.

$\Diamond$ The diameter of a curve is equal to the diameter of its convex hull. The area of this hull is equivalent to its diameter multiplied by its breadth (Appendix C, §4). Therefore $\mathcal{Q}3$ can also be written as:

for every point x_0 of Γ, $\displaystyle\limsup_{\epsilon \to 0} \frac{\mathrm{breadth}(\mathcal{K}(x_0{}^\frown x))}{\mathrm{diam}\,(x_0{}^\frown x)} > 0$.

This is a good translation of the fact that the curve Γ is not flat in the neighborhood of any of its points; it preserves a *perturbed* structure (with respect to a straight line) at any scale.

Relations between the properties $\mathcal{Q}i$ Here are the relations that exist between these properties: for any given curve Γ:

$$\mathcal{Q}3 \Longrightarrow \mathcal{Q}2 \Longrightarrow \mathcal{Q}1 .$$

▶   a) Assume that $\mathcal{Q}3$ holds on a curve Γ. For every $x_0 \in \Gamma$, there exists a non-null real number h and a point x_n of Γ whose distance to x_0 is $\leq 1/n$ such that

$$\mathcal{A}(\mathcal{K}(x_0 \frown x_n)) \geq h \left(\text{diam}\,(x_0 \frown x_n)\right)^2 .$$

Let $\epsilon_n = \inf\{\, r \text{ such that } x_0 \frown x_n \subset B_r(x_0) \,\}$, the largest distance between a point of $x_0 \frown x_n$ and x_0. The sequence ϵ_n tends to 0. The angle $\theta_{\epsilon_n}(x_0)$ belongs to the interval $[0, \pi]$. If this angle is different from π, $x_0 \frown x_n$ is then included in a semicone of vertex x_0 and angle $\theta_{\epsilon_n}(x_0)$.

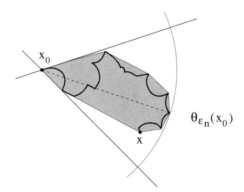

Fig. 11.1. *The convex hull $\mathcal{K}(x_0 \frown x_n)$ is included in a semicone of angle $\theta_{\epsilon_n}(x_0)$.*

The convex hull $\mathcal{K}(x_0 \frown x_n)$ is then included in a circular sector of $B_{\epsilon_n}(x_0)$ (Fig. 11.1) whose area is $\frac{1}{2}\theta_{\epsilon_n}(x_0)\,\epsilon^2$. If $\theta_{\epsilon_n}(x_0) = \pi$, then $\mathcal{K}(x_0 \frown x_n)$ is included in $B_{\epsilon_n}(x_0)$ whose area is $\pi\,\epsilon_n{}^2$.

In all cases we deduce that:

$$\theta_{\epsilon_n}(x_0)\,\epsilon_n{}^2 \geq h\,\epsilon_n{}^2 ,$$

and therefore

$$\theta_{\epsilon_n}(x_0) \geq h .$$

Since h is not null, we can conclude that $\theta_\epsilon(x_0)$ does not converge to 0. Thus $\mathcal{Q}3$ implies $\mathcal{Q}2$.

b) $\mathcal{Q}2$ implies $\mathcal{Q}1$ because if $\mathcal{Q}1$ does not hold, then there is a subarc Γ^* of Γ whose length is finite. Almost everywhere on Γ^*, we shall have: $\lim_{\epsilon \to 0} \theta_\epsilon(x_0) = 0$ (Chap. 7, §4). This contradicts $\mathcal{Q}2$. ◀

11.3 Diameter, size

Keeping the method used in §1, we shall look for a characterization of fractal curves using a very simple geometrical parameter: the **diameter**

$$\text{diam}\,(E) = \sup\{\,\text{dist}(x,y)\,,\ \text{where}\ x,\ y \in E\,\}\,.$$

However, though it is simple to define the diameter, it is not always easy to evaluate it, and it is sometimes helpful to define other parameters that are of the same order and that will be grouped under the same general notion: **size**.

We call **size** *any set function, generally denoted* size(E)*, that satisfies the following three conditions:*

1. size(E) *is equivalent to diameter: there are two constants* c_1 *and* $c_2 > 0$ *such that for every bounded set* E:

$$c_2\,\text{diam}\,(E) \leq \text{size}(E) \leq c_1\,\text{diam}\,(E)\,.$$

2. size(E) *is an increasing function:*

$$E_1 \subset E_2 \Longrightarrow \text{size}(E_1) \leq \text{size}(E_2)\,.$$

3. size(E) *is continuous with respect to the Hausdorff distance:*

$$\lim_{n\to\infty}\text{dist}(E_n, E) = 0 \Longrightarrow \lim_{n\to\infty}\text{size}(E_n) = \text{size}(E)\,.$$

An evident example of the size function is the diameter itself. But we can, depending on the situation, use others.

• If Ox_1 and Ox_2 are two orthogonal axes in the plane, then to each set E we can associate the smallest rectangle containing E whose sides are parallel to the axis. We call it the *circumscribing rectangle* of E. We may then define

$$\text{size}(E) = \text{length of the circumscribing rectangle of}\ E\,.$$

This is a parameter equivalent to the diameter, with $c_1 = 1$ and $c_2 = 1/\sqrt{2}$, as is

$$\text{size}(E) = \text{perimeter of the circumscribing rectangle of}\ E\,,$$

with $c_1 = 2\sqrt{2}$ and $c_2 = 2$.

• The circumscribing rectangle is a convex set. We may also use the convex hull of E and write that:

$$\text{size}(E) = L(\partial\mathcal{K}(E))\,.$$

This is the length of the boundary of the convex hull (Appendix C). This function is equivalent to the diameter with $c_1 = \pi$ and $c_2 = 2$.

• Let $C(E)$ be the circumcircle of E (the smallest circle containing E). Its diameter is not always equal to that of E, as can be seen in the case of an equilateral triangle. However, since $C(E)$ is also the circumcircle of $\mathcal{K}(E)$, its diameter is then equivalent to diam(E) (Appendix C). We can write

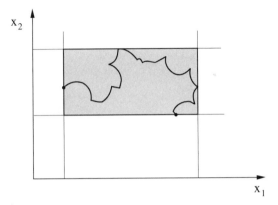

Fig. 11.2. *The smallest rectangle containing a given curve whose sides are parallel to the axis. Its length is equivalent to the diameter.*

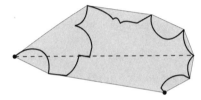

Fig. 11.3. *The convex hull of a curve. Its perimeter is equivalent to the diameter of the curve.*

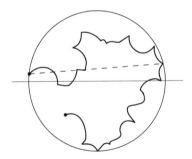

Fig. 11.4. *The circumcircle of a curve. Its diameter is an equivalent function to the diameter of the curve.*

$$\text{size}(E) = \text{diam}\,(C(E)) \,,$$

with $c_1 = 2/\sqrt{3}$ and $c_2 = 1$.

◇   We note the similarity between this section and §2 in Appendix C, which is devoted to the diameter of the convex set. In every case, the size of E is equal to the size of its convex hull $\mathcal{K}(E)$.

11.4 Characterization of a fractal curve

Suppose that a size function (diameter or other) has been chosen. Let Γ be a parameterized curve, the image of the interval $[a, b]$ by a continuous function γ. We call a *local arc* around a point $\gamma(t)$ of Γ the image of an interval $[t - \tau, t + \tau]$. It depends on τ, half the length of the *time window* (Fig. 11.5).

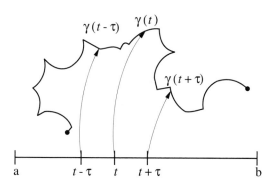

Fig. 11.5. *The local arc $\gamma(t - \tau) ^\frown \gamma(t + \tau)$ constructed around the point $\gamma(t)$ is the image by the parameterization γ of the time window $[t - \tau, t + \tau]$.*

With respect to this parameterization, we define a *local size function* analogous to the distance function defined in (Chap. 6, §5 and Chap. 10, §1). It is constructed so that it is continuous on the interval $[a, b]$, and, in view of its numerical applications, it does not have any important variations near the endpoints (on some neighborhood of a and of b). Therefore, for all t in $[a, b]$ and all τ in $]0, (b - a)/2]$, we write

$$T(t, \tau) = \begin{cases} \text{size}(\gamma(a) ^\frown \gamma(a + 2\tau)), & \text{if } t - \tau \leq a; \\ \text{size}(\gamma(t - \tau) ^\frown \gamma(t + \tau)), & \text{if } a \leq t - \tau < t + \tau \leq b; \\ \text{size}(\gamma(b - 2\tau) ^\frown \gamma(b)), & \text{if } b \leq t + \tau. \end{cases}$$

The function

$$\overline{T}_\tau = \frac{1}{b - a} \int_a^b T(t, \tau)\, dt$$

is the "average local size," and the ratio $\overline{T}_\tau / 2\tau$ is an evaluation of the average local velocity.

When the curve is of finite length, this ratio tends to a finite limit $L(\Gamma)/(b-a)$. When the curve is of infinite length, $\overline{T}_\tau / \tau$ tends to infinity when τ tends to 0. Thus we have the following characterization:

We say that Γ is **fractal** *if*
(Q4) The ratio

$$\frac{T(t, \tau)}{\tau}$$

tends to infinity uniformly with respect to t when τ tends to 0.

The property $\mathcal{Q}4$ can also be expressed as:

For all real A as large as we wish, we can find $\tau_0 > 0$ such that for all $\tau \leq \tau_0$ and all $t \in [a, b]$,

$$T(t, \tau) \geq A\tau \, .$$

In fact, this indicates that the average of $T(t, \tau)$ on any subinterval of $[a, b]$ tends to infinity. Therefore any subarc of Γ is of infinite length, and Γ is nowhere rectifiable. Also, when $\tau \leq \tau_0$, all the arcs of measure 2τ satisfy the same inequality:

$$\text{size}(\gamma(t - \tau)^\frown\gamma(t + \tau)) \geq A\tau \, .$$

In this sense the curve Γ is homogeneous.

$\diamond$ We see that what was suggested is a loose definition of a fractal curve. We may make it a bit more restrictive by assuming (while following the same order of ideas) that there exists a function $f(\tau)$ and two positive constants c_1, c_2 such that

$$\lim_{\tau \to 0} \frac{f(\tau)}{\tau} = +\infty \, ,$$

and for all $t \in [a, b]$,

$$c_2 f(\tau) \leq T(t, \tau) \leq c_1 f(\tau) \, .$$

This, in fact, allows us to say that arcs of the same measures have equivalent sizes.

$\diamond$ An advantage of this type of characterization of fractal curves is that it will lead us to natural algorithms to compute their dimensions (Chaps. 12 to 15).

12 Graphs of Nondifferentiable Functions

12.1 Curves parameterized by the abscissa

A set of numerical data is often given as a set of points in a plane with two orthogonal axes in such a way that two distinct points have distinct abscissas. This is the case of most *time signals*, but there are plenty of other examples: profiles of rough surfaces, distance from a fixed point to a parameterized curve or to a brownian motion on the line. We shall consider such curves to be the graph of a continuous function $z(t)$ defined on the interval $[a, b]$. It is naturally parameterized by its abscissa t, and the parameterization is the function $\gamma(t)$ that assigns to each value of t the point $(t, z(t))$ of the plane. We assume continuity in this model; if it is not likely in some cases, we must find other models with jumps, which is not the topic of this chapter.

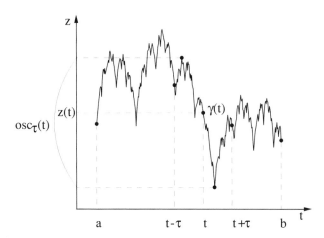

Fig. 12.1. *The curve Γ is the graph of a continuous function $z(t)$. The parameterization γ is the function that associates to every t the point $(t, z(t))$ of the plane. The projection of the local arc $\gamma(t - \tau)^\frown \gamma(t + \tau)$ on Ot is the interval $[t - \tau, t+\tau]$, and its projection on Oz is the interval $[\inf_{|t-t'|\leq\tau} z(t'), \sup_{|t-t'|\leq\tau} z(t')]$.*

For this particular type of parameterized curve what we call *local arc* $\gamma(t - \tau)^\frown \gamma(t + \tau)$ is the part of the graph that corresponds to all values of t in the subinterval $[t - \tau, t + \tau]$.

◇   When the abscissa t represents time, the curve Γ is a trajectory. In the case where Γ is of finite length, the function $z(t)$ is almost everywhere differentiable and the instantaneous velocity is given by $v(t) = \sqrt{1 + z'(t)^2}$. The length is obtained by

$$L(\Gamma) = \int_a^b \sqrt{1 + z'(t)^2}\, dt$$

(Chap. 6, §3). Thus, the velocity is larger whenever the oscillation of z at t is stronger. In the case where the curve is of infinite length the notion of derivative disappears, and it is the local oscillation that should be measured.

12.2 Size of local arcs

Figure 12.1 shows how a local arc, which is the image of the interval $[t - \tau, t + \tau]$, would be inscribed inside a rectangle of bases $2\,\tau$. Its height, which will be called the τ–**oscillation** of the function z at t, is equal to:

$$\operatorname{osc}_\tau(t) = \sup_{|t-t'|\le\tau} z(t') - \inf_{|t-t'|\le\tau} z(t')$$

$$= \sup\{\, z(t') - z(t'')\,,\ \text{where } t' \text{ and } t'' \text{ belong to } [t - \tau, t + \tau]\,\}\ .$$

When $a \le t \le a + 2\tau$, we take $\operatorname{osc}_\tau(t) = \operatorname{osc}_\tau(a + \tau)$; when $b - 2\tau \le t \le b$, we take $\operatorname{osc}_\tau(t) = \operatorname{osc}_\tau(b - \tau)$. It is convenient to measure the *size* of a local arc by the length of this rectangle (Chap. 11, §3):

$$T(t, \tau) = \max\{\, 2\,\tau\,,\ \operatorname{osc}_\tau(t)\,\}\ .$$

By reconsidering the ($\mathcal{Q}4$-characterization) of fractal curves (Chap. 11, §4), we can write:

The graph Γ is fractal if

$$\frac{\operatorname{osc}_\tau(t)}{\tau} \longrightarrow_{\tau \to 0} +\infty\ ,$$

uniformly with respect to t.

Thus for a fractal graph, the circumscribing rectangles $2\tau \times \operatorname{osc}_\tau(t)$ become higher and narrower as the precision τ of the observation gets smaller. In practice, a sufficiently precise printer will draw such curves as successive vertical segments. This phenomenon is related to the fact that the continuous function $z(t)$ is nowhere differentiable.

12.3 Variation of a function

The τ–variation of $z(t)$ on $[a, b]$ is the integral of the oscillations. We denote it by $\mathrm{Var}_\tau(z)$, or Var_τ when the function is clear. Thus

$$\mathrm{Var}_\tau = (b-a)\overline{\mathrm{osc}}_\tau$$

$$= \int_a^b \mathrm{osc}_\tau(t)\, dt .$$

This quantity is evidently related to the average of local size $\overline{T}_\tau$ (Chap. 11, §4). However, this relation is not a direct one. If for some values of t, $\mathrm{osc}_\tau(t) < 2\tau$, then the size of the local arc is 2τ and not $\mathrm{osc}_\tau(t)$. At these points the graph Γ has a flattened profile, when seen at a precision τ. If such a phenomenon occurs almost everywhere, then it is possible to obtain the following inequality: $\mathrm{Var}_\tau < (b-a)\overline{T}_\tau$. Nevertheless these two functions are always equivalent:

   If the function $z(t)$ is not constant, then

$$\boxed{\mathrm{Var}_\tau \simeq \overline{T}_\tau .}$$

▶   By integrating both sides of the following inequality:

$$\mathrm{osc}_\tau(t) \le T(t, \tau) \le \mathrm{osc}_\tau(t) + 2\tau ,$$

we can deduce that:

$$\mathrm{Var}_\tau \le (b-a)\overline{T}_\tau \le \mathrm{Var}_\tau + 2\tau(b-a) .$$

If $z(t)$ is not constant, then let $c = \sup_{t\in[a,b]} z(t) - \inf_{t\in[a,b]} z(t)$ be the total oscillation of $z(t)$ on $[a, b]$. The inequality $\mathrm{Var}_\tau \ge c\tau$ is always true (a geometrical proof of this fact will be given later on in this section). Then

$$\mathrm{Var}_\tau \le (b-a)\overline{T}_\tau \le (2\,\frac{b-a}{c} + 1)\mathrm{Var}_\tau . \quad ◀$$

Thus we can deduce that when the curves are of infinite length, the ratios Var_τ/τ and $\overline{T}_\tau/\tau$ have the same order of growth to $+\infty$ when τ tends to 0. We shall see that this order of growth is directly related to the fractal dimension of Γ (§4). We prefer to use the function Var_τ, which is simpler to analyze. Here are some of its properties.

Geometry of Var_τ We have denoted (Chap. 9, §4) $\Gamma(\tau, \pi/2)$ the surface scanned by Γ when it is horizontally translated by $+\tau$ and $-\tau$; this is also the union of all horizontal segments centered on Γ whose length is 2τ. Since the interval $[a, b]$ is the domain of definition of $z(t)$, we truncate this surface on the left of the line

$t = a$ and on the right of the line $t = b$ in such a way that we keep the points whose abscissa is between a and b. Let U_τ be this surface (Fig. 12.2):

$U_\tau = \{ (t_1, t_2)$ such that

   $(i)\ a \le t_1 \le b$

   (ii) there exists $t_0 \in [a, b]$ such that $|t_1 - t_0| \le \tau$ and $t_2 = z(t_0) \}$.

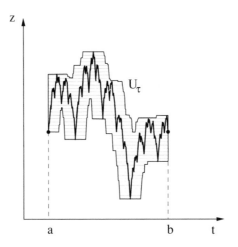

Fig. 12.2. *To construct the surface U_τ, we take the union of all the horizontal segments centered on Γ whose length is 2τ. We truncate the union on the left and on the right to keep just the points whose abscissa is between a and b.*

This surface U_τ is nothing but a sausage around Γ. Its area is precisely the τ–variation of $z(t)$:

$$\mathrm{Var}_\tau = \mathcal{A}(U_\tau) \ .$$

▶   If (t, z) is any point of U_τ, then there exists t_0 in $[a, b]$ such that $|t - t_0| \le \tau$ and $z = z(t_0)$. We deduce that U_τ is the union of vertical segments (not necessarily centered on Γ!) of abscissa $t \in [a, b]$ and of length $\mathrm{osc}_\tau(t)$. When integrating this length along $[a, b]$ with respect to the variable t, we obtain the area of the surface; therefore $\mathcal{A}(U_\tau) = \int_a^b \mathrm{osc}_\tau(t)\, dt$. ◀

◇   If $z(t)$ is not constant, we denote the total oscillation of $z(t)$ on $[a, b]$ by $c = \sup_{t \in [a,b]} z(t) - \inf_{t \in [a,b]} z(t)$. The horizontal sections of U_τ are all of length $\ge \tau$, whenever $\tau \le b - a$. By integrating the length of these horizontal sections on the interval $[\inf_{t \in [a,b]} z(t), \sup_{t \in [a,b]} z(t)]$, we obtain the area of U_τ. Therefore

$$\mathrm{Var}_\tau = \mathcal{A}(U_\tau) \ge c\tau \ .$$

Analytical properties of Var_τ

1. *For all continuous functions $z(t)$, we have $\text{Var}_\tau(z) \geq 0$ and*

$$\text{Var}_\tau(z) = 0 \iff \text{osc}_\tau(t) = 0 \text{ for all } t$$
$$\iff z(t) \text{ is a constant function .}$$

2. *For all constants c_1, c_2,*

$$\text{Var}_\tau(c_1\, z + c_2) = |c_1|\, \text{Var}_\tau(z) .$$

3. *For all pairs of functions z_1, z_2 defined on the same interval,*

$$\text{Var}_\tau(z_1) - \text{Var}_\tau(z_2) \leq \text{Var}_\tau(z_1 + z_2) \leq \text{Var}_\tau(z_1) + \text{Var}_\tau(z_2) .$$

▶   1. This is immediate, because the integral of a positive continuous function is null if and only if the function itself is null.

2. It suffices to verify that $\text{Var}_\tau(c\,z) = c\,\text{Var}_\tau(z)$ if $c \geq 0$, $\text{Var}_\tau(-z) = \text{Var}_\tau(z)$, and $\text{Var}_\tau(z + c) = \text{Var}_\tau(z)$.

3. First, we note the following inequalities

$$\inf_{|t-t'|\leq\tau} z_1(t') + \inf_{|t-t'|\leq\tau} z_2(t') \leq \inf_{|t-t'|\leq\tau} (z_1 + z_2)(t')$$
$$\leq \sup_{|t-t'|\leq\tau} (z_1 + z_2)(t') \leq \sup_{|t-t'|\leq\tau} z_1(t') + \sup_{|t-t'|\leq\tau} z_2(t') ,$$

With evident notations, this implies the following: for all t,

$$\text{osc}_\tau(z_1 + z_2) \leq \text{osc}_\tau(z_1) + \text{osc}_\tau(z_2) .$$

By integrating both sides we get one of the desired inequalities:

$$\text{Var}_\tau(z_1 + z_2)(t) \leq \text{Var}_\tau(z_1)(t) + \text{Var}_\tau(z_2)(t) .$$

To prove the other inequality, we note that:

$$\text{Var}_\tau(z_1) = \text{Var}_\tau(z_1 + z_2 - z_2) \leq \text{Var}_\tau(z_1 + z_2) + \text{Var}_\tau(-z_2)$$
$$= \text{Var}_\tau(z_1 + z_2) + \text{Var}_\tau(z_2) . \quad ◀$$

◇   From these properties, we can deduce that for all τ, $\text{Var}_\tau(z)$ is a **norm** on the set of continuous functions of mean zero defined on $[a, b]$.

12.4 Fractal dimension of a graph

If $z(t)$ is a constant function defined on $[a, b]$, $\mathrm{Var}_\tau(z) = 0$ for all τ. Otherwise, $\mathrm{Var}_\tau(z)$ is a non-null quantity that converges to 0 and whose order of growth is directly related to the dimension of the graph.

THEOREM   *Let $z(t)$ be a non constant continuous function, and let Γ be its graph, then*

$$\Delta(\Gamma) = \lim_{\tau \to 0} (2 - \frac{\log \mathrm{Var}_\tau(z)}{\log \tau}) \, .$$

In particular, if z is differentiable on $[a, b]$, then $\mathrm{Var}_\tau \simeq \tau$ and the limit is 1. This corresponds to the dimension of a curve of finite length.

When $\mathrm{Var}_\tau(z)$ does not have a well-defined order of growth, the formula will not be convergent. In this case, we replace lim by lim sup, and we obtain

$$\Delta(\Gamma) = \limsup_{\tau \to 0}(2 - \frac{\log \mathrm{Var}_\tau(z)}{\log \tau}) \, .$$

▶   To prove this theorem, we reconsider the crosses covering domain $X_\tau(x)$ defined in (Chap. 10, §7), where $x = (t, z(t))$ runs through Γ. We take $\theta_1 = \pi/2$, $\theta_2 = 0$, so that the branches of any cross are parallel to the axis. The union

$$X_\tau = \bigcup_{x \in \Gamma} X_\tau(x)$$

is a sausage, which is in fact the union of two sausages: the first is made by the horizontal segments of length 2τ, noted by $\Gamma(\pi/2, \tau)$; the second by the vertical segments of length 2τ, noted by $\Gamma(0, \tau)$ (Chap. 9, §4). Therefore

$$\mathcal{A}(\Gamma(\pi/2, \tau)) \le \mathcal{A}(X_\tau) \le \mathcal{A}(\Gamma(\pi/2, \tau)) + \mathcal{A}(\Gamma(0, \tau)) \, .$$

The area $\mathcal{A}(\Gamma(0, \tau))$ is the integral along $[a, b]$ of the lengths of the vertical sections, which are all equal to 2τ. We obtain $\mathcal{A}(\Gamma(0, \tau)) = 2(b - a)\tau$.

If we truncate the surface $\Gamma(\pi/2, \tau)$ on the left by the line $t = a$ and on the right by $t = b$, it becomes equal to the surface U_τ (§3). Let $c = \sup_{t \in [a,b]} z(t) - \inf_{t \in [a,b]} z(t)$, which is not null because $z(t)$ is not constant. Recalling that $\mathcal{A}(U_\tau) = \mathrm{Var}_\tau$, we can write:

$$\mathrm{Var}_\tau \le \mathcal{A}(\Gamma(\pi/2, \tau)) \le \mathrm{Var}_\tau + 2\,c\,\tau \, .$$

Finally, $c\tau \le \mathrm{Var}_\tau$. Putting all these results together we get:

$$\mathrm{Var}_\tau \le \mathcal{A}(X_\tau) \le \mathrm{Var}_\tau + 2\,(c + b - a)\,\tau$$

$$\le (3 + 2\frac{b - a}{c})\mathrm{Var}_\tau \, .$$

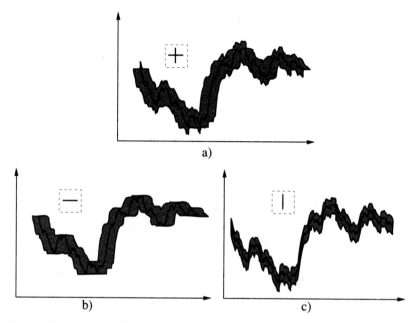

Fig. 12.3. *The surface X_τ formed by crosses whose centers run through Γ (a) is the union of the surface $\Gamma(\pi/2, \tau)$ formed by the horizontal segments (b) and by the surface $\Gamma(0, \tau)$ formed by the vertical segments (c).*

Therefore,

$$\limsup_{\tau \to 0}(2 - \frac{\log \mathrm{Var}_\tau(z)}{\log \tau}) = \limsup_{\tau \to 0}(2 - \frac{\log \mathcal{A}(X_\tau)}{\log \tau}) \ .$$

The right-hand side is exactly $\Delta(\Gamma)$ (Chap. 11, §4).   ◄

◊   Let z_1 and z_2 be two continuous functions on $[a, b]$ whose respective graphs are Γ_1 and Γ_2. Let Γ be the graph of $z_1 + z_2$. The inequality

$$\mathrm{Var}_\tau(z_1 + z_2) \le \mathrm{Var}_\tau(z_1) + \mathrm{Var}_\tau(z_2) \ ,$$

proved in §3, allows us to say that the order of growth to 0 of $z_1 + z_2$ is higher than the maximum order of growth of z_1 and of z_2, and therefore:

$$\Delta(\Gamma) \le \max\{\Delta(\Gamma_1), \Delta(\Gamma_2)\} \ .$$

An inequality in the other direction is more difficult to obtain because the oscillations of the functions z_1 and z_2 may cancel. In the case where $\Delta(\Gamma_1) \ge \delta(\Gamma_1) > \Delta(\Gamma_2)$, then $\mathrm{Var}_\tau(z_2)$ is negligible relative to $\mathrm{Var}_\tau(z_1)$, and the inequality

$$\mathrm{Var}_\tau(z_1 + z_2) \ge \mathrm{Var}_\tau(z_1) - \mathrm{Var}_\tau(z_2)$$

of §3 implies that $\Delta(\Gamma) = \Delta(\Gamma_1)$. In general we may say that:

When the graphs of two functions have different dimensions, the dimension of their sum is the larger of the two dimensions.

We note that *in most cases*, if $\Delta(\Gamma_1) = \Delta(\Gamma_2)$, the same equality $\Delta(\Gamma) = \Delta(\Gamma_1)$ holds. This follows because the functions z_1 and z_2 are *independent* and their oscillations are not directly opposed. This notion of independence has to be strictly defined.

12.5 Hölder exponent

If a function $z(t)$ is differentiable at t, the ratio $\mathrm{osc}_\tau(t)/2\tau$ tends to $z'(t)$ when τ tends to 0, and the graph Γ of z can, at a neighborhood of the point $(t, z(t))$, be amalgamated with the tangent. But if the lim sup of $\mathrm{osc}_\tau(t)/\tau$ is infinite, there is no derivative at t. To measure the irregularity of Γ at this point, we define the "Hölder exponent" H, with $0 < H \leq 1$:

Hölder   *The function $z(t)$ is Holderian of exponent H at t if there exists a constant c such that for all t',*

$$|z(t) - z(t')| \leq c\,|t - t'|^H .$$

In terms of oscillation, this condition can be written as follows:

Hölder   *The function $z(t)$ is Holderian of exponent H at t, with $0 < H \leq 1$, if there exists a constant c such that for all τ,*

$$\mathrm{osc}_\tau(t) \leq c\,\tau^H .$$

▶   In fact, $\mathrm{osc}_\tau(t) \leq c\,\tau^H$ straightforwardly implies that $|z(t) - z(t')| \leq c\,|t - t'|^H$. Conversely, assume that $|z(t) - z(t')| \leq c\,|t - t'|^H$ for all t'. Let t_1 satisfy $|t - t_1| \leq \tau$ and $z(t_1) = \sup_{|t-t'|\leq\tau} z(t')$. Let t_2 satisfy $|t - t_2| \leq \tau$ and $z(t_2) = \inf_{|t-t'|\leq\tau} z(t')$. We have:

$$\mathrm{osc}_\tau(t) = z(t_1) - z(t_2)$$
$$= z(t_1) - z(t) + z(t) - z(t_2) \leq 2\,c\,\tau^H . \quad ◀$$

◇   Hölder's condition indicates that on a neighborhood of the point $(t, z(t))$, the graph Γ will be inscribed between the graphs of the two functions:

$$z(t') = \pm c\,|t - t'|^H .$$

◇   In the typical case of fractal curves, this Hölder exponent is the same for all t. Moreover, if the constant c is independent from t ($z(t)$ is uniformly Holderian), we obtain by integration over the domain of definition $[a, b]$:

$$\mathrm{Var}_\tau \leq c\,\tau^H ,$$

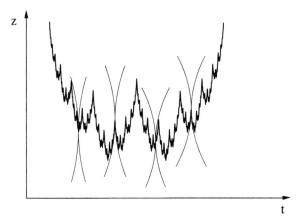

Fig. 12.4. *Graph of the continuous function* $z(t) = \sum_0^\infty 2^{-n/2} \cos(2^n t)$, *which is Holderian of exponent* $1/2$ *for all* t *(§7). On the neighborhood of* t, *the graph oscillates inside a hull formed by the two parabolas of type* $z(t') = \pm c \sqrt{|t - t'|}$.

and we deduce the following result on the dimension (§4):

$$\Delta(\Gamma) \le 2 - H .$$

To transform this inequality into equality we need a condition on the oscillation that goes in the other direction. This is the "anti–Hölder" condition.

Anti–Hölder   *The function* $z(t)$ *is anti–Holderian of exponent* H *at* t *if there is a constant* $c > 0$ *such that for all* τ,

$$\mathrm{osc}_\tau(t) \ge c\,\tau^H .$$

Hence we have the following result on the dimension:

If there exist two constants H *and* $c > 0$, *such that* $\mathrm{osc}_\tau(t) \ge c\tau^H$ *for all* t, *then*

$$\Delta(\Gamma) \ge 2 - H .$$

The **Hölder** and **anti–Hölder** conditions are opposed but are not negations of each other. Many well-known mathematical functions—functions of Knopp, Weierstrass, etc.—satisfy both of these properties for all t and for the same value of H. These are functions with well defined local geometry for which we have that $\Delta(\Gamma) = 2 - H$. Here we note that the **anti–Hölder** condition is usually more difficult to prove. In §6 we shall present an easy case (function of Knopp). In the subsequent sections, we shall see that the fractal dimension can be evaluated without using this condition.

◇ Even if the Hölder exponent H is constant along the curve, it might happen that the constant c varies with t. Then the equality $\Delta(\Gamma) = 2 - H$ may not hold any longer. Other more sophisticated notions of dimension must then be used.

12.6 Functions defined by series

A method used quite often to construct graphs of continuous nowhere differentiable functions consists of putting:

$$z(t) = \sum_{n=0}^{\infty} a_n g(\omega_n t) \,,$$

where g is a continuous periodic function, (a_n) a sequence that tends to 0, and (ω_n) a sequence that tends to $+\infty$ such that $|a_n|\omega_n$ tends to $+\infty$. In this manner, the proper frequencies of the signal tend to infinity faster than the convergence to 0 of their amplitudes. This is the reason behind the local irregularities of the graph of $z(t)$. To insure the absolute convergence of the series and hence the continuity of $z(t)$, we may assume that $\sum |a_n|$ converges.

Here is a heuristic method to compute the dimension: we fix an integer n and let $\tau = 1/\omega_n$. At each t, the τ–oscillation of $z(t)$ has the same order as $|a_n|$ which can be written as $\tau^{\log |a_n|/\log \omega_n}$. Therefore, at each point we can expect the function to be Holderian of exponent

$$H = \lim_{n \to \infty} \frac{\log |a_n|}{\log \omega_n}$$

and

$$\Delta(\Gamma) = 2 - H \,.$$

In most cases, the sequences (a_n) and (ω_n) are geometric progressions: there exists a real number $\omega > 1$ such that $\omega_n = \omega^n$, and $a_n = \omega^{-nH}$, $0 < H < 1$. However, even in this simple case the expected result about the dimension will depend on the function $g(t)$. Here is an example.

The Knopp function   Let $g(t)$ be a periodic function of period 1, defined on $[0, 1]$ by

$$g(t) = \begin{cases} 2t, & \text{if } 0 \leq t \leq 1/2; \\ 2 - 2t, & \text{if } 1/2 \leq t \leq 1. \end{cases}$$

Let $\omega = 2$ and choose a parameter $0 < H < 1$. Then the Knopp function is defined as follows:

$$z(t) = \sum_{n=0}^{\infty} 2^{-nH} g(2^n t) \,.$$

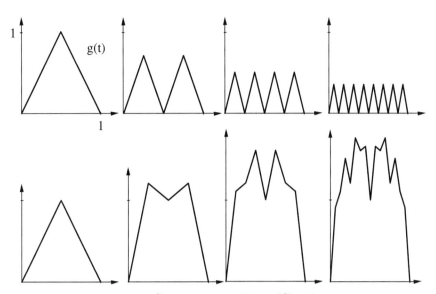

Fig. 12.5. *The functions* $2^{-n/2}g(2^n t)$ *and* $\sum_{i=0}^{n} 2^{-i/2}g(2^i t)$ *for* $n = 0$ *to* 3.

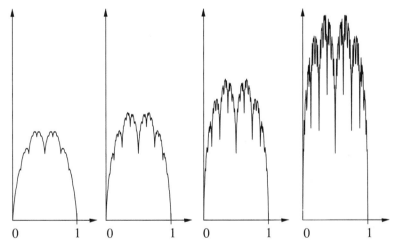

Fig. 12.6. *Knopp functions* $\sum_{n=0}^{\infty} 2^{-nH}g(2^n t)$ *for* $H = 1$, $3/4$, $1/2$, $1/4$; *the fractal dimension* $2 - H$ *measures the global irregularity of the curve.*

We shall show that there are two non-null constants c_1 and c_2, such that for all t and τ,

$$c_1 \tau^H \le \mathrm{osc}_\tau(t) \le c_2 \tau^H ,$$

thus proving that $z(t)$ is at the same time Holderian and anti–Holderian with exponent H. Thus the dimension of its graph Γ is

$$\Delta(\Gamma) = 2 - H .$$

▶   We consider the particular values $\tau = 2^{-n}$. Let $z_n(t) = \sum_{i=0}^{n} 2^{-iH} g(2^i t)$.
a) The graph of the function $2^{-iH} g(2^i t)$ is formed by segments of slope $2^{1+i(1-H)}$.
The graph of z_n is therefore formed by line segments of slope less than
$\sum_{0}^{n} 2^{1+i(1-H)} \leq d_1 2^{n(1-H)}$, where d_1 depends only on H. The τ-oscillation of
z_n at each point can be bounded from above by $2\tau d_1 2^{n(1-H)} = 2 d_1 \tau^H$. On the
other hand, the τ-oscillation of $z - z_n = \sum_{i=n+1}^{\infty} 2^{-iH} g(2^i t)$ is less than the sum
of amplitudes of this series, namely, $\sum_{n+1}^{\infty} 2^{-iH} \leq d_2 2^{-nH} = d_2 \tau^H$, where d_2
that depends only on H. Finally, we find that:

$$\mathrm{osc}_\tau(t)(z) \leq \mathrm{osc}_\tau(t)(z_n) + \mathrm{osc}_\tau(t)(z - z_n) \leq c_2 \tau^H \ ,$$

with $c_2 = 2d_1 + d_2$.
b) To prove the other inequality, we use the fact that at each dyadic point of
type $t = k\, 2^{-n}$, where k is an integer, the value of z is the same as the value
of z_n. In fact, the functions $2^{-iH} g(2^i t)$ take the value 0 at these points whenever
$i > n$. Since z_n is linear on the interval $[k\, 2^{-n}, (k+1)\, 2^{-n}]$, it is either increasing
or decreasing. Assume it is an increasing function, and consider the median point
$t_0 = (2k+1)2^{-(n+1)}$, where $z(t_0) = z_{n+1}(t_0)$. The value $z(t_0)$ is at least equal to
$z_n(k\, 2^{-n}) + 2^{-(n+1)H}$. It follows that the τ-oscillation of z at t is at least equal
to $2^{-(n+1)H} = 2^{-H} \tau^H$. Since all intervals of length 2τ contains a dyadic interval
of this type, we obtain

$$\mathrm{osc}_\tau(t) \geq 2^{-H} \tau^H$$

for all t. To end the proof, take $c_1 = 2^{-H}$.   ◀

12.7 Weierstrass function

Here, the basic periodic function is $g(t) = \cos t$, and the Weierstrass function is
defined as follows:

$$W(t) = \sum_{n=0}^{\infty} \omega^{-nH} \cos(\omega^n t) \ ,$$

where $\omega > 1$ and $0 < H < 1$. To obtain a more realistic rough profile, we may
introduce phases:

$$W(t) = \sum_{n=0}^{\infty} \omega^{-nH} \cos(\omega^n t + \phi_n) \ ,$$

where (ϕ_n) is any sequence. The sequence (ϕ_n) often consists of random numbers
(Fig. 12.7).
    We shall prove the following result concerning the dimension of the graph of
this function. This result is similar to the one proved about Knopp functions:

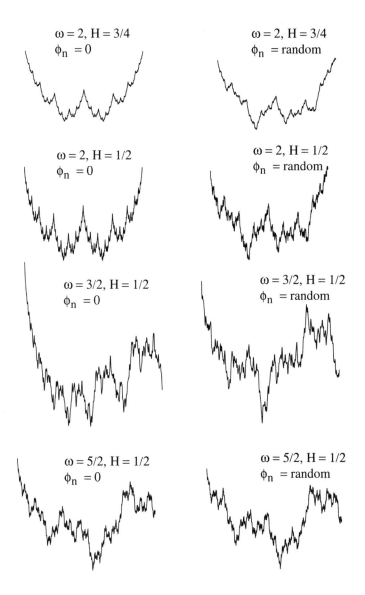

Fig. 12.7. *The Weierstrass functions for different values of ω, H, and ϕ_n. The fractal dimension depends only on H. When the sequence (ϕ_n) is chosen at random, it has a uniform distribution over $[0, \pi[$.*

Let $W(t)$ be the function defined on the interval $[a, b]$. The dimension of its graph Γ is

$$\Delta(\Gamma) = 2 - H .$$

▶   a) The proof of the inequality

$$\Delta(\Gamma) \leq 2 - H$$

is similar to the one in §6. We show that at each point $W(t)$ is Holderian of exponent H. We put $W_n(t) = \sum_{i=0}^{n} \omega^{-iH} \cos(\omega^i t + \phi_i)$. This is a differentiable function whose derivative at any point satisfies:

$$|W_n'(t)| \leq \sum_{0}^{n} \omega^{i(1-H)} \leq d_1 \, \omega^{n(1-H)} ,$$

with d_1 depending only on H and ω. If $\tau = \omega^{-n}$, then the τ–oscillation of W_n is bounded by $2 \tau \, d_1 \, \omega^{n(1-H)} = 2 \, d_1 \, \tau^H$. On the other hand, the τ–oscillation of $W - W_n = \sum_{i=n+1}^{\infty} \omega^{-iH} \cos(\omega^i t + \phi_i)$ is less than $2 \sum_{n+1}^{\infty} \omega^{-iH} \leq d_2 \, \omega^{-nH} = d_2 \, \tau^H$, where d_2 depends only on H and ω. It follows that for all t and τ,

$$\mathrm{osc}_\tau(t)(W) \leq \mathrm{osc}_\tau(t)(W_n) + \mathrm{osc}_\tau(t)(W - W_n) \leq (2d_1 + d_2)\tau^H .$$

By §5, this proves the desired inequality.

b) Now that we have proved that W is Holderian, it would be natural to show that it is anti–Holderian at each point. This is true, but its proof is complicated. Here, we prefer to give a more direct proof of the inequality

$$\Delta(\Gamma) \geq 2 - H .$$

If $H = 1$, it is clear that $\Delta(\Gamma) \geq 2 - 1 = 1$. Therefore we assume that $H < 1$. To evaluate the τ–variation

$$\mathrm{Var}_\tau = \int_a^b \mathrm{osc}_\tau(t) \, dt ,$$

we shall make use of the following approximation:

$$\mathrm{osc}_\tau(t) \geq |W(t + \tau) - W(t - \tau)| .$$

We fix an integer n, and put $\tau = \omega^{-n}$:

$$W(t + \tau) - W(t - \tau) = -2 \omega^{-nH} \sin(1) \sin(\omega^n t + \phi_n)$$

$$- 2 \sum_{i \neq n} \omega^{-iH} \sin(\omega^i \tau) \sin(\omega^i t + \phi_i) .$$

Let $h(t)$ be a continuous function such that for all t $|h(t)| < 1$ (later, we shall give the precise expression of this function). We write

$$\text{Var}_\tau \geq \int_a^b |h(t)(W(t+\tau) - W(t-\tau))| \, dt$$

$$\geq \left| \int_a^b h(t)(W(t+\tau) - W(t-\tau)) \, dt \right|$$

$$\geq 2\omega^{-nH} \sin(1) \left| \int_a^b h(t) \sin(\omega^n t + \phi_n) \, dt \right|$$

$$- 2 \sum_{i \neq n} \omega^{-iH} \left| \int_a^b h(t) \sin(\omega^i t + \phi_i) \, dt \right| .$$

We choose $h(t) = \sin(\omega^n t + \phi_n)$. All that we have to do now is to evaluate the following integrals:

$$\int_a^b (\sin(\omega^n t + \phi_n))^2 \, dt \simeq \frac{b-a}{2} \, ;$$

$$\left| \int_a^b \sin(\omega^n t + \phi_n) \sin(\omega^i t + \phi_i) \, dt \right| \leq \frac{2}{|\omega^n - \omega^i|} + \frac{2}{\omega^n + \omega^i}$$

$$\leq \frac{4}{|\omega^n - \omega^i|} \, ,$$

if $i \neq n$. This is where the fact that the sequence of frequencies of $W(t)$ is a geometric progression will prove to be crucial. In fact,

$$\frac{1}{|\omega^n - \omega^i|} \leq \frac{1}{\omega^n - \omega^{n-1}} = \frac{\omega^{-n}}{1 - \omega^{-1}} \, .$$

Putting all these results together we get:

$$\text{Var}_\tau \geq c_1 \omega^{-nH} - c_2 \omega^{-n} = c_1 \tau^H - c_2 \tau \, ,$$

where $c_1 = (b-a) \sin(1)$ and $c_2 = (8/(1-\omega^{-1})) \sum \omega^{-iH}$. But $H < 1$, and τ is negligible compared to τ^H. Therefore

$$\text{Var}_\tau \succeq \tau^H \, ,$$

which proves the desired inequality; namely, $\Delta(\Gamma) \geq 2 - H$ (§4). ◀

12.8 Fractal dimension and the structure function

The **structure function** of a continuous function $z(t)$ defined on the interval $[a, b]$ is written as follows:

$$S_2(\tau) = \int_a^b (z(t+\tau) - z(t-\tau))^2 \, dt \, .$$

To define it well, we need to define $z(t - \tau)$ when $t - \tau < a$ and $z(t + \tau)$ when $t + \tau > b$. We can, for example, put $z(t) = z(a)$ if $t < a$, and $z(t) = z(b)$ if $t > b$, or we may give any other possible value. In what follows we leave aside this boundary problem.

What is the utility of the structure function? When $z(t)$ is periodic of period T_0, $S_2(\tau)$ is null for $\tau = T_0/2$. Conversely, this function can be used to indicate the proper frequencies of a signal by the values of τ that nullify the function. We can generalize this by putting:

$$S_\alpha(\tau) = \int_a^b |z(t + \tau) - z(t - \tau)|^\alpha \, dt \ ,$$

where α is a parameter ≥ 1. The function $S_1(\tau)$ was already used in the proof of the previous section; it gave an evaluation by default of Var_τ. In general $S_\alpha(\tau)$ can serve to evaluate the fractal dimension of the graph. But we need some more conditions.

THEOREM   Let $z(t)$ be a continuous function defined on $[a, b]$. We assume that there exist two parameters $\alpha \geq 1$, $0 < H < 1$ and two constants c_1 and c_2 such that

$$(i) \quad \sup_{t \in [a,b]} |z(t + \tau) - z(t - \tau)| \leq c_1 \, \tau^H \ ;$$

$$(ii) \quad \left[\int_a^b |z(t + \tau) - z(t - \tau)|^\alpha \, dt \right]^{1/\alpha} \geq c_2 \, \tau^H \ .$$

Then

$$\boxed{\Delta(\Gamma) = 2 - H \ .}$$

▶   The hypothesis (i) implies that $z(t)$ is a Holderian function with exponent H at all t. Therefore (§5)

$$\Delta(\Gamma) \leq 2 - H \ .$$

On the other hand, we fix τ and let $f(t) = |z(t + \tau) - z(t - \tau)|$ giving:

$$\text{Var}_\tau(z) \geq \int_a^b f(t) \, dt = \int_a^b \frac{f(t)^\alpha}{f(t)^{\alpha - 1}} \, dt \ .$$

Using (i) once more: $f(t)^{\alpha - 1} \leq c_1^{\alpha - 1} \, \tau^{H(\alpha - 1)}$. Therefore

$$\text{Var}_\tau(z) \geq c_1^{\alpha - 1} \, \tau^{H(\alpha - 1)} \int_a^b f(t)^\alpha \, dt \ .$$

Finally (ii) implies that $\int_a^b f(t)^\alpha \, dt \geq c_2^\alpha \, \tau^{\alpha H}$. This gives

$$\text{Var}_\tau(z) \geq c_1^{1 - \alpha} \, c_2^\alpha \, \tau^H \ .$$

Therefore $\Delta(\Gamma) \geq 2 - H.$ ◀

The expression

$$\left[\int_a^b |f(t)|^\alpha \, dt \right]^{1/\alpha}$$

is called the "α–norm of $f(t)$." It satisfies the usual properties of a norm. Furthermore it satisfies the following:

$$\lim_{\alpha \to \infty} \left[\int_a^b |f(t)|^\alpha \, dt \right]^{1/\alpha} = \sup_{t \in [a,b]} f(t) .$$

This gives the idea that the fractal dimension of a graph can be computed by replacing the oscillations by the averages of local differences. For example:

For every continuous function $z(t)$ defined on $[a, b]$ with graph Γ:

$$\Delta(\Gamma) \geq \lim_{\tau \to 0} \left(2 + \frac{1}{\alpha} - \frac{\log \left(\int_a^b \left[\int_0^\tau |z(t + s) - z(t - s)|^\alpha \, ds \right]^{1/\alpha} dt \right)}{\log \tau} \right) .$$

▶ In fact, the (integral) mean value theorem allows us to say that for all t, there exists an s_0, $0 < s_0 \leq \tau$, such that

$$\int_0^\tau |z(t + s) - z(t - s)|^\alpha \, ds = \tau \, |z(t + s_0) - z(t - s_0)|^\alpha .$$

Since the second member is less than $\tau \operatorname{osc}_\tau(t)^\alpha$, we obtain

$$\int_a^b \left[\int_0^\tau |z(t + s) - z(t - s)|^\alpha \, ds \right]^{1/\alpha} dt \leq \int_a^b (\tau \operatorname{osc}_\tau(t)^\alpha)^{1/\alpha} \, dt$$

$$= \tau^{1/\alpha} \operatorname{Var}_\tau ,$$

and therefore

$$\operatorname{Var}_\tau \succeq \tau^{-1/\alpha} \int_a^b \left[\int_0^\tau |z(t + s) - z(t - s)|^\alpha \, ds \right]^{1/\alpha} dt .$$ ◀

We can find infinitely many types of such relations. This one has the advantage of containing local and global averages that regulate the results and allow a more precise evaluation of the dimension.

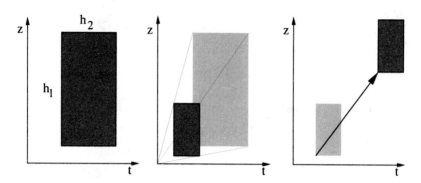

Fig. 12.8. *The affine transformation*

$$F(x) = \begin{pmatrix} \rho_1 & 0 \\ 0 & \rho_2 \end{pmatrix} x + \begin{pmatrix} b_1 \\ b_2 \end{pmatrix}$$

transforms a rectangle $h_1 \times h_2$ with sides parallel to the axis, onto a rectangle $\rho_1 h_1 \times \rho_2 h_2$ (change of the scale), then translates the new rectangle according to a translation vector $B = \begin{pmatrix} b_1 \\ b_2 \end{pmatrix}$.

12.9 Functions constructed by diagonal affinities

In a plane with two orthogonal axes Ot, Oz, a *diagonal affinity* is a plane transformation consisting of a translation and two changes of scale: one on Ot, the second on Oz.

Such a transformation can be written in its matrix form as follows:

$$F\begin{pmatrix} t \\ z \end{pmatrix} = \mathcal{M}\begin{pmatrix} t \\ z \end{pmatrix} + B \,,$$

where $\mathcal{M}$ is a diagonal matrix:

$$\mathcal{M} = \begin{pmatrix} \rho_1 & 0 \\ 0 & \rho_2 \end{pmatrix} \,.$$

This matrix multiplies the abscissas by ρ_1 and the ordinates by ρ_2. Here $B = \begin{pmatrix} b_1 \\ b_2 \end{pmatrix}$ is the translation vector. This is a particular case of affine transformations, where, in general, the matrix $\mathcal{M}$ is of any kind.

In this section we are interested in graphs Γ of continuous functions $z(t)$ for which there exist an integer $N \geq 2$ and N distinct diagonal affinities $F_1, \ldots, F_N$, such that

$$\Gamma = F_1(\Gamma) \cup \ldots \cup F_N(\Gamma) \,.$$

We then say that Γ has a **self–affine** structure. Such a curve is the union of N affine copies of itself. To construct a function of this type so that it is defined on the interval $[a, b]$, we may proceed as follows. Given

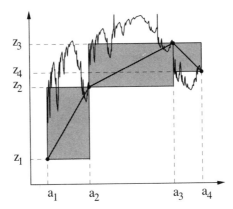

Fig. 12.9. *The graph Γ is the union of three graphs $F_1(\Gamma)$, $F_2(\Gamma)$, and $F_3(\Gamma)$. The affine applications F_1, F_2, and F_3 transform the rectangle $[a_1, a_4] \times [z_1, z_4]$ into the three shaded rectangles.*

(i) $N + 1$ numbers in $[a, b]$ such that: $a = a_1 < a_2 < \ldots < a_{N+1} = b$;

(ii) $N + 1$ numbers on Oz: $z_1, \ldots, z_{N+1}$ such that $z_1 \neq z_{N+1}$, and $|z_{i+1} - z_i| < |z_{N+1} - z_1|$ for all $i = 1, \ldots, N$;

(iii) For all $i = 1, \ldots, N$: a diagonal affinity F_i that transforms the segment whose endpoints are (a, z_1) and (b, z_{N+1}), into the segment of endpoints (a_i, z_i) and (a_{i+1}, z_{i+1}).

The curve Γ is entirely determined by these N transformations. It can be constructed with the help of the set function

$$\mathcal{T}(E) = F_1(E) \cup \ldots \cup F_N(E) \,.$$

Starting with any domain E (for example, the initial rectangle $R = [a, b] \times [z_1, z_{N+1}]$ whose diagonal has endpoints (a, z_1) and (b, z_{N+1})), we successively construct the sets $E_1 = \mathcal{T}(E)$, $E_2 = \mathcal{T}(E_1)$, $\ldots$, $E_n = \mathcal{T}(E_{n-1})$, and so forth. We get a sequence (E_n) that converges, in the sense of Hausdorff distance, to a curve Γ. This curve satisfies the desired property $\Gamma = F_1(\Gamma) \cup \ldots \cup F_N(\Gamma)$, and it passes through the points (a_i, z_i). We do not give the proof of this fact. It is not different from the one given in Chapter 13 on self–similar curves. A classical proof can be found in any topological manual on fractal sets. Its main argument rests on the fact that the F_i are *contractions*, because $|a_{i+1} - a_i| < b - a$ and $|z_{i+1} - z_i| < |z_{N+1} - z_1|$.

Particular case Let ρ_1 and ρ_2 be two real numbers, $0 < \rho_1 < \rho_2 < 1$. We shall assume that all the transformations F_i make the same change of scale defined by $\pm\rho_1$ on the abscissa and by $\pm\rho_2$ on the ordinate. The number ρ_1 is then necessarily equal to $1/N$, and $a_i = a + \frac{(i-1)(b-a)}{N}$. The rectangles $F_i(R)$ are all of equal dimensions $(b - a)/N \times \rho_2|z_{N+1} - z_1|$ (Fig. 12.10). The arc of Γ that corresponds to the abscissas $a_i \leq t \leq a_{i+1}$ is equal to $F_i(\Gamma)$. Therefore, if $c = \sup_{t \in [a,b]} z(t) - \inf_{t \in [a,b]} z(t)$ is the total oscillation of $z(t)$ on $[a, b]$, the oscillation of $z(t)$ on $[a_i, a_{i+1}]$ is exactly $c\rho_2$. Similarly, if we divide $[a, b]$ into N^k

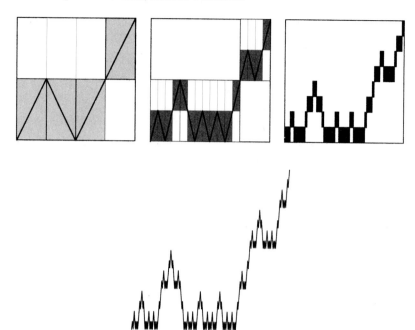

Fig. 12.10. *Construction of a fractal curve (steps 1, 2, 3, and 5) by four diagonal affinities. The affine transformations map the square $[0,1] \times [0,1]$ to the four shaded rectangles. In matrix form the affine applications F_1, F_2, F_3, F_4 are defined by*

$$F_1(x) = \begin{pmatrix} 1/4 & 0 \\ 0 & 1/2 \end{pmatrix} x, \qquad\qquad F_2(x) = \begin{pmatrix} -1/4 & 0 \\ 0 & 1/2 \end{pmatrix} x + \begin{pmatrix} 1/2 \\ 0 \end{pmatrix},$$

$$F_3(x) = \begin{pmatrix} 1/4 & 0 \\ 0 & 1/2 \end{pmatrix} x + \begin{pmatrix} 1/2 \\ 0 \end{pmatrix}, \quad and \quad F_4(x) = \begin{pmatrix} 1/4 & 0 \\ 0 & 1/2 \end{pmatrix} x + \begin{pmatrix} 3/4 \\ 1/2 \end{pmatrix}.$$

In this example, $N = 4$, $\rho_1 = 1/4$, and $\rho_2 = 1/2$; therefore $H = 1/2$ and $\Delta(\Gamma) = 3/2$.

equal intervals, the oscillation of $z(t)$ on each interval is $c\,\rho_2^k$. We deduce that at each t, $z(t)$ is Holderian of exponent

$$H = \log \rho_2 / \log \rho_1 = |\log \rho_2| / \log N .$$

The dimension of Γ is therefore

$$\Delta(\Gamma) = 2 - H .$$

12.10 Invariance under change of scale

In this section we shall discuss the continuous function $z(t)$ defined for all $t \geq 0$ that satisfy the following relation:

$$z(\rho_1 \, t) = \rho_2 \, z(t).$$

Here ρ_1 and ρ_2 are constant, with

$$0 < \rho_1 \leq \rho_2 < 1 \, .$$

For every integer n, we therefore obtain

$$z(\rho_1^n) = \rho_2^n \, z(1) \, , \text{ and } z(\rho_1^{-n}) = \rho_2^{-n} \, z(1) \, .$$

These equalities prove that $z(t)$ tends to 0 when t tends to 0 and that if $z(t)$ is not a null function, then $|z(t)|$ tends to $+\infty$ when t tends to $+\infty$. These growths are ruled by the coefficients ρ_1 and ρ_2. At 0, the function is Holderian of exponent $H = \log \rho_2 / \log \rho_1$.

Let Γ be the graph of $z(t)$ on $[0, +\infty)$, and let F be the affine transformation defined by $F(x) = \begin{pmatrix} \rho_1 & 0 \\ 0 & \rho_2 \end{pmatrix} x$. The invariance relation can be formulated as:

$$F(\Gamma) \subset \Gamma \, .$$

Example 1 We construct a graph Γ on the interval $[0, 1]$ with the help of N diagonal affinities $F_1, \ldots, F_N$ as in the previous section. The curve Γ satisfies the equality $\Gamma = F_1(\Gamma) \cup \ldots \cup F_N(\Gamma)$. In particular,

$$F_1(\Gamma) \subset \Gamma \, .$$

If $F_1(x)$ is given by $F_1(x) = \begin{pmatrix} \rho_1 & 0 \\ 0 & \rho_2 \end{pmatrix} x$, then we have $z(\rho_1 \, t) = \rho_2 \, z(t)$ for all $t \in [0, 1]$. It is easy, thanks to this relation, to extend the domain of definition to $[0, +\infty)$ (Fig. 12.11). For each t, we find an integer n such that $\rho_1^n \, t \leq 1$, and we put

$$z(t) = \rho_2^{-n} \, z(\rho_1^n \, t) \, .$$

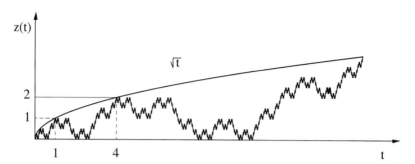

Fig. 12.11. *On* $[0,1]$, *this curve is identical to the curve of Figure* 12.10. *It is extended to* $[0,+\infty)$ *by applying the relation* $z(t) = z(4\,t)/2$.

The value of $z(t)$ does not depend on the choice of the integer n. We remark that in this extension, the only transformation that plays any role is F_1.

Example 2 The function

$$z(t) = t^H$$

also satisfies an invariance relation for every pair (ρ_1, ρ_2) with $H = \log\rho_2/\log\rho_1$. This relation can be written as:

$$z(t) = \rho_1^{-H} z(\rho_1\,t) \ .$$

Thus, we see that the invariance relation can be satisfied by functions whose graphs are locally rectifiable.

Example 3 See the Weierstrass–Mandelbrot function in §11.

Geometrical properties When

$$z(\rho_1\,t) = \rho_2 z(t) \ ,$$

the growth to $+\infty$ of $|z(t)|$ is ruled by the Hölder exponent $H = \log\rho_2/\log\rho_1$. In fact, if (t_0, z_0) is any point of the graph, then all the points $(\rho_1^{-n}\,t_0, \rho_1^{-nH}\,z_0)$ also belong to the graph and to the intersection of the graph of $z(t)$ with the graph of the function $(z_0/t_0)t^H$. In particular, if z_0 represents the maximum of $|z(t)|$ on $[0,1]$, then the curve Γ will lie between the two graphs $\pm(z_0/t_0)t^H$. The general growth of Γ is the same as the growth of t^H.

The associated periodic function We observe that Γ contains parts that are repeated infinitely many times, but not in a periodic manner; their scale increases with t^H. We shall obtain a true periodic function after we divide $z(t)$ by t^H and draw it on a logarithmic coordinate system. That is,

Let ρ and H be two parameters in $]0,1[$, and let $z(t)$ be a function such that for all $t > 0$, $z(t) = \rho^{-H} z(\rho\,t)$. Let

$$z^*(t) = \rho^{Ht} z(\rho^{-t}) \ :$$

then $z^*(t)$ is a periodic function of period 1.

▶ The invariance relation implies:

$$z^*(t+1) = \rho^{Ht+H} z(\rho^{-t-1}) = \rho^{Ht} \rho^H z(\rho^{-1}\rho^{-t})$$
$$= \rho^{Ht} z(\rho^{-t}) .$$ ◀

If $z(t) = t^H$, we simply obtain $z^*(t) = 1$.

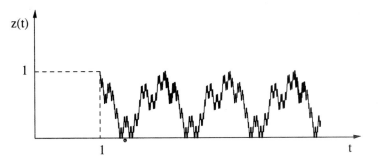

Fig. 12.12. *The graph of the periodic function* $z^*(t) = 2^t z(4^{-t})$ *associated with the function* $z(t)$ *of Figure* 12.11.

The inverse transformation Take any periodic function z^* of period 1 and two parameters ρ and H between 0 and 1. If we put, for all $t > 0$,

$$z(t) = t^H z^*\left(\frac{\log t}{|\log \rho|}\right) ,$$

then z satisfies the invariance relation $z(\rho t) = \rho^H z(t)$. Thus, starting with the periodic functions, we can obtain all the functions that satisfy the relation.

Comparison of the two associated functions The functions $z(t)$ and $z^*(t)$ have different global behaviors, but their local behavior is similar. Here is the reason: the graph of z^* is induced by the graph of z via the following plane transformation

$$(t, z) \longrightarrow \left(\frac{\log t}{|\log \rho|}, t^{-H} z\right) .$$

This is a bijection defined on $]0, \infty) \times (-\infty, +\infty)$ whose image is the entire plane. It is continuous with partial derivatives with respect to t and to z at every point of its domain of definition. That means that this function is a local *diffeomorphism*. We have seen in Chapter 10, §3 that such a transformation does not change the value of the dimension. Thus $z^*(t)$ has all the local characteristics of $z(t)$. We can observe this by comparing Figures 12.11 and 12.12.

Conclusion The relation

$$z(\rho_1 t) = \rho_2 z(t)$$

characterizes some fractal curves whose dimension is related to the exponent $H = \log \rho_2 / \log \rho_1$. However, this relation is not unique to fractal curves.

12.11 The Weierstrass–Mandelbrot function

The function of Weierstrass (§7) $W(t) = \sum_0^\infty w^{-nH} \cos(w^n t)$ satisfies the relation $W(w\,t) = w^H(W(t) - \cos t)$, which is not an invariance relation like the one studied in §10. The Weierstrass–Mandelbrot function:

$$WM(t) = \sum_{-\infty}^{+\infty} w^{-nH}(1 - \cos(w^n t))$$

was constructed from $W(t)$ so that it satisfies the invariance relation

$$WM(w\,t) = w^H WM(t) \;.$$

In this case, $\rho_1 = w^{-1}$ and $\rho_2 = w^{-H}$, $w > 1$. It is worth verifying that this series converges.

▶ The sum $\sum_0^\infty w^{-nH}(1 - \cos(w^n t))$ converges absolutely. On the other hand, when n tends to $-\infty$, $1 - \cos(w^n t)$ is equivalent to $w^{2n} t^2$. Therefore $w^{-nH}(1 - \cos(w^n t))$ is equivalent to $w^{n(2-H)} t^2$, whose series for $n \le 0$ converges. ◀

The graph of $WM(t)$ grows like t^H. This is due to the introduction of low frequencies in the Weierstrass series.

◇ These functions WM do not present a self–affine structure as defined in §9.

◇ The dimension of every bounded part of the graph of WM is $2 - H$, as is the dimension of the graph of WM^*.

▶ Write $WM(t)$ as

$$WM(t) = \sum_{-\infty}^{-1} w^{-nH}(1 - \cos(w^n t)) + \sum_0^{+\infty} w^{-nH} - \sum_0^{+\infty} w^{-nH} \cos(w^n t) \;.$$

This is the sum of three functions. The first is differentiable because the series of the derivatives of the terms

$$\sum_{-\infty}^{-1} w^{n(1-H)} \sin(w^n t)$$

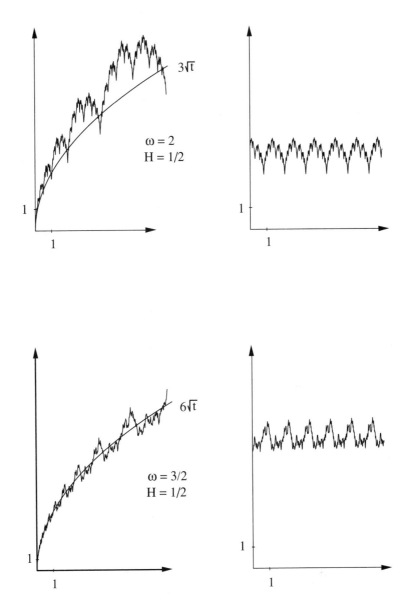

Fig. 12.13. *On the left, the graphs of Weierstrass–Mandelbrot functions for different values of parameters ω and H. On the right, the graph of the associated periodic functions $WM^*(t) = \omega^{-Ht}WM(\omega^t) = \sum_{-\infty}^{+\infty} \omega^{-H(n+t)}(1 - \cos(\omega^{n+t})).$*

is absolutely convergent. Its graph on any interval $[a, b]$ $a > 0$, is of dimension 1. The second term is constant, and the dimension of its graph is 1. Finally, the third term is $W(t)$ itself; the dimension of its graph is $2 - H$. ◀

◇ Every series of the form $z(t) = \sum_{n=0}^{\infty} \omega^{-nH} g(\omega^n t)$ can provide an analogous example if H is < 1 and if the function $g(t)$ is periodic and continuously differentiable at 0. We can then put:

$$z(t) = \sum_{n=-\infty}^{+\infty} \omega^{-nH} (g(0) - g(\omega^n t)) \ .$$

This function presents, like the Weierstrass–Mandelbrot function (which corresponds to the case of $g(t) = \cos t$), a scale invariance.

▶   The series that defines $z(t)$ converges. In fact, the first three terms of the Taylor series of $g(t)$ in the neighborhood of 0 exist and

$$|g(t) - g(0)| \leq c t \ ,$$

where c is a constant associated to the value of the derivative. We deduce that when n tends to $-\infty$, $|g(0) - g(\omega^n t)|$ is less than $c\omega^n t$. Thus we find that $c\omega^{n(1-H)}t$ is an upper bound for each term of the series. Finally, $z(t)$ satisfies the same invariance relation, namely:

$$z(\omega t) = \omega^H z(t) \ . ◀$$

12.12 The spectrum of invariant functions

Assume that $z(t)$ satisfies the invariance relation

$$z(\rho t) = \rho^H z(t) \ ,$$

where $0 < \rho < 1$ and $0 < H < 1$. We shall study the effect of this relation on the spectrum of $z(t)$.
• If the spectrum of $z(t)$ is *discrete*, we consider the **power spectral density**

$$G_z(s) = \lim_{T \to \infty} \frac{1}{T^2} \left| \int_0^T e^{-ist} z(t) \, dt \right|^2 \ .$$

Assume that this limit exists, and let us calculate $G_z(\rho s)$. With a change of variable, the integral $\int_0^T e^{-i\rho st} z(t) \, dt$ can be written as:

$$\rho^{-1} \int_0^{\rho T} e^{-isu} z(\rho^{-1} u) \, du = \rho^{-(H+1)} \int_0^{\rho T} e^{-isu} z(u) \, du \ .$$

Therefore,

$$G_z(\rho s) = \lim_{T \to \infty} \frac{\rho^{-2(H+1)}}{T^2} \left| \int_0^{\rho T} e^{-ist} z(t) \, dt \right|^2$$

$$= \rho^{-2H} \lim_{T \to \infty} \frac{1}{(\rho T)^2} \left| \int_0^{\rho T} e^{-ist} z(t) \, dt \right|^2 .$$

Finally,

$$G_z(\rho s) = \rho^{-2H} G_z(s) .$$

Therefore, this spectral density presents a scale invariance. Every fundamental frequency s_0 of the spectrum (a value for which $G_z(s)$ is not null) is at the origin of a sequence of frequencies $\rho^{-1} s_0$, $\rho^{-2} s_0$, ..., $\rho^{-k} s_0$, ... tending to infinity for which G_z is not null. Moreover,

$$G_z(\rho^{-k} s_0) = \rho^{-2kH} G_z(s_0) .$$

The parameter $2H$ controls the growth to 0 of $G_z(s)$ when s tends to infinity.
- If the spectrum of $z(t)$ is *continuous*, we evaluate its **power spectrum**

$$P_z(s) = \lim_{T \to \infty} \frac{1}{T} \left| \int_0^T e^{-ist} z(t) \, dt \right|^2 ,$$

provided this limit exists. The same type of computation proves that

$$P_z(\rho s) = \rho^{-2H-1} P_z(s) .$$

When s tends to $+\infty$, the power spectrum tends to 0 with a growth dominated by s^{-2H-1}.

◇ In the case of the curve $WM(t)$ whose spectrum is discrete, the decrease of the power spectral density is controlled by the parameter $2H$, though the dimension of its graph is $2 - H$. But we should not conclude that the behavior of the spectrum of $z(t)$ is always related to the value of its graph dimension. We give a counterexample.

Example The Weierstrass function

$$W(t) = \sum_0^\infty 2^{-nH} \cos(2\pi \, 2^n \, t)$$

has a graph of dimension $2 - H$ on any finite interval. This is a periodic function whose period is 1. Taking the "inverse transformation" of W (as defined in §10), for all parameters ρ and H' that are between 0 and 1, we can associate a function $w(t)$ that is invariant under the change of scale:

$$w(t) = t^{H'} \sum_0^\infty 2^{-nH} \cos\left(\frac{2\pi \, 2^n}{|\log\rho|} \log t\right) .$$

This new function satisfies the relation

$$w(\rho\, t) = \rho^{H'} \, w(t) \, ,$$

and the decrease of its spectrum is related to the parameter H'. On the other hand, the dimension of its graph is $2 - H$. These two parameters are independent.

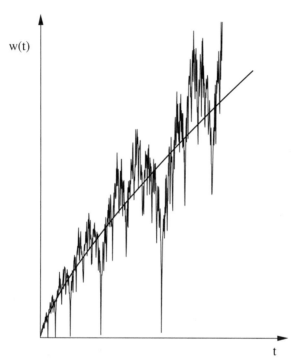

Fig. 12.14. *The graph of the function* $w(t) = t^{3/4} \sum_0^\infty 2^{-n/4} \cos\left(\frac{2\pi \, 2^n}{\log 2} \log t\right)$. *Its dimension is* $2 - 1/4 = 7/4$. *It satisfies the invariance relation* $w(2\, t) = 2^{3/4}\, w(t)$. *We have drawn the parabolic curve* $z(t) = 6\, t^{3/4}$ *that indicates the growth of* $w(t)$.

12.13 Computing the dimensions of the graphs

As for sets of null measure (Chap. 3, §5), we shall look for a quantity $Q(\epsilon)$ such that, theoretically, the limit of the ratio

$$\frac{\log Q(\epsilon)}{|\log \epsilon|}$$

is equal to the fractal dimension when ϵ tends to 0. Given a decreasing sequence $\epsilon_1, \epsilon_2, \ldots, \epsilon_N$, we draw all the points

$$(|\log \epsilon_n| \, , \log Q(\epsilon_n))$$

that form the *logarithmic diagram* (or log-log plot). The slope of the least square line is the approximate value of the dimension.

The result, and above all the correlation of the diagram, depends a lot on the measured quantity $Q(\epsilon)$. For some choices of $Q(\epsilon)$, the diagram presents a systematic concavity or important dispersion; the value of the slope is then meaningless. Other choices of $Q(\epsilon)$, that is, different algorithms, will give a better correlation.

To test the efficiency of such algorithms, we can digitize into sets of 500, 1000, or 2000 points the graphs of functions with known dimension. The functions of Knopp and Weierstrass, or any other function of the same type, is often used. We consider a method to be "good" if the error of approximating the theoretical value of the dimension is not higher than 5%. In fact, this numerical approximation will depend on the number of points considered. A good method gives a good approximation for a minimum number of points, approximately 1000.

A good choice of the sequence (ϵ_n) is important and so is the choice of the extreme values $\epsilon_N < \epsilon_1$. In fact, if ϵ is too large we "see" the curve from afar, and the dimension will be near 0, the dimension of a point. If ϵ is too small, we see the curve from nearby, and the dimension is either 0 or 1, depending on whether the data are discrete or joined by line segments (linear interpolation). The fractal hypothesis ideally consists of constructing the curve for infinitesimal values of ϵ in such a way that the small arcs of the curve are images, in some sense, of the entire curve. To proceed with such a construction is to do a *fractal interpolation*. It is unusual to do this in practice before computing the dimension, since the dimension is unknown. Such an interpolation will be meaningful if it does not depend on any previously set value of this dimension. This domain of research needs more development.

Minkowski sausage   We evaluate the quantity

$$Q(\epsilon) = \frac{1}{\epsilon^2}\mathcal{A}(\Gamma(\epsilon))$$

and construct the logarithmic diagram

$$\left(|\log \epsilon_n| \, , \log \frac{1}{\epsilon_n^2}\mathcal{A}(\Gamma(\epsilon_n))\right) .$$

This method is a direct result of the theoretical definition of $\Delta(\Gamma)$. Experience shows that it is greatly ineffectual. The experimental profiles are usually defined with a multiplicity factor. The profile of rough surfaces, for example, contains oscillation of the order of micron, which must be multiplied by 10^3 or 10^4 to be visible and measurable. This multiplication is in fact a plane transformation that transforms a circle into a long ellipse. The sausages of the "real" and "multiplied" profiles are not similar. This explains why, in the case of profiles, the method of the Minkowski sausage is sensitive to the multiplication of data by even a factor of 2. The quantities $Q(\epsilon)$, which are invariant under the multiplication, will give better and more stable results. This is the case with the "variation" method, which we shall discuss later.

Boxes   Here we compute the quantity

$$Q(\epsilon) = \omega_\epsilon(\Gamma) \, ,$$

the number of boxes of a network covering the curve (Chap. 10, §5). We construct the logarithmic diagram

$$(|\log \epsilon_n| \, , \, \log \omega_{\epsilon_n}(\Gamma))$$

and calculate the slope of the least square line. Though it is an easy method to implement, it presents the same inconveniences as the previous one: it does not present any form of invariance. The squares are transformed by affine transformations into rectangles. Thus the method is sensitive to the scaling of data. Moreover, the quantity $Q(\epsilon)$ takes only integer values, and therefore it undergoes unpredictable jumps when ϵ varies. This usually widens the dispersion of the results. To improve it, it is interesting to present the results in a different way. Let $\omega_{\epsilon,k}$ be the number of boxes of side ϵ that intersect the graph of $z(t)$ and can be projected into the same interval $[k\epsilon, (k+1)\epsilon]$. Thus, $\omega_\epsilon(\Gamma) = \sum_k \omega_{\epsilon,k}$. The quantity $\epsilon \omega_{\epsilon,k}$ can be considered to be an approximation of the oscillation of $z(t)$ on the interval $[k\epsilon, (k+1)\epsilon]$. If we replace $\epsilon \omega_{\epsilon,k}$ by the real oscillation

$$\mathrm{osc}_{\epsilon,k} = \sup_{k\epsilon \le t \le (k+1)\epsilon} z(t) \; - \inf_{k\epsilon \le t \le (k+1)\epsilon} z(t) \, ,$$

we find a new method, where we have to evaluate the quantity

$$Q(\epsilon) = \frac{1}{\epsilon} \sum_k \mathrm{osc}_{\epsilon,k} \, .$$

We can still do better: we write the previous quantity as $(1/\epsilon^2)\epsilon \sum_k \mathrm{osc}_{\epsilon,k}$, and we replace $\epsilon \sum_k \mathrm{osc}_{\epsilon,k}$ by an integral that is the average of the oscillations of $z(t)$ on all the intervals of length ϵ. This is precisely the variation method that follows.

Variation   Let τ, instead of ϵ, denote the precision of the measurement of the curve. We compute the quantity

$$Q(\tau) = \frac{1}{\tau^2} \mathrm{Var}_\tau \, ,$$

where Var_τ is the integral of the τ-oscillations of $z(t)$ (§3 and §4). We draw the logarithmic diagram

$$\left(|\log \tau_n|, \log(\frac{1}{\tau_n^2} \mathrm{Var}_{\tau_n}) \right) \, .$$

Here are some advantages of this method:

(i) Fast computation: we evaluate the τ-oscillations of $z(t)$ for small values of τ; then we increase this value by using previous results. This algorithm is even faster than the box-counting method.

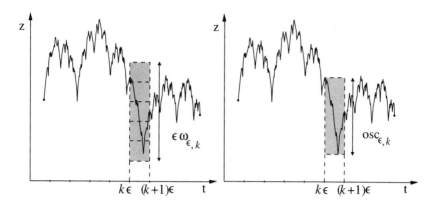

Fig. 12.15. *The box-counting method amounts to measuring the oscillation of $z(t)$ on $[k\epsilon, (k+1)\epsilon]$ by counting the number $\omega_{\epsilon,k}$ of boxes intersecting the curve. But the total height $\epsilon\,\omega_{\epsilon,k}$ can be advantageously replaced by the real oscillation $osc_{\epsilon,k}$.*

(ii) Effectiveness: the obtained results are good approximations of the theoretical values of the dimension when applied to test-curves (Weierstrass, Knopp, brownian motion). This is due in part to the following reason:

(iii) The type of invariance under change of scale. As we have already seen, Var_τ is essentially the area of the surface obtained by horizontal segments whose form does not change when the data are multiplied by a constant factor. Algebraically this can be written as follows (§3):

$$Var_\tau(c\,z) = |c|\,Var_\tau(z)\,,$$

for all constant c. In principle, the two functions $z(t)$ and $c\,z(t)$ give the same slope of logarithmic diagram.

◇   We can find an inconvenience in this method; the measure of the oscillations depends on the extreme values of $z(t)$. In experimental cases, these extreme values are often erroneous.

Other methods Therefore, we may prefer other methods that take more care with all the values of the function $z(t)$. Generally speaking, the larger the number of averages taken on the computed measure, the better aligned the logarithmic diagram will be (the correlation coefficient approaches or equals 1). However, it should be noted that some methods may give, at least theoretically, a result other than the fractal dimension $\Delta(\Gamma)$. For example, the structure function

$$\int_a^b (z(t+\tau) - z(t-\tau))^2\,dt$$

does not give the real value of Δ except for some categories of functions (§8). The function

$$\left(\int_a^b \left[\int_0^\tau |z(t+s) - z(t-s)|^\alpha \, ds \right]^{1/\alpha} dt \right),$$

introduced in §8, gives in principle a value less than or equal to the value of the dimension. The corresponding logarithmic diagram is:

$$\left(|\log \tau_n| \, , \, \log \left(\frac{1}{\tau_n^{2+(1/\alpha)}} \int_a^b \left[\int_0^{\tau_n} |z(t+s) - z(t-s)|^\alpha \, ds \right]^{1/\alpha} dt \right) \right).$$

12.14 Bibliographical notes

In this chapter, one should distinguish between the study of the characteristic parameters of a profile (oscillation, variation, dimension) that constitutes the *signal analysis* and the construction of mathematical models (Weierstrass, Knopp, ...) which is done with the help of *recursive systems*. These are two different domain s of research. In the spirit of this book, we subordinate the second to the first; theoretical curves interest us in as much as they provide us with test–functions that can be compared to real data.

The *signal analysis* point of view, for example, finds its application in the topographic study of rough materials. Such a study was carried out in [C. Tricot, et al.], where the variation method was introduced. The method of data acquisition will have an effect on the choice of algorithm. For example, if a rough surface profile is obtained at the micron scale, then it must be magnified by a power of 10 so that its irregularities become *visible*. We see the importance of choosing a stable method, vis–à–vis the multiplication of data.

The question of the power spectral density was discussed in [C. Tricot 6].

Where models are concerned, the Weierstrass curve goes back to 1875, but the first exhaustive study is to be found in [G.H. Hardy 2] (1916). It was Hardy who proved that this is a unformly Hölderian and anti–Hölderian function of exponent H (to use the vocabulary of this chapter). In 1903, [T. Takagi] gave a much more simple example of a nowhere differentiable function that exactly corresponds to the Knopp function of §6, with $H = 1$. The dimension of the graph of such a funct ion is 1. The "function of [K. Knopp]" (1920) is a bit more general since H can be any real number between 0 and 1; some authors call it the "Takagi function." On this subject, the reader may also consult [E.W. Hobson] (1926) and [M. Hata & M. Yamaguti]. Other functions defined by series and the dimension of their graphs are given in [J. Kaplan et al.]. The Weierstrass function satisfies a functional relation that was studied and generalized in [M. Yamaguti & M. Hata]. Our references are by no means exhaustive, but we should salute the progress made by the Japanese school under the influence of M. Yamaguti. See also the survey article of [M. Hata].

There exists another way to construct fractal curves as attractors of a system of iterated affinities: the example in Figure 12.10 is taken from [B. Mandelbrot 3]. This article contains a general study of *self-affine* structures, and the relation

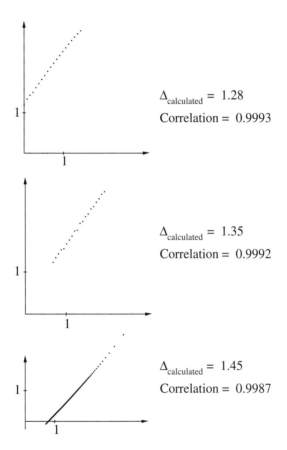

Fig. 12.16. *A comparison of different methods for computing the dimension with logarithmic diagrams: in each case the slope of the regression line and the correlation coefficient are computed:*

   a) the Minkowski sausage method $Q(\epsilon) = (1/\epsilon^2)\mathcal{A}(\Gamma(\epsilon))$;
   b) the box–counting method $Q(\epsilon) = \omega_\epsilon(\Gamma)$; *and*
   c) the variation method $Q(\tau) = (1/\tau^2)\mathrm{Var}_\tau$.

These methods are all applied to the same test–function: a Weierstrass function with parameters $\omega = 2$ *and* $H = 1/2$. *The theoretical value of the dimension is* 1.5.

between the Hölder exponent and the fractal dimension. On this subject the reader may also consult [N. Kôno]. The Weierstrass–Mandelbrot function can be found in [B. Mandelbrot 1].

13 Curves Constructed by Similarities

13.1 Similarities

On the line, a similarity is a function $f(t)$ that can be written as:

$$f(t) = \rho t + b ,$$

where ρ and b are real parameters, $\rho \neq 0$. It is a **translation** if $\rho = 1$, a **symmetry** if $\rho = -1$, and a **homothety** of center O if $b = 0$. These notions can be generalized to the plane.

The plane similarity transformations are compositions of translations, rotations, symmetries, and homotheties, as summarized in Fig. 13.1. Every transformation that is a composition of the first three will be called a **displacement**, because it conserves the forms of the objects. It displaces the object without altering its shape. If F is a displacement, then for all pairs of points x, y in the plane, the distance between the image of the points is the same as the distance between the points themselves:

$$\text{dist}(F(x), F(y)) = \text{dist}(x, y) .$$

In a plane with a Cartesian coordinate system Ox_1, Ox_2, every displacement can be algebraically noted by:

$$F(x) = \mathcal{M}x + B ,$$

where B is a translation vector and $\mathcal{M}$ is an *orthonormal* 2×2-matrix (its column vectors are orthogonal and of length 1). The determinant of $\mathcal{M}$ is ± 1. It is 1 if $\mathcal{M}$ is a rotation matrix of type $\mathcal{R}_\theta$; it is -1 if $\mathcal{M}$ can be written as the product of a rotation matrix and a symmetry matrix $\mathcal{R}_\theta \mathcal{S}_1$ or $\mathcal{S}_1 \mathcal{R}_\theta$.

A **similarity** is a displacement followed by a homothety or a homothety followed by a displacement. It can be written as:

$$F(x) = \rho \mathcal{M}x + B ,$$

where ρ is a real number > 0, $\mathcal{M}$ is an orthogonal matrix, and B is a translation vector. If $\rho < 1$, F is a *contraction*. If $\rho > 1$, F is a *dilation*. If $\rho = 1$, it is a displacement. The number ρ is the *ratio of the similarity*. Therefore, a similarity of ratio $\neq 1$ displaces the object and changes its size in an isotropic manner. In particular, if F is a similarity, then for all pairs of points x, y in the plane,

$$\text{dist}(F(x), F(y)) = \rho \, \text{dist}(x, y) .$$

Figure 13.1 *A similarity is composed of transformations of the following type:*

Translation : *defined by a fixed vector* **B** $=$ $b_1\mathbf{i} + b_2\mathbf{j}$. *To each point* $x = \begin{pmatrix} x_1 \\ x_2 \end{pmatrix}$, *this transformation associates the point* $F(x) = \begin{pmatrix} x_1 + b_1 \\ x_2 + b_2 \end{pmatrix}$.

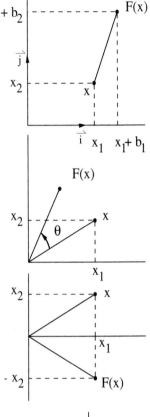

Rotation *of center 0: defined by an angle* θ. *The matrix of this transformation is*

$$\mathcal{R}_\theta = \begin{pmatrix} \cos\theta & -\sin\theta \\ \sin\theta & \cos\theta \end{pmatrix}.$$

To each point x, *this transformation associates the point* $F(x) = \begin{pmatrix} x_1\cos\theta - x_2\sin\theta \\ x_1\sin\theta + x_2\cos\theta \end{pmatrix}$.

Symmetry *with respect to* Ox_1: *defined by the matrix*

$$S_1 = \begin{pmatrix} 1 & 0 \\ 0 & -1 \end{pmatrix}.$$

To each x, *it associates the point* $F(x) = \begin{pmatrix} x_1 \\ -x_2 \end{pmatrix}$.

Symmetry *with respect to* Ox_2: *defined by the matrix*

$$S_2 = \begin{pmatrix} -1 & 0 \\ 0 & 1 \end{pmatrix}.$$

Thus , $F(x) = \begin{pmatrix} -x_1 \\ x_2 \end{pmatrix}$. *We remark that* $S_2 = \mathcal{R}_\pi S_1$.

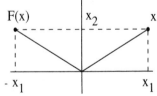

Homothety *of center O: defined by a real* ρ. *To each* x, *it associates the point* $F(x) = \begin{pmatrix} \rho x_1 \\ \rho x_2 \end{pmatrix}$.

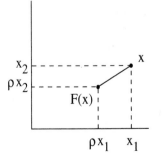

◇   Given four points of the plane, $A \neq B$, $C \neq D$, there exist two similarities in the plane such that $F(A) = C$ and $F(B) = D$. One is of determinant > 0, the other is of determinant < 0. There are also two similarities of the plane such that $F(A) = D$ and $F(B) = C$. All in all there are four distinct similarities F such that $F(AB) = CD$ (Fig. 13.2).

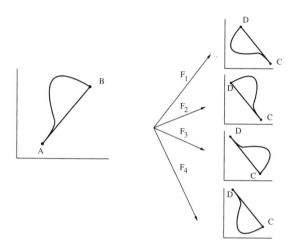

Fig. 13.2. *Given two segments AB and CD of length different from 0, there exist four similarities such that the image of AB is equal to CD. Here F_1 and F_3 have determinant 1; F_2 and F_4 have determinant -1. All have the similarity ratio:*

$$\rho = \frac{\operatorname{dist}(C, D)}{\operatorname{dist}(A, B)} \ .$$

13.2 Self–similar structure

The following theorem gives the necessary conditions to construct a curve in the plane using similarities. The proof is contained in Sections 3 to 6.

THEOREM   *Given $N + 1$ distinct points of the plane, $N \geq 2$:*

$$A = A_1, A_2, \ldots, A_{N+1} = B \,,$$

satisfying $\operatorname{dist}(A_i, A_{i+1}) < \operatorname{dist}(A, B)$ *for all* $i = 1, \ldots, N$; *and given N similarities $F_1, \ldots, F_N$ such that*

$$F_i(AB) = A_i A_{i+1} \,,$$

there exists a unique curve Γ satisfying

$$\Gamma = \bigcup_{i=1}^{N} F_i(\Gamma) \,.$$

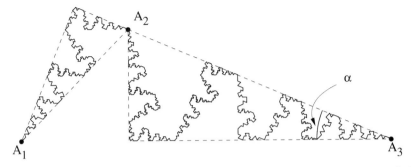

Fig. 13.3. *Example of a self–similar curve Γ. The two arcs $A_1 \frown A_2$ and $A_2 \frown A_3$ are the images of Γ by two similarities F_1 and F_2. In this particular case, the curve Γ is inscribed in a right–angled triangle whose smallest angle is α, $0 < \alpha \le \pi/4$, which passes through the points $A = A_1 = (0,0)$, $A_2 = (1 - \cos 2\alpha, \tan \alpha \cos 2\alpha)$, $A_3 = B = (1,0)$. The transformations F_1 and F_2 contain a symmetry so that they are of negative determinant; they are defined algebraically by*

$$F_1(x) = \tan \alpha \begin{pmatrix} -\sin 2\alpha & -\cos 2\alpha \\ -\cos 2\alpha & \sin 2\alpha \end{pmatrix} x + \begin{pmatrix} 1 - \cos 2\alpha \\ \tan \alpha \cos 2\alpha \end{pmatrix}$$

$$F_2(x) = \frac{\cos 2\alpha}{\cos \alpha} \begin{pmatrix} \cos \alpha & -\sin \alpha \\ -\sin \alpha & -\cos \alpha \end{pmatrix} x + \begin{pmatrix} 1 - \cos 2\alpha \\ \tan \alpha \cos 2\alpha \end{pmatrix} .$$

The curve of this figure corresponds to $\alpha = \pi/8$.

This curve has A and B as endpoints and passes through all the points A_i.

   This *equality of similarity* $\Gamma = \cup_{i=1}^{N} F_i(\Gamma)$ shows that the curve is the union of N copies of itself. Here, the word "copy" means "an image by a similarity." This is a typical case of fractal curves, provided that Γ is of infinite length.

   We denote by

$$\rho_i = \frac{\text{dist}(A_i, A_{i+1})}{\text{dist}(A, B)}$$

the similarity ratio of F_i, and by

$$\rho_{\max} = \max\{\rho_1, \ldots, \rho_N\}$$

the largest of these ratios. The F_i are contractions, and $\rho_{\max}$ is strictly less than 1.

13.3 Generator

Using the notation of the previous theorem, we call the polygonal curve $\mathbf{P}_1$ whose vertices are: $A_1, \ldots, A_{N+1}$ the **generator** of Γ. It is formed by the N segments $A_i A_{i+1}$; this is the first polygonal approximation of the curve Γ whose existence will be proved in §6.

We note that $\mathbf{P}_1$ is not by itself sufficient to completely determine the curve Γ. In fact, for all i there exist four similarities F_i such that $F_i(AB) = A_i A_{i+1}$, as we have already seen in §1. Thus, in principle, $\mathbf{P}_1$ corresponds to 4^N families of different similarities $\{F_1, \ldots, F_N\}$. Except in some cases of symmetry, these will provide us with 4^N different curves.

We continue the construction of finer polygonal approximations of our curve.

The polygonal curve $\mathbf{P}_2$ is obtained by replacing each of the segments $A_i A_{i+1}$ constituting $\mathbf{P}_1$ with its copy $F_i(\mathbf{P}_1)$, which has the same endpoints A_i and A_{i+1}. Thus, $\mathbf{P}_2$ contains N^2 segments, and

$$\mathbf{P}_2 = \bigcup_{i=1}^{N} F_i(\mathbf{P}_1) \,,$$

and so forth. Assume that the polygonal curve $\mathbf{P}_{k-1}$ has been constructed; its endpoints are A and B. It passes through all the points A_i, and it contains N^{k-1} segments. Since the copy $F_i(\mathbf{P}_{k-1})$ has A_i and A_{i+1} as endpoints, we replace each segment $A_i A_{i+1}$ of $\mathbf{P}_1$ by $F_i(\mathbf{P}_{k-1})$, this produces a polygonal curve $\mathbf{P}_k$, made of N^k segments. It may be written as:

$$\mathbf{P}_k = \bigcup_{i=1}^{N} F_i(\mathbf{P}_{k-1}) \,.$$

When we draw these curves with ordinary precision we quickly realize that $\mathbf{P}_k$ and $\mathbf{P}_{k+1}$ are indistinguishable. This is due to the fact that these curves will converge to a limit curve, which is Γ. In §5 we give parameterizations γ_k of $\mathbf{P}_k$; in §6, we will show that they converge to a limit parameterization γ, whose image is the curve Γ.

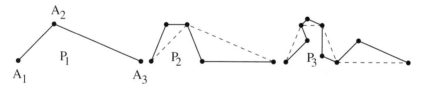

Fig. 13.4. *Construction of the polygonal approximations $\mathbf{P}_k$ corresponding to the two similarities of Figure 13.3.*

$\diamondsuit$ The segments of $\mathbf{P}_1$ have length $\leq \rho_{\max}\mathrm{dist}(A, B)$. Similarly, those of $\mathbf{P}_2$ have length $\leq \rho_{\max}^2\mathrm{dist}(A, B)$, and those of $\mathbf{P}_k$ have length $\leq \rho_{\max}^k\mathrm{dist}(A, B)$. This number tends to 0 when k tends to infinity. If the limit curve Γ exists, the sequence $(\mathbf{P}_k)$ is a sequence of polygonal approximations of Γ (Chap. 5, §3).

13.4 Self–similar structure on $[0, 1]$

Let $\mathbf{I}$ be the interval $[0, 1]$ of the real line. To define parameterizations of the curves $\mathbf{P}_k$, it is necessary to construct a self–similar structure on $\mathbf{I}$. We do it as follows:

We choose $N + 1$ points in $\mathbf{I}$, such that

$$t_1 = 0 < t_2 < \ldots < t_{N+1} = 1 .$$

For all i, $i = 1, \ldots, N$, there exist two similarities f_i such that $f_i(\mathbf{I}) = [t_i, t_{i+1}]$. We choose the one that goes in the same direction as F_i. More precisely:

Case 1 If $F_i(A) = A_i$ and $F_i(B) = A_{i+1}$, f_i must be chosen so that $f_i(0) = t_i$ and $f_i(1) = t_{i+1}$. Therefore,

$$f_i(t) = (t_{i+1} - t_i)t + t_i .$$

Case 2 If $F_i(A) = A_{i+1}$ and $F_i(B) = A_i$, f_i must be chosen so that $f_i(0) = t_{i+1}$ and $f_i(1) = t_i$. Therefore,

$$f_i(t) = (t_i - t_{i+1})t + t_{i+1} .$$

Having thus defined the family $\{f_1, \ldots, f_N\}$ of similarities on the line, we observe that $\mathbf{I}$ can be written as:

$$\mathbf{I} = \bigcup_{i=1}^{N} [t_i, t_{i+1}] = \bigcup_{i=1}^{N} f_i(\mathbf{I}) .$$

This interval has a self–similar structure.

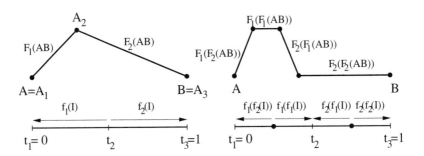

Fig. 13.5. *The construction of similarities f_1 and f_2 on $\mathbf{I} = [0, 1]$ corresponding to the similarities of the plane F_1 and F_2 of Figures 13.3 and 13.4. The function f_1 is decreasing, because $F_1(A) = A_2$ and $F_1(B) = A_1$ (arrow to the left). The function f_2 is increasing, because $F_2(A) = A_2$ and $F_2(B) = A_3$ (arrow to the right).*

13.5 Parameterization of the generator

We shall parameterize the generator $\mathbf{P}_1$ and then all of the polygonal approximations $(\mathbf{P}_k)$.
- We start by parameterizing the segment AB. The coordinates of the point A
are (a_1, a_2); those of the point B are (b_1, b_2). For example, we may put:

$$\gamma_0(t) = \begin{pmatrix} (1-t)\, a_1 + t\, b_1 \\ (1-t)\, a_2 + t\, b_2 \end{pmatrix} .$$

This parameterization corresponds to motion with a constant velocity on AB,
with $\gamma_0(0) = A$, $\gamma_0(1) = B$.
- To construct a parameterization γ_1 of the generator $\mathbf{P}_1$, we start by defining γ_1
on the interval $[t_i, t_{i+1}]$ in the following manner:

$$\text{If } t_i \le t \le t_{i+1} , \ \gamma_1(t) = F_i(\gamma_0(f_i^{-1}(t))) .$$

The function f_i defined in §4 is a bijective application. We remark that in case 1
studied in §4,

$$\gamma_1(t_i) = F_i(\gamma_0(f_i^{-1}(t_i))) = F_i(\gamma_0(0)) = F_i(A) = A_i ,$$

while in case 2,

$$\gamma_1(t_i) = F_i(\gamma_0(1)) = F_i(B) = A_i .$$

The result is the same. We can verify that in both cases $\gamma_1(t_{i+1}) = A_{i+1}$.
Moreover,

$$\gamma_1([t_i, t_{i+1}]) = F_i(\gamma_0(f_i^{-1}([t_i, t_{i+1}]))) = F_i(\gamma_0(\mathbf{I})) = F_i(AB) = A_i A_{i+1} .$$

Taking the union of all the intervals $[t_i, t_{i+1}]$, we obtain a continuous function γ_1
defined on $\mathbf{I}$, such that

$$\gamma_1(\mathbf{I}) = \bigcup_{i=1}^{N} A_i A_{i+1} = \mathbf{P}_1 .$$

Yet γ_1 has the following property:
  For all $t \in \mathbf{I}$ and all $i = 1, \ldots, N$,

$$\gamma_1(f_i(t)) = F_i(\gamma_0(t)) .$$

- We continue in the same manner. Assume that the parameterization γ_{k-1}
of $\mathbf{P}_{k-1}$ has been constructed with $\gamma_{k-1}(t_i) = A_i$ for all i. We define γ_k on
$[t_i, t_{i+1}]$ as follows:

$$\text{If } t_i \le t \le t_{i+1} , \ \gamma_k(t) = F_i(\gamma_{k-1}(f_i^{-1}(t))) .$$

We verify that $\gamma_k(t_i) = A_i$, $\gamma_k(t_{i+1}) = A_{i+1}$, and $\gamma_k([t_i, t_{i+1}]) = F_i(\mathbf{P}_{k-1})$. This allows us to consider the union of the intervals $[t_i, t_{i+1}]$ with γ_k defined and continuous on $\mathbf{I}$, and

$$\gamma_k(\mathbf{I}) = \mathbf{P}_k .$$

Moreover,

for all $t \in \mathbf{I}$ and for all $i = 1, \ldots, N$,

$$\gamma_k(f_i(t)) = F_i(\gamma_{k-1}(t)) .$$

$\diamond$ Now that we have constructed a sequence (γ_k), we can prove the following result about its convergence:

The sequence of the functions (γ_k) uniformly converges to a continuous function, which we shall denote by γ.

▶   a) We first show that for some constant C and for all $t \in \mathbf{I}$,

$$\mathrm{dist}(\gamma_k(t), \gamma_{k+1}(t)) \leq C \, \rho_{\mathrm{max}}^k .$$

Like most of the proofs in this chapter, this is to be proved by induction. Let

$$C = \max_{1 \leq i \leq N+1, 1 \leq j \leq N+1} \mathrm{dist}(A_i, A_j)$$

be the largest distance between any two vertices of the generator. Since every point of AB is at a distance $\leq C$ from the point A, we have $\mathrm{dist}(\gamma_0(t), \gamma_1(t)) \leq C$ for all t. Thus the result is proved for $k = 0$.

Assume it is true for $k - 1$: for every $t \in \mathbf{I}$,

$$\mathrm{dist}(\gamma_{k-1}(t), \gamma_k(t)) \leq C \, \rho_{\mathrm{max}}^{k-1} .$$

Now, every $t \in \mathbf{I}$ belongs to an interval $[t_i, t_{i+1}]$ for some i; therefore it can be written as $t = f_i(s)$ where $s \in \mathbf{I}$. Thus

$$\begin{aligned}
\mathrm{dist}(\gamma_k(t), \gamma_{k+1}(t)) &= \mathrm{dist}(\gamma_k(f_i(s)), \gamma_{k+1}(f_i(s))) \\
&= \mathrm{dist}(F_i(\gamma_{k-1}(s)), F_i(\gamma_{k-1}(s))) \\
&= \rho_i \, \mathrm{dist}(\gamma_{k-1}(s), \gamma_k(s)) \\
&\leq C \, \rho_{\mathrm{max}}^k ,
\end{aligned}$$

proving the desired inequality.

b) We deduce that for all integers $k > l$,

$$\text{dist}(\gamma_k(t), \gamma_l(t)) \leq \sum_{i=k}^{l-1} \text{dist}(\gamma_i(t), \gamma_{i+1}(t))$$

$$\leq \sum_{i=k}^{l-1} C\, \rho_{\max}^i$$

$$\leq \frac{C}{1 - \rho_{\max}}\, \rho_{\max}^k \; .$$

For sufficiently large k, the right-hand side can be made as small as we wish. We deduce that for each value of t, the sequence $(\gamma_k(t))$ is a Cauchy sequence, and therefore it converges to a limit, which we shall denote by $\gamma(t)$. Moreover, this convergence is uniform (independent of t), because

$$\text{dist}(\gamma_k(t), \gamma(t)) \leq \frac{C}{1 - \rho_{\max}}\, \rho_{\max}^k \; .$$

Since the uniform limit of continuous functions is continuous, we deduce that γ is continuous. ◄

13.6 The limit curve Γ

We have just proved the existence of a continuous function $\gamma(t)$, defined on $\mathbf{I} = [0, 1]$. We put

$$\Gamma = \gamma(\mathbf{I}) \; .$$

This is a parameterized curve. According to previous computations the sequence of polygonal curves $(\mathbf{P}_k)$ converges to Γ in the Hausdorff distance. This distance is evaluated by

$$\text{dist}(\Gamma, \mathbf{P}_k) \leq \frac{C}{1 - \rho_{\max}}\, \rho_{\max}^k \; .$$

For smaller $\rho_{\max}$ the convergence is faster (the minimal value of $\rho_{\max}$ is $1/N$, where N is the number of similarities). Here are some properties of Γ:
- We have seen in §5 that for all $t \in \mathbf{I}$,

$$\gamma_k(f_i(t)) = F_i(\gamma_{k-1}(t)) \; .$$

Since all these functions are continuous, we can pass to the limit when k tends to infinity, and we find that:

$$\gamma(f_i(t)) = F_i(\gamma(t)) \; .$$

This is true for any integer i between 1 and N. We can symbolize this result by the following *commutative diagram*:

$$\begin{array}{ccc} \mathbf{I} & \xrightarrow{\ f_i\ } & f_i(\mathbf{I}) \\ \downarrow{\scriptstyle\gamma} & & \downarrow{\scriptstyle\gamma} \\ \Gamma & \xrightarrow{\ F_i\ } & F_i(\Gamma) \end{array}$$

- Since $\mathbf{I} = \cup_{i=1}^{N} f_i(\mathbf{I})$, we deduce from the previous equality that:

$$\Gamma = \gamma(\mathbf{I}) = \gamma\Big(\bigcup_{i=1}^{N} f_i(\mathbf{I})\Big) = \bigcup_{i=1}^{N} \gamma(f_i(\mathbf{I}))$$

$$= \bigcup_{i=1}^{N} F_i(\gamma(\mathbf{I})) = \bigcup_{i=1}^{N} F_i(\Gamma) \, .$$

This equality

$$\boxed{\ \Gamma = \cup_{i=1}^{N} F_i(\Gamma)\ }$$

proves that Γ is the union of N copies of itself; thus Γ is **self–similar**.
- The curve Γ passes through all the vertices A_i of the generator because all the polygonal approximations $\mathbf{P}_k$ pass through these points. In fact, $A_i = \gamma(t_i)$, with the time t_i defined in §4.
- Finally, Γ is the only closed set that satisfies the similarity equality.

▶   To prove this we need a lemma on the Hausdorff metric:

Let E_1, E_2, E_3, E_4 be any sets:

$$\operatorname{dist}(E_1 \cup E_2, E_3 \cup E_4) \leq \max\{\operatorname{dist}(E_1, E_3), \operatorname{dist}(E_2, E_4)\} \, .$$

Because if $\epsilon \geq \max\{\operatorname{dist}(E_1, E_3), \operatorname{dist}(E_2, E_4)\}$, then $E_1 \subset E_3(\epsilon)$ and $E_2 \subset E_4(\epsilon)$. Therefore, $E_1 \cup E_2 \subset E_3(\epsilon) \cup E_4(\epsilon) = (E_3 \cup E_4)(\epsilon)$. By symmetry, $E_3 \cup E_4 \subset (E_1 \cup E_2)(\epsilon)$. Therefore, $\epsilon \geq \operatorname{dist}(E_1 \cup E_2, E_3 \cup E_4)$. This proves the lemma. Now if E is a closed set such that $E = \cup_{i=1}^{N} F_i(E)$:

$$\operatorname{dist}(\Gamma, E) = \operatorname{dist}(\cup_{i=1}^{N} F_i(\Gamma), \cup_{i=1}^{N} F_i(E))$$

$$\leq \max_{1 \leq i \leq N} \operatorname{dist}(F_i(\Gamma), F_i(E)) \leq \rho_{\max} \operatorname{dist}(\Gamma, E) \, .$$

Since $\rho_{\max} < 1$, this inequality is impossible unless $\operatorname{dist}(\Gamma, E) = 0$. ◀

This proof ends the proof of the theorem in §2 concerning the existence and the uniqueness of the curve Γ.

◇   The examples in Figures 13.6 and 13.7 prove that the concept of *curve* is too general. We shall talk in the rest of this chapter about *simple curves*.

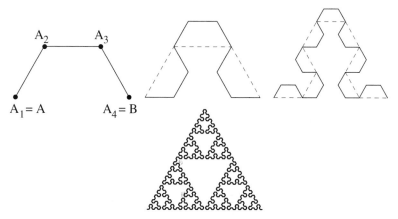

Fig. 13.6. *We see that the well-known "Sierpinski triangle" is a self–similar curve. If $A = (0,0)$ and $B = (1,0)$, we can define it with the help of the following three similarities (with the notations of §1):*

$$F_1(x) = \frac{1}{2}\mathcal{R}_{\pi/3}\,\mathcal{S}_1\,x\ ,\ \ F_2(x) = \frac{1}{2}x + \begin{pmatrix} 1/4 \\ \sqrt{3}/4 \end{pmatrix}\ ,\ \ F_3(x) = \frac{1}{2}\mathcal{R}_{2\pi/3}\,x + \begin{pmatrix} 1 \\ 0 \end{pmatrix}.$$

We show the generator $\mathbf{P}_1$, and $\mathbf{P}_2$, $\mathbf{P}_3$, $\mathbf{P}_6$.

13.7 Simplicity criterion

The construction of $\varGamma$ by its generator, described in §3, allows us to obtain good representations of a self–similar curve. But it does not allow us to predict some of the properties of the curve. For example, we cannot tell whether or not this curve has a double point. Many models, like those representing geographical lines, must be simple curves. Given a system of N similarities satisfying the conditions described in §2, it is interesting to have a simplicity criterion that can be applied to the initial data. In this section, we shall give a criterion that we shall apply to the many examples in §9. This criterion is associated with a method for **covering** the curve, which allows a visual comprehension of the curve that is better than drawing it using its polygonal approximations.

CLOSED SET CRITERION   *Given $N + 1$ points A_1, ..., A_{N+1} and N similarities F_1, ..., F_N as in §2, let $\varGamma$ be the self–similar curve defined by these points and similarities. Assume that there exists a closed bounded set D of non-null area satisfying: 1) $F_i(D) \subset D$ for all $i = 1,\dots,N$; 2) $F_i(D) \cap F_{i+1}(D) = \{A_{i+1}\}$ for all $i = 1,\dots,N-1$; 3) $F_i(D)$ and $F_j(D)$ are disjoint, whenever $|i - j| \geq 2$. Then $\varGamma$ is simple.*

In what follows, when we talk about a *self–similar simple curve*, we always mean a curve that satisfies this criterion. The proof will consist of two parts: *a*) We show that the curve is included in D; *b*) then we show that the parameterization as defined in §6 is one-to-one.

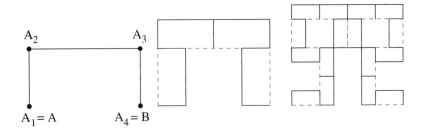

A_2 A_3

$A_1 = A$ $A_4 = B$

Fig. 13.7. *The square* $\mathbf{I} \times \mathbf{I}$ *is also a self–similar curve, with* $A = (0,0)$, $B = (1,0)$,

$$F_1(x) = \frac{1}{2} \begin{pmatrix} 0 & 1 \\ 1 & 0 \end{pmatrix} x \;,\quad F_2(x) = \frac{1}{2}x + \begin{pmatrix} 0 \\ 1/2 \end{pmatrix} \;,$$

$$F_3(x) = \frac{1}{2}x + \begin{pmatrix} 1/2 \\ 1/2 \end{pmatrix} \;,\quad F_4(x) = \frac{1}{2} \begin{pmatrix} 0 & -1 \\ 1 & 0 \end{pmatrix} x + \begin{pmatrix} 1 \\ 0 \end{pmatrix} \;.$$

In fact, this square is the union of four squares of sides $1/2$. *The parameterization* γ *induced by these similarities is an example of the Peano curve. We show the generator* $\mathbf{P}_1$, *and* $\mathbf{P}_2$, $\mathbf{P}_3$, $\mathbf{P}_5$.

▶   a) Let $D_0 = D$, $D_1 = \cup_{i=1}^{N} F_i(D)$, ..., $D_k = \cup_{i=1}^{N} F_i(D_{k-1})$. Condition 1) implies that $D_1 \subset D_0$, so that by induction, $D_k \subset D_{k-1}$. The sequence (D_k) is a sequence of embedded closed sets. We want to show that $\Gamma \subset D_k$ for all k. We make the following estimate:

$$\mathrm{dist}(\Gamma, D_k) = \mathrm{dist}(\cup_i F_i(\Gamma), \cup_i F_i(D_{k-1}))$$
$$\leq \max_i \left\{ \mathrm{dist}(F_i(\Gamma), F_i(D_{k-1})) \right\} \quad (\S\,6)$$
$$\leq \rho_{\max}\, \mathrm{dist}(\Gamma, D_{k-1}) \;,$$

where $\rho_{\max}$ is the largest similarity ratio of $F_1, \ldots, F_N$. We deduce that:

$$\mathrm{dist}(\Gamma, D_k) \leq \rho_{\max}^k \mathrm{dist}(\Gamma, D) \;.$$

But D is bounded and so is Γ, so $\mathrm{dist}(\Gamma, D)$ is finite. The right-hand-side term then tends to 0. Thus,

$$\lim_{k \to \infty} \mathrm{dist}(\Gamma, D_k) = 0 \;.$$

Since $D_k \subset D_{k-1}$, we deduce that:

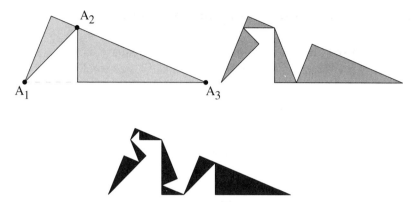

Fig. 13.8. *We reconsider the curve Γ of Figures 13.3, 13.4, and 13.5. It is included in a right–angled triangle D. The images of D by the similarities F_1 and F_2 have the unique common point A_2 and are included in D. This suffices to show (closed-set criterion) that Γ is simple. By iterating the procedure, we obtain covers of Γ by "chains" whose elements are smaller and smaller.*

$$\Gamma = \bigcap_{k=1}^{\infty} D_k \ .$$

b) Conditions 1) and 2) show that the families of intervals $\{f_i(\mathbf{I})\}$ and of domains $\{F_i(D)\}$ are geometrically structured in the same manner: two intervals $f_i(\mathbf{I})$ and $f_j(\mathbf{I})$, $i \neq j$, are in fact either disjoint ($|i - j| \geq 2$) or adjacent ($|i - j| = 1$). In the first case $F_i(D)$ and $F_j(D)$ are disjoint; in the second case they have a unique common point. Without going through all the details, we claim that this will be the same at rank k; the N^k intervals of rank k, which can be written as $f_{i_1}(\dots(f_{i_k}(\mathbf{I}))\dots)$, $1 \leq i_j \leq N$, are either disjoint or adjacent. Then the corresponding domains $F_{i_1}(\dots(F_{i_k}(\mathbf{I}))\dots)$ have either 0 or 1 point in common, respectively.

Let $p_{\max} = \max\{t_{i+1} - t_i\}$. This is a number < 1, and all the intervals of rank k are of length $\leq p_{\max}^k$. If we take any two reals t and t' in $\mathbf{I}$, we can always find a large enough integer k so that t and t' belong to two disjoint intervals of rank k. Their images $\gamma(t)$ and $\gamma(t')$ belong to two disjoint domains of rank k. Therefore, $\gamma(t) \neq \gamma(t')$. This proves that γ is one-to-one. That is, Γ has no double point. ◀

◇ Figure 13.6 shows a curve with infinitely many double points, though its polygonal approximations $\mathbf{P}_k$ are simple. Indeed Γ does not satisfy the closed-set criterion.

◇ In all the examples in this chapter, the domain D used to show the simplicity of Γ is a convex set. This implies that the segment AB joining the endpoints of Γ is included in D. Similarly, $\mathbf{P}_k$ is included in D_k. But the convexity is not necessary to prove the simplicity of Γ; it will be used in Chapter 14, §6.

13.8 Similarity and dimension exponent

We consider a self–similar curve Γ defined by the similarities $F_1, \ldots, F_N$ whose similarity ratios are $\rho_1, \ldots, \rho_N$. We are specifically interested in these ratios:

The inequality

$$\sum_{i=1}^{N} \rho_i \geq 1 \,,$$

is always true. Moreover, if the curve satisfies the closed-set criterion (§7), then

$$\sum_{i=1}^{N} \rho_i^2 \leq 1 \,.$$

▶ If A and B are the endpoints of Γ, they are also the endpoints of the generator $\mathbf{P}_1 = \cup_i F_i(AB)$. By similarity, $L(F_i(AB)) = \rho_i \, L(AB)$. Therefore $L(AB) \leq L(\mathbf{P}_1) = L(AB) \sum_i \rho_i$. The limit case $\sum_i \rho_i = 1$ corresponds to the case where AB is a generator. Thus every polygonal curve $\mathbf{P}_k$ is equal to AB. It follows that Γ is itself equal to AB.

If the curve is simple, the conditions of the closed set criterion imply that:

$$\sum_i \mathcal{A}(F_i(D)) \leq \mathcal{A}(D) \,,$$

where $\mathcal{A}(F_i(D)) = \rho_i^2 \, \mathcal{A}(D)$. Since $\mathcal{A}(D) \neq 0$, we get the second inequality. ◀

Since $\sum_i \rho_i^x$ is a strictly decreasing and continuous function when x varies from 1 to 2, there is a unique real number e (an "exponent") between 1 and 2 such that:

$$\sum_{i=1}^{N} \rho_i^e = 1 \,.$$

This is the "similarity exponent" of G. Bouligand.

◊ In the case where all the similarity ratios are equal to ρ the above equation can be written as: $N \rho^e = 1$. Thus

$$e = \frac{\log N}{|\log \rho|} \,.$$

This similarity exponent can be interpreted to be the dimension of the curve.

THEOREM *Let Γ be a self–similar simple curve defined by N similarities of ratio $\rho_1, \ldots, \rho_N$. Then*

$$\Delta(\Gamma) = \delta(\Gamma) = e \,.$$

▶   We recall the following notation: $\Gamma(\epsilon)$ is the ϵ–Minkowski sausage of Γ, and D is the closed domain of the simplicity criterion. To shorten the notation, let $D_i = F_i(D)$ and $\Gamma_i = F_i(\Gamma)$. We define the following two functions of ϵ:

$$f(\epsilon) = \epsilon^{e-2} \mathcal{A}(\Gamma(\epsilon) \cap D)$$
$$g(\epsilon) = \epsilon^{e-2} \mathcal{A}(\Gamma(\epsilon)) \ .$$

We shall show that on a neighborhood of 0, they are always bounded by two non-null constants. The fact that the D_i are all included in D and disjoint, except eventually on the boundaries, implies that:

$$\mathcal{A}(\Gamma(\epsilon) \cap D) \geq \sum_i \mathcal{A}(\Gamma(\epsilon) \cap D_i) \geq \sum_i \mathcal{A}(\Gamma_i(\epsilon) \cap D_i) \ .$$

The set $\Gamma_i(\epsilon) \cap D_i$ is the image by the similarities F_i of the set $\Gamma(\epsilon/\rho_i) \cap D$. Consequently,

$$f(\epsilon) \geq \epsilon^{e-2} \sum_i \mathcal{A}(\Gamma_i(\epsilon) \cap D_i) = \epsilon^{e-2} \sum_i \rho_i^2 \mathcal{A}(\Gamma(\epsilon/\rho_i) \cap D)$$
$$= \sum_i \rho_i^e \, f(\epsilon/\rho_i) \ .$$

Assume that $\rho_1 \leq \ldots \leq \rho_N$. The function f is continuous at $\epsilon > 0$. It has a non-null lower bound on $[1, 1/\rho_N]$. Let c_1 be this bound.

If $\epsilon \in [\rho_N, 1]$, then $\epsilon/\rho_i \in [1, 1/\rho_N]$, and therefore $f(\epsilon) \geq \sum \rho_i^e \, c_1 = c_1$.

By induction we get: if $\epsilon \in [\rho_N^{k+1}, 1]$, then $\epsilon/\rho_i \in [\rho_N^k, 1/\rho_N]$, and $f(\epsilon) \geq c_1$. We deduce that for all $\epsilon > 0$,

$$f(\epsilon) \geq c_1 > 0 \ .$$

Similarly,

$$g(\epsilon) \leq \epsilon^{e-2} \sum_i \mathcal{A}(\Gamma_i(\epsilon)) = \epsilon^{e-2} \sum_i \rho_i^2 \, \mathcal{A}(\Gamma(\epsilon/\rho_i))$$
$$= \sum_i \rho_i^e \, g(\epsilon/\rho_i) \ .$$

Putting $c_2 = \sup_{[1,1/\rho_N]} g(\epsilon)$, we prove by induction that for all $\epsilon > 0$,

$$g(\epsilon) \leq c_2 \ .$$

In conclusion,
$$0 < c_1 \leq f(\epsilon) \leq g(\epsilon) \leq c_2 \ .$$

These inequalities show that $\lim_{\epsilon \to 0} \log g(\epsilon)/\log \epsilon = 0$, and therefore

$$\lim_{\epsilon \to 0} \left(2 \frac{\log \mathcal{A}(\Gamma(\epsilon))}{\log \epsilon} \right) = e \ .$$

This is also the value of $\Delta(\Gamma)$ and $\delta(\Gamma)$ (Chap. 10, §3). ◀

13.9 Examples

Example 1 The curve in Figures 13.3, 13.4, 13.5, and 13.8 depends on a parameter α, $0 < \alpha < \pi/4$. It is constructed by two similarities of ratios $\rho_1 = \tan \alpha$ and $\rho_2 = \cos 2\alpha / \cos \alpha$. Its similarity exponent is therefore obtained by solving the equation:

$$(\tan \alpha)^e + (\cos 2\alpha / \cos \alpha)^e = 1 .$$

We remark that if α tends to 0 while the distance between A_1 and A_3 remains fixed, the point A_2 tends to A_1, and the curve tends to the segment $A_1 A_3$. But at the same time, e tends to 2 in the equation. If α tends to $\pi/4$, the point A_2 tends to A_3, the curve tends to $A_1 A_3$ and e tends to 1. The two similarity ratios are equal when $\alpha = \pi/6$; their common value is $1/\sqrt{3}$. In this case $e = \log 4 / \log 3$.

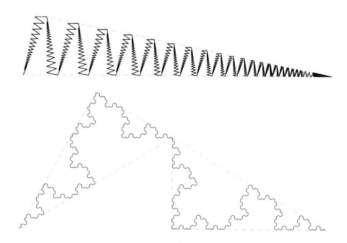

Fig. 13.9. *A picture of the curve defined in Figure 13.3 for two values of the parameter α: $\alpha = \pi/16$ (dimension $1.51508\ldots$) and $\alpha = \pi/6$ (dimension $1.26186\ldots$).*

Example 2 Usually the Von Koch curve is defined by four similarities of equal ratio, their common value is $1/3$. But we can define it by two similarities of ratio $1/\sqrt{3}$. Figure 13.10 shows this, starting with a convenient domain D (an isosceles triangle has an angle of $\pi/6$). We can prove the simplicity of Γ by using the criterion of §7. This domain is in fact the convex hull of the curve. The dimension, or similarity exponent (§8), equals

$$\frac{\log 4}{\log 3} = \frac{\log 2}{\log \sqrt{3}} = 1.26186\ldots .$$

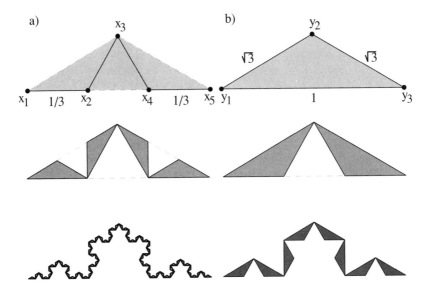

Fig. 13.10. *The construction of the Von Koch curve starting with an initial domain D. In the following two cases: a) Γ is defined by four similarities of ratio $1/3$; b) Γ is defined by two similarities of ratio $1/\sqrt{3}$. In each case we present the step 0 (domain D), step 1 (domain $D_1 = \cup_i F_i(D)$), and step 3 (domain $D_3 = \cup_i F_i(D_2)$) of the construction.*

$\diamond$ Surely, we can start by choosing a domain other than the isosceles triangle. If the chosen domain D is too small, the images $F_i(D)$ are not included in D, and neither is the curve Γ. If it is too big, the images $F_i(D)$ are no longer disjoint, and it will be impossible to predict the simplicity of Γ. In Figure 13.11 we start with a square.

Example 3 If we vary the angle $\phi = \pi/6$ in the previous example, the shape and dimension of the curve will also vary. The curve Γ is determined by two similarities of ratio $\rho = 1/(2\cos(\phi/2))$. The curve is simple if $0 < \phi < \pi/2$. The dimension

$$\frac{\log 2}{\log(2\cos\frac{\phi}{2})}$$

varies between two limits: 1 (for $\phi \to 0$) and 2 (for $\phi \to \pi/2$).

Example 4 The Gosper curve has a generator formed by three equal segments. Because of the symmetries, the curve is completely determined by $\mathbf{P}_1$ and $\mathbf{P}_2$ (Fig. 13.12). With the same notation as in this figure, the dimension is:

$$\frac{\log 3}{|\log r|} = \frac{2\log 3}{\log(5 + 4\cos\phi)} ,$$

where $0 < \phi < \pi$. The curve is simple whenever ϕ is small enough, but it is not when ϕ gets close to π. To obtain a condition for simplicity, we use the domain D

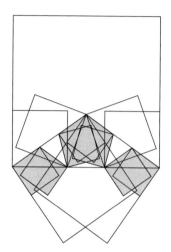

Fig. 13.11. *Construction of the Von Koch curve starting with a square as its initial domain D. The successive images of D are mixed up, and it is hard to predict the simplicity of Γ.*

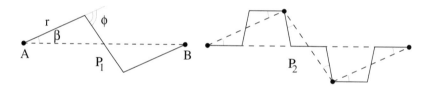

Fig. 13.12. *Generators* $\mathbf{P}_1$ *and* $\mathbf{P}_2$ *of the Gosper curve. If the two endpoints A and B are at a distance 1, the parameters r, φ, and β are related by the relations*

$$r = \frac{1}{\sqrt{5 + 4\cos\phi}} \, , \quad \cos\beta = \frac{1}{4r} + \frac{3r}{4} = \frac{2 + \cos\phi}{\sqrt{5 + 4\cos\phi}} \, .$$

The angle φ can take all the values between 0 and π. Knowing $\mathbf{P}_2$ *is a necessary condition to determine the curve.*

in Figure 13.13, which is a parallelogram such that $F_i(D) \subset D$ for $i = 1, 2, 3$. For Γ to be simple, we need that the $F_i(D)$ have either zero or one point in common (§7). In the notation of Figure 13.13, this is assured when

$$2\phi < \pi + \beta \, ,$$

which is equivalent to

$$\cos\phi > -\frac{1}{4} \, .$$

This is a *sufficient* condition for simplicity: we have not proved that Γ has double points for the values of φ which are less than arccos(−1/4). When $\phi \to 0$, the

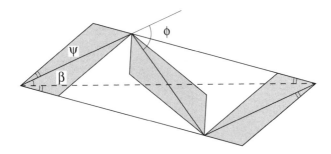

Fig. 13.13. *We can prove the simplicity of the Gosper curve with the help of a domain D, which is a parallelogram as above. The angle ψ satisfies $2(\psi + \beta) = \phi$. The three parallelograms, images of D by the three similarities, are included in D; their interiors are disjoint if $\beta + 2\psi < \pi - \phi$, or if $2\phi < \pi + \beta$.*

dimension of the curve tends to 1; when $\phi \to \arccos(-1/4)$, the dimension tends to $\log 3 / \log 2$.

Example 5 The quadratic curve of Mandelbrot has a generator formed by eight equal segments (Fig. 13.14). The similarity ratios are all equal to $1/4$. The dimension is equal to $3/2$.

13.10 The natural parameterization

Let Γ be a self–similar simple curve whose endpoints are A and B, and assume that Γ is parameterized by the continuous function $\gamma : \mathbf{I} = [0,1] \to \Gamma$. We have constructed this parameterization in §6, but it was not unique because the function γ depends on the choice of the real numbers $t_1, \ldots, t_{N+1}$ in $\mathbf{I}$, with

$$0 = t_1 < t_2 < \ldots < t_{N+1} = 1 \, .$$

Let us denote by $p_i = t_{i+1} - t_i$, $i = 1, \ldots, N$. These numbers are not null, and $\sum_i p_i = 1$. To every choice of the family $\{p_i\}$, there corresponds a family $\{f_i\}$ of similarities with ratio p_i on $\mathbf{I}$ (§4) and therefore a parameterization γ such that for all t (§6):

$$\gamma(f_i(t)) = F_i(\gamma(t)) \, .$$

Probabilistic interpretation If we want to "randomly" choose a point x of Γ, we must first put x in the form

$$x = \bigcap_{k=1}^{\infty} F_{i_1}(\ldots(F_{i_k}(\Gamma))\ldots) \, ,$$

where $(F_{i_1}(\ldots(F_{i_k}(\Gamma))\ldots))_{k \geq 1}$ is a sequence of embedded arcs whose diameters tend to 0. To every infinite sequence $(i_k)_{k \geq 1}$ of integers between 1 and N there corresponds a unique point of Γ.

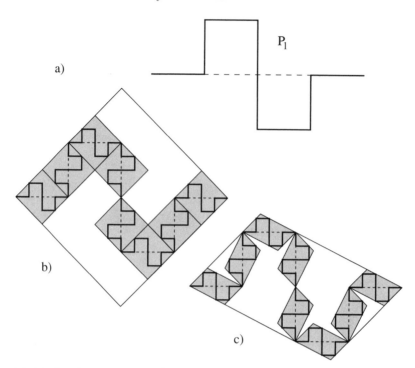

Fig. 13.14. *In a) a generator of the quadratic curve of Mandelbrot; in b) we can include the curve in a square. But to show the simplicity of Γ it is better to take the parallelogram of c).*

We start by randomly choosing the integer i_1 with a probability p_{i_1}, and this determines the arc $F_{i_1}(\Gamma)$. Then we choose the integer i_2 with probability p_{i_2}, and this determines the arc $F_{i_1}(F_{i_2}(\Gamma))$. We repeat this process. Since Γ is simple, the p_i define a **probability measure** on Γ. For all k, the measure of the arc $F_{i_1}(\ldots(F_{i_k}(\Gamma))\ldots)$, that is, the probability of finding x in this arc, is exactly $p_{i_1}\ldots p_{i_k}$.

Let us call this measure μ. We know (Chap. 6, §2) that μ induces a parameterization, which is, in this case, γ. The correspondence between μ and γ is given by the following relation

$$\mu(A, \gamma(t)) = t \ ,$$

which means, in terms of trajectory, that the measure of the arc is the time needed for the object to run through this arc. Here the time spent in running through the arc $F_{i_1}(\ldots(F_{i_k}(\Gamma))\ldots)$ is equal to $p_{i_1}\ldots p_{i_k}$.

Natural choice of the p_i There exists a choice of the p_i that seems to be more natural than the others. It is inspired by the equation:

$$\sum_1^N \rho_i^e = 1 \ ,$$

where the ρ_i are the ratios of the similarities F_1, ..., F_N defining Γ. We put:

$$p_i = \rho_i^e .$$

The following result shows the importance of this choice.

THEOREM   *Let Γ be a self–similar simple curve defined by the similarities F_i, $i = 1, \ldots, N$, with ratios ρ_i and similarity exponent e. Also assume $\sum_i \rho_i^e = 1$. Let γ be the parameterization $\mathbf{I} \to \Gamma$ such that the measure of the arc $F_i(\Gamma)$ is ρ_i^e and the measure of the arc $F_{i_1}(\ldots(F_{i_k}(\Gamma))\ldots)$ is $(\rho_{i_1} \ldots \rho_{i_k})^e$.*

Let $T(t, \tau)$ be the local size function that corresponds to this parameterization (Chap. 11, §4). Then

$$T(t, \tau) \simeq \tau^{1/e}$$

uniformly with respect to t.

In other words, there exist two nonzero constants c_1 and c_2 such that for all $t \in \mathbf{I}$:

$$c_1\, \tau^{1/e} \leq T(t, \tau) \leq c_2\, \tau^{1/e} .$$

The order of growth of $T(t, \tau)$ when τ tends to 0 is then equal to $1/e$.

If $e = 1$, then $\sum_i \rho_i = 1$, and Γ is the segment AB.

If $e > 1$, then $T(t, \tau)/\tau \simeq \tau^{(1/e)-1}$ for all t. This tends to $+\infty$ when τ tends to 0. The curve Γ is a fractal curve in the sense in Chapter 11, §4.

◇   The theorem shows that two arcs of the same measure are of equivalent sizes. That is why the parameterization γ is said to be "natural."

▶   We prove the theorem. Assume that Γ satisfies the closed-set criterion (§7). Let Γ^* be an arc of Γ. We would like to find two constants c_1 and c_2 independent of Γ^*, such that:

$$0 < c_1 \leq \frac{\mu(\Gamma^*)^{1/e}}{\operatorname{diam}(\Gamma^*)} \leq c_2 ,$$

where the measure $\mu(\Gamma^*)$ is the time spent to run through Γ^*. This suffices because if $\Gamma^* = \gamma([t - \tau, t + \tau])$, then $\operatorname{diam}(\Gamma^*) = T(t, \tau)$ and $\mu(\Gamma^*) = 2\,\tau$. The previous inequalities prove that $T(t, \tau) \simeq \tau^{1/e}$.

a)   First, assume that Γ^* is an arc of rank k; that is, it can be written as $F_{i_1}(\ldots(F_{i_k}(\Gamma))\ldots)$. Then $\operatorname{diam}(\Gamma^*) = \rho_{i_1} \ldots \rho_{i_k} \operatorname{diam}(\Gamma)$ by similarity, and $\mu(\Gamma^*) = (\rho_{i_1} \ldots \rho_{i_k})^e$. Therefore

$$\frac{\mu(\Gamma^*)^{1/e}}{\operatorname{diam}(\Gamma^*)} = \frac{1}{\operatorname{diam}(\Gamma)} .$$

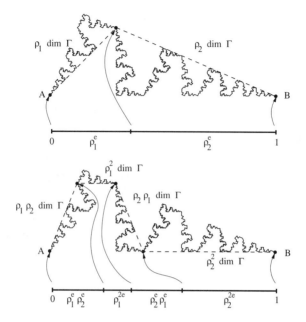

Fig. 13.15. *The curve Γ of Figures 13.3, 4, 5, and 8 is defined by two similarities F_1 and F_2 of ratios ρ_1 and ρ_2. The similarity exponent satisfies the equation $\rho_1^e + \rho_2^e = 1$. The two weights $p_1 = \rho_1^e$ and $p_2 = \rho_2^e$ induce a probability measure on Γ: the length of an interval is equal to the measure of its image by γ. This measure is equivalent to the diameter of the arc to the power e.*

b)   If Γ^* is any arc, there exists a smallest integer k such that Γ^* contains an arc Γ_1 of rank k. Let Γ_2 be the arc of rank $k-1$ that contains Γ_1. If, for example, $\Gamma_1 = F_{i_1}(\dots(F_{i_k}(\Gamma))\dots)$, then $\Gamma_2 = F_{i_1}(\dots(F_{i_{k-1}}(\Gamma))\dots)$, and

$$\frac{\mu(\Gamma_1)^{1/e}}{\mu(\Gamma_2)^{1/e}} = \frac{\operatorname{diam}(\Gamma_1)}{\operatorname{diam}(\Gamma_2)} \geq \rho_{\min},$$

where $\rho_{\min}$ is the smallest among the ratios of similarity ρ_i. We distinguish two cases.

b1) $\Gamma^* \subset \Gamma_2$: then $\operatorname{diam}(\Gamma_1) \leq \operatorname{diam}(\Gamma^*) \leq \operatorname{diam}(\Gamma_2)$, and $\mu(\Gamma_1) \leq \mu(\Gamma^*) \leq \mu(\Gamma_2)$. This gives:

$$\frac{\operatorname{diam}(\Gamma_1)}{\operatorname{diam}(\Gamma_2)} \frac{\mu(\Gamma_1)^{1/e}}{\operatorname{diam}(\Gamma_1)} \leq \frac{\mu(\Gamma^*)^{1/e}}{\operatorname{diam}(\Gamma^*)} \leq \frac{\operatorname{diam}(\Gamma_2)}{\operatorname{diam}(\Gamma_1)} \frac{\mu(\Gamma_2)^{1/e}}{\operatorname{diam}(\Gamma_2)}.$$

Using a), we find

$$\boxed{\frac{\rho_{\min}}{\operatorname{diam}(\Gamma)} \leq \frac{\mu(\Gamma^*)^{1/e}}{\operatorname{diam}(\Gamma^*)} \leq \frac{1}{\rho_{\min}\operatorname{diam}(\Gamma)}.}$$

b2) Γ^* is not included in Γ_2. In this case $\Gamma^* - \Gamma_2$ (the set of points of Γ^* that do not belong to Γ_2) is an arc, and there exists a smallest integer $l \geq k$ such that $\Gamma^* - \Gamma_2$ contains an arc Γ_3 of rank l. Let Γ_4 be the arc of rank $l - 1$ that contains Γ_3. Γ_4 is not included in Γ_2 but is adjacent to Γ_2 (because otherwise there will be another arc between Γ_2 and Γ_4 that is of rank $l - 1$ and included in Γ^*, which is impossible by the definition of the integer l). We deduce the following:

$$\Gamma_1 \cup \Gamma_3 \subset \Gamma^* \subset \Gamma_2 \cup \Gamma_4 .$$

Therefore,

$$\mathrm{diam}\,(\Gamma_1) + \mathrm{diam}\,(\Gamma_3) \leq \mathrm{diam}\,(\Gamma^*) \leq \mathrm{diam}\,(\Gamma_2) + \mathrm{diam}\,(\Gamma_4)$$

and

$$\mu(\Gamma_1) + \mu(\Gamma_3) \leq \mu(\Gamma^*) \leq \mu(\Gamma_2) + \mu(\Gamma_4) .$$

This gives

$$\frac{(\mu(\Gamma_1) + \mu(\Gamma_3))^{1/e}}{\mathrm{diam}\,(\Gamma_2) + \mathrm{diam}\,(\Gamma_4)} \leq \frac{\mu(\Gamma^*)^{1/e}}{\mathrm{diam}\,(\Gamma^*)} \leq \frac{(\mu(\Gamma_2) + \mu(\Gamma_4))^{1/e}}{\mathrm{diam}\,(\Gamma_1) + \mathrm{diam}\,(\Gamma_3)} .$$

Using the results of a) and the inequalities $\max\{a, b\} \leq a + b \leq 2\max\{a, b\}$, which are true for any two positive real numbers a, b, we finally find that

$$\frac{\rho_{\min}}{2\,\mathrm{diam}\,(\Gamma)} \leq \frac{\mu(\Gamma^*)^{1/e}}{\mathrm{diam}\,(\Gamma^*)} \leq \frac{2^{1/e}}{\rho_{\min}\,\mathrm{diam}\,(\Gamma)} .$$

◀

13.11 The algorithm of local sizes

To numerically estimate the fractal dimension of a curve, some universal algorithms such as box counting or the Minkowski sausage, are often used. We have already discussed them in the case of continuous graphs (Chap. 12, §13). Experience shows that the results obtained by these methods are not trustworthy. For the graphs of functions or curves of a very particular type, we have proposed more efficient methods based on the particular geometry of these curves.

We can do the same with self-similar curves. Assume that such a curve is defined by its natural parameterization. Section 10 suggests a very efficient method, known as the **method of diameters**, to compute the fractal dimension. We evaluate the function of local diameter $T(t, \tau)$ at each t and deduce their mean

$$\overline{T}_\tau = \int_0^1 T(t, \tau)\, dt .$$

It is the order of growth at 0 of this function $\overline{T}_\tau$ when τ tends to 0, which will inform us about the value of the dimension. This is not a new idea. This function $\overline{T}_\tau$ was already used under its form Var_τ (Chap. 12, §3), which gave the **variation method** of a graph. We shall have the occasion to recall this in Chapter 15, where we shall be looking for a unified approach to algorithms used to compute the dimension of curves with a particular geometry. We return to self–similarity. We distinguish two types of curves.

Strict self–similarity These curves have been the topic of the present chapter. They are mathematically defined by N similarities (Von Koch, Gosper,...). We assume that all of these similarities have the same ratio ρ. To construct such a curve we start with a generator $\mathbf{P}_1$ (§3) and continue by constructing the usual polygonal approximations $\mathbf{P}_2, \ldots, \mathbf{P}_k$. We obtain a good approximation of Γ for $k = 5$ or 6. The curve $\mathbf{P}_k$ has in fact $N^k + 1$ vertices that are recorded. The set of these vertices constitutes a *digitization* of Γ. The advantage of this procedure is that it exactly follows the natural parameterization of Γ. We know that an ideal arc of Γ located between two vertices is of measure N^{-k}. The measure of any arc located between two vertices of $\mathbf{P}_k$ is therefore obtained by simply counting the points of $\mathbf{P}_k$ that belong to this arc. For example, all arcs of Γ containing $M + 1$ vertices of $\mathbf{P}_k$, $1 \le M \le N$, have a measure $\tau = M N^{-k}$. We calculate the size (for example, the diameter) of *all* the arcs of measure τ and compute their mean. We thus obtain $\overline{T}_\tau$. We have proved in §10 that $\overline{T}_\tau \simeq \tau^{1/e}$, where e is the similarity exponent, that is, the fractal dimension of the curve.

The **method of the diameters** for computing the dimension consists of using the following formula:

$$\Delta(\Gamma) = \lim_{\tau \to 0} \frac{\log \tau}{\log \overline{T}_\tau} \, .$$

But we know that an exact theoretical computation of the dimension can be done, since Γ is self–similar and the N similarities have the same ratio ρ. Thus $\Delta(\Gamma)$ is equal to $\log N / |\log \rho|$ (§8). We can compare this number with the numerical result gotten by the limit and realize how good the obtained results are.

We are left with a question. Is finding the dimension of such curves (self–similar whose similarities are of the same ratio) more than a mere scholarly exercise? In fact, strict self–similarity does not exist in nature. Thus when it is "strict," the model is not "plausible." The diameter method will not be interesting unless it can be applied to a wider family of curves. Which family then?

Similarities in experimental data There are many examples of curves to which a notion of similarity is attached: geographical coasts, level curves of a rough surface, diffusion frontier, and aggregate boundary. Digitized data form a finite number of points, and we have subjective reasons to believe that these points belong to a self-similar curve. We do not really mean a strict self-similarity, because we will not really find the same figure zoomed all along the curve.

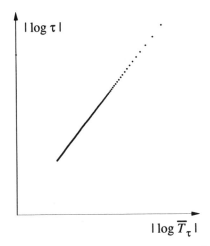

Fig. 13.16. *Estimation by the method of the diameters of the dimension of the Von Koch curve (Fig. 13.10) calculated up to rank 6 with four similarities of ratio 1/3. We thus have $4^6 + 1 = 4097$ points. The values of τ are between 1 and 3^{-6} (distance between two consecutive points). For each τ, we calculate the mean $\overline{T}_\tau$ of the diameters of all the arcs of the curve whose measure is τ. The slope of the regression line of the points $(|\log \overline{T}_\tau|, |\log \tau|)$ is 1.2647... (correlation coefficient 0.999987). The exact dimension is $\log 4/\log 3 = 1.26186....$*

One used to speak of "statistical self–similarity" with reference to some curves obtained by a random procedure. In practice, this seems to mean that every subarc is "similar" to the whole curve. But what is meant by the word "similar" in this context? It is important to note that one of the best successes of fractal models is due in fact to statistical self–similarity. Though we cannot find any justification in either the vocabulary or the mathematical hypothesis attached to this expression. We believe that this is an important question. We shall discuss it in Chapter 15, §6, where we propose a general criterion for self–similarity. The method of the diameters can be applied to all curves satisfying this criterion.

13.12 Bibliographical notes

The first self–similar perfect set was constructeu by Cantor, while the first self–similar curve seems to be the Von Koch curve (1904). Similar independent constructions appeared, for example, in [G. Bouligand 2] (1927). Since the notion of self–similarity in its general form is sufficiently discussed in [B. Mandelbrot 2], we feel that there is no need to add to the history or bibliographical references of this book. In revenge, nothing in the literature seems to establish a link between the following two different notions: strict self–similarity constructed via a recursive system whose main interest is purely theoretical, on one side; and the *statistical* or *generalized* self–similarity on the other. This latter notion is

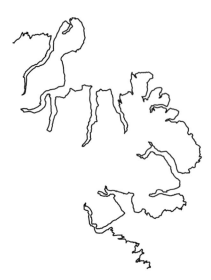

Fig. 13.17. *Boundary of a part of the coast of Baffin globe. Is it self–similar?*

chiefly based on the idea that *every part of the curve is similar to the entire curve* without specifying the meaning of the word similar.

The notion of *generator*, the first polygonal approximation, also comes from [B. Mandelbrot 2], as well as some of the examples of §9 (Ex. 2 (Von Koch), Ex. 3 (Lévy), Ex. 4 (Gosper), Ex. 5).

Another point of view, i.e., of a "dynamical system," can be found in [J.E. Hutchinson]: the self–similar sets are considered as the *attractors* of a system of contractive similarities. There exists an "open set" criterion, which permits one in a general manner to determine the dimension of this set. Our "closed-set" criterion is very similar; it allows us to deduce that the resultant curve is nothing but a curve.

Finally, we find in [G. Bouligand 5] a sketch of the notion of the *similarity exponent*, or the *similarity dimension* as it is known nowadays, and its relation to the dimension of the set.

14 Deviation, and Expansive Curves

14.1 Introducing new notions

To analyze a curve, we often use the notion of **size**, which is a generalization of the notion of **diameter**. The evaluation of local sizes along a given curve allows us to give an estimate of their length in the particular case of curves of finite length or of their dimension in the infinite-length case. But the relation between local size and dimension depends on the type of curve: it is not the same for graphs of continuous functions (Chap. 12) and for self–similar curves (Chap. 13). We can in fact provide a unified approach (Chap. 15), but not without introducing a new geometrical notion concerning curves; the notion of **deviation** follows from the notion of **breadth** of convex sets. These notions allow us to measure how distant a curve is from a straight line. The two complementary notions of deviation and size suffice to characterize the geometry of a curve where its dimensional properties are concerned.

The notion of deviation will be discussed in §2 and §3. But we will not really be able to apply it, except to the family of **expansive** curves to which we devote the rest of this chapter. Intuitively speaking, an expansive curve is a curve that does not come back near itself too often, and therefore its Minkowski sausage can be estimated as the union of convex hulls. This family is quite large. It includes the well-known mathematical models and all experimental curves.

14.2 Deviation of a set

We shall often use definitions and explicit results from Appendix C. Here we simply recall the following definition:

> The **breadth** of a convex set K is the smallest distance between two parallel lines enclosing K.

Recall also that the convex hull $\mathcal{K}(E)$ of a set E is the smallest closed convex set containing E. To increase the field of applications, we now generalize the notion of breadth to **deviation**. Even though this word is more appropriate for curves, it is still interesting to define the deviation of any bounded set. Here is how we do it:

> A **deviation** is a set function $\mathrm{dev}(E)$ satisfying the following three conditions:

1 . dev(E) *is* **equivalent** *to* breadth($\mathcal{K}(E)$). *There exist two constants* c_1 *and* $c_2 > 0$ *such that for all bounded sets* E:

$$c_2 \, \text{breadth}(\mathcal{K}(E)) \leq \text{dev}(E) \leq c_1 \, \text{breadth}(\mathcal{K}(E)) \,.$$

2 . dev(E) *is an* **increasing** *function:*

$$E_1 \subset E_2 \Longrightarrow \text{dev}(E_1) \leq \text{dev}(E_2) \,.$$

3 . dev(E) *is* **continuous** *with respect to Hausdorff distance:*

$$\lim_{n\to\infty} \text{dist}(E_n, E) = 0 \Longrightarrow \lim_{n\to\infty} \text{dev}(E_n) = \text{dev}(E) \,.$$

An example of the deviation function is the breadth of $\mathcal{K}(E)$ itself. But other deviation functions can be defined and used depending on the situation.

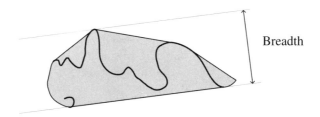

Fig. 14.1. *The quantity* breadth($\mathcal{K}(\Gamma)$), *the breadth of the convex hull, is a measure of the deviation of* Γ.

●   We can take:
$$\text{dev}(E) = \text{diam int}(\mathcal{K}(E)) \,,$$

where diam int($K(E)$) is the diameter of the largest circle inscribed in $\mathcal{K}(E)$. This is a deviation function with $c_1 = 1$ and $c_2 = 2/3$.

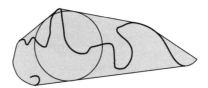

Fig. 14.2. *The diameter of the largest circle inscribed in the convex hull of the curve* Γ *is a measure of the deviation of* Γ *with respect to a straight line.*

- For *every* size function, the ratio

$$\frac{\mathcal{A}(\mathcal{K}(E))}{\text{size}(E)}$$

is a deviation function, provided it is increasing, e.g., the ratio area/perimeter.

Property 1 $\text{dev}(E) = 0$ if and only if the set E is included in a straight line. This property justifies the use of the word "deviation" for curves; a nondeviated curve is a line segment.

Property 2 For all size and deviation functions,

$$\mathcal{A}(\mathcal{K}(E)) \simeq \text{size}(E)\,\text{dev}(E) \ .$$

▶   See the evaluation of the area of a convex set, Appendix C, §4.   ◀

Property 3 Recall that $\mathcal{K}(E)(\epsilon)$ is the set of points whose distance to $\mathcal{K}(E)$ is $\leq \epsilon$. Then we have:

$$\textit{If } \epsilon \simeq \text{dev}(E) \ , \ \textit{ then } \mathcal{A}(\mathcal{K}(E)(\epsilon)) \simeq \epsilon\,\text{size}(E).$$

▶   In fact, $\mathcal{K}(E)(\epsilon)$ is itself a convex set of diameter $\text{diam}\,(E) + 2\epsilon$ and breadth breadth$(\mathcal{K}(E)) + 2\epsilon$. Since $\text{diam}\,(E) \geq$ breadth$(\mathcal{K}(E)) \simeq \text{dev}(E)$, the quantity $\text{diam}\,(E) + 2\epsilon$ is equivalent to $\text{diam}\,(E)$. Thus, $\mathcal{A}(\mathcal{K}(E)(\epsilon)) \simeq \text{diam}\,(E)\,\text{breadth}(\mathcal{K}(E)) \simeq \epsilon\,\text{size}(E)$.   ◀

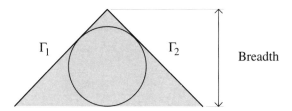

Fig. 14.3. *The segments Γ_1 and Γ_2 are each of length 1. Each is of deviation 0. However, the deviation of $\Gamma = \Gamma_1 \cup \Gamma_2$, taken in the sense of the breadth of $\mathcal{K}(\Gamma)$, is $1/\sqrt{2}$. In the sense of the inner diameter, it is equal to $2 - \sqrt{2}$.*

◇   These notions, size and deviation, have common properties. The deviation induces a particular difficulty: if the curve Γ is divided into two arcs Γ_1 and Γ_2, then there is no relation among $\text{dev}(\Gamma)$, $\text{dev}(\Gamma_1)$, and $\text{dev}(\Gamma_2)$. See, for example, Figure 14.3. For the diameter, however, we can always say that:

If the curve Γ is the union of two curves Γ_1 and Γ_2, then

$$\max\{\,\text{diam}\,(\Gamma_1), \text{diam}\,(\Gamma_1)\,\} \leq \text{diam}\,(\Gamma_1 \cup \Gamma_2) \leq \text{diam}\,(\Gamma_1) + \text{diam}\,(\Gamma_2)\ .$$

▶   The first inequality is a consequence of the fact that $\text{diam}\,(E)$ is an increasing function. The second can be verified by choosing two points y and z

in $\Gamma_1 \cup \Gamma_2$ such that $\operatorname{dist}(y, z) = \operatorname{diam}(\Gamma_1 \cup \Gamma_2)$. If y and z belong to Γ_1, then $\operatorname{diam}(\Gamma_1 \cup \Gamma_2) = \operatorname{diam}(\Gamma_1)$. If y is in Γ_1 and z in Γ_2, let x be a point in $\Gamma_1 \cap \Gamma_2$:

$$\operatorname{diam}(\Gamma_1 \cup \Gamma_2) = \operatorname{dist}(y, z) \le \operatorname{dist}(y, x) + \operatorname{dist}(x, z)$$
$$\le \operatorname{diam}(\Gamma_1) + \operatorname{diam}(\Gamma_2) . \quad \blacktriangleleft$$

14.3 Constant deviation along a curve

Let $\operatorname{dev}(\cdot)$ be a deviation function. Let Γ be a curve parameterized by γ, a continuous function defined on $[a, b]$. Let $A = \gamma(a)$ and $B = \gamma(b)$ be the endpoints of Γ. We shall show that we can construct a sequence of arcs along Γ that have the same deviation.

First, given a point x in Γ and a real number ϵ less than $\operatorname{dev}(x^\frown B)$, the function $\operatorname{dev}(x^\frown y)$ increases when y runs through the arc $x^\frown B$. Since it is continuous, it takes all values between 0 and $\operatorname{dev}(x^\frown B)$. In particular, it takes the value ϵ at least once; there exists a point y_1 of $x^\frown B$ such that $\operatorname{dev}(x^\frown y_1) = \epsilon$. Since the function $\operatorname{dev}(x^\frown y)$ is not a strictly increasing one, there might be more than one point y_1 that satisfies this. In this case we choose the *first* such point. Thus we have fixed y_1 without any ambiguity. If $x = \gamma(t)$, we define the following continuous function

$$\tau_1(t, \epsilon) = \sup \{ \tau \, : \, \operatorname{dev}(\gamma([t, t + \tau])) < \epsilon \}$$
$$= \inf \{ \tau \, : \, \operatorname{dev}(\gamma([t, t + \tau])) \ge \epsilon \} .$$

We put

$$y_1 = \gamma(t + \tau_1(t, \epsilon)) .$$

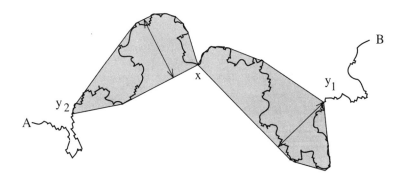

Fig. 14.4. *Starting with a point x of the curve, we can construct two arcs $x^\frown y_1$ and $x^\frown y_2$; both of them have deviation ϵ.*

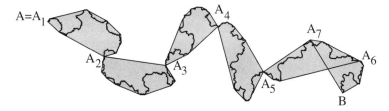

Fig. 14.5. *Construction of a cover of Γ by arcs of the same deviation ϵ. They are either disjoint or adjacent, except the last two. The "step" (segment) $A_k A_{k+1}$ is large in the regular parts and small in the chaotic parts.*

In the other direction, if ϵ is less than $\mathrm{dev}(x^\frown A)$, we can always find a point y_2 of $x^\frown A$ such that $\mathrm{dev}(x^\frown y_2) = \epsilon$. If there is more than one such point, we define the function $\tau_2(t, \epsilon)$ to be

$$\tau_2(t, \epsilon) = \sup \{ \tau \; : \; \mathrm{dev}(\gamma([t - \tau, t])) < \epsilon \} \,,$$

which is continuous. If $x = \gamma(t)$, we put

$$y_2 = \gamma(t - \tau_2(t, \epsilon)) \,.$$

We use these two functions τ_1 and τ_2 to construct a *path* along the curve, formed by steps of deviation ϵ. This can be done in the following manner.

We choose ϵ such that $\epsilon < \mathrm{dev}(\Gamma)$. We define the sequence (t_k) in $[a, b]$ and a sequence (A_k) of points of Γ as follows:

$$t_1 = a \,, \qquad\qquad A_1 = \gamma(t_1) = A$$
$$t_2 = t_1 + \tau_1(t_1, \epsilon) \,, \qquad A_2 = \gamma(t_2)$$
$$t_3 = t_2 + \tau_1(t_2, \epsilon) \,, \qquad A_3 = \gamma(t_3)$$
$$\cdots$$
$$t_k = t_{k-1} + \tau_1(t_{k-1}, \epsilon) \,, \quad A_k = \gamma(t_k) \,.$$

This procedure stops at the first integer N satisfying $\mathrm{dev}(A_N{}^\frown B) < \epsilon$. Thus, we obtain an increasing sequence (t_k) and a sequence of points of the curve (A_k) such that:

$$\mathrm{dev}(A_k{}^\frown A_{k+1}) = \epsilon \,.$$

If $A_N = B$, we stop. Else, we have to go a step backward and put:

$$t_{N+1} = b - \tau_2(b, \epsilon) \,, \quad A_{N+1} = \gamma(t_{N+1}) \,.$$

We thus have:

$$\mathrm{dev}(A_N{}^\frown A_{N+1}) = \epsilon \,.$$

The fact that $\mathrm{dev}(\cdot)$ is an increasing function implies that:

$$t_{N-1} \leq t_{N+1} < t_N < b \,.$$

Let $\Gamma_k = A_k{}^\frown A_{k+1}$, $k = 1, \ldots, N - 1$, and $\Gamma_{N+1} = A_{N+1}{}^\frown B$ if $A_N \neq B$.

The family of arcs $(\Gamma_k)_{1 \leq k \leq N \text{ or } N+1}$ is a cover of Γ. All these arcs are either disjoint or adjacent except Γ_N and Γ_{N+1} (if it exists).

◇ The successive convex hulls $\mathcal{K}(\Gamma_k)$ constitute a good approximation of Γ at ϵ–precision. This approximation follows the particular geometry of the curve better than the ϵ–Minkowski sausage. Thus, we may be able to do a finer analysis of the curve.

14.4 Definition of an expansive curve

Intuitively, the curve Γ is expansive if it can be covered by arcs of the same deviation whose convex hulls do not "overlap too much." We translate this condition to the following inequality on the areas:

We shall say that Γ is an **expansive** *curve if there exists a constant $c > 1$ such that for all $\epsilon \le \mathrm{dev}(\Gamma)$ there exists a cover $\Gamma = \cup_k \Gamma_k^\epsilon$ by arcs Γ_k^ϵ satisfying the following conditions:*

$$(i)\quad \frac{\epsilon}{c} \le \mathrm{dev}(\Gamma_k^\epsilon) \le \epsilon ;$$

$$(ii)\quad \sum_k \mathcal{A}(\mathcal{K}(\Gamma_k^\epsilon)) \le c\,\mathcal{A}(\cup_k \mathcal{K}(\Gamma_k^\epsilon)) .$$

Since the inequality:

$$\sum_k \mathcal{A}(\mathcal{K}(\Gamma_k^\epsilon)) \ge \mathcal{A}(\cup_k \mathcal{K}(\Gamma_k^\epsilon))$$

is always true, condition (ii) implies that the two sides of the inequality have equivalent sizes when $\epsilon \to 0$.

◇ When Γ is a segment of a straight line, $\mathrm{dev}(\Gamma)$ is 0 and conditions (i) and (ii) cannot be applied. Nevertheless, segments are considered to be expansive curves.

Fig. 14.6. *For every n, the Von Koch curve (Fig. 13.10) is covered by 4^n arcs whose convex hulls are triangles with breadth (the smallest height) $(1/2\sqrt{3})\,3^{-n}$. If ϵ is any real between 0 and $\sqrt{3}/2$ and n is an integer such that $(1/2\sqrt{3})\,3^{-n} \le \epsilon < (1/2\sqrt{3})\,3^{-n+1}$, then the curve is covered by 4^n arcs whose deviation is between ϵ and $\epsilon/3$. The convex hulls of these arcs have disjoint interiors. Conditions (i) and (ii) of §4 are both satisfied, and the curve is expansive.*

14.5 Expansivity criterion

Let Γ be a simple curve. Here is an expansivity criterion that can be formulated in different manners.

THEOREM   *The following three conditions are equivalent, and they imply the expansivity of Γ:*

1. *For every point x in Γ there exists a straight line $\mathbf{D}_x$ passing through x that does not contain any point of Γ other than x:*

$$\mathbf{D}_x \cap \Gamma = \{x\} \ .$$

2. *For every subarc Γ^* of Γ, the convex hull $\mathcal{K}(\Gamma^*)$ does not contain any points of Γ other than those of Γ^*:*

$$\mathcal{K}(\Gamma^*) \cap \Gamma = \Gamma^* \ .$$

3. *For every partition of Γ into two adjacent subarcs Γ_1 and Γ_2 whose common endpoint is x, the convex hulls of Γ_1 and Γ_2 have only x as a common point:*

$$\mathcal{K}(\Gamma_1) \cap \mathcal{K}(\Gamma_2) = \{x\} \ .$$

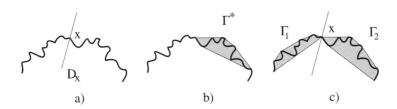

a)   b)   c)

Fig. 14.7. *Three representations of the expansivity criterion: a) $\mathbf{D}_x \cap \Gamma = \{x\}$; b) $\mathcal{K}(\Gamma^*) \cap \Gamma = \Gamma^*$; c) $\mathcal{K}(\Gamma_1) \cap \mathcal{K}(\Gamma_2) = \{x\}$.*

▶ $(1 \Rightarrow 2)$ Assume that condition **1** holds for all x in Γ. Take an arc Γ^* of Γ. Assume, for example, that the endpoints A_1 and A_2 of Γ^* are distinct from the endpoints of Γ A and B. The curve Γ is divided into three arcs:

$$\Gamma = A \frown A_1 \cup A_1 \frown A_2 \cup A_2 \frown B \ ,$$

the middle one is Γ^*. Draw the lines $\mathbf{D}_{A_1}$ and $\mathbf{D}_{A_2}$. Since, by hypothesis, they contain no other point of Γ, they divide the plane into four parts (three if they are parallel). Three of these parts contain $A \frown A_1$, $A_1 \frown A_2$, and $A_2 \frown B$, respectively (Fig. 14.8). These regions, which are not necessarily bounded, are convex; thus they contain the convex hulls of these arcs. Therefore $\mathcal{K}(A_1 \frown A_2)$ is included in

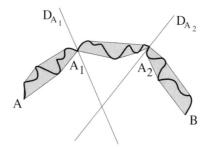

Fig. 14.8. *The lines* $\mathbf{D}_{A_1}$ *and* $\mathbf{D}_{A_2}$ *divide the plane into four convex regions.*

one of them. Its unique common point with $\mathbf{D}_{A_1}$ is A_1, and its unique common point with $\mathbf{D}_{A_2}$ is A_2. We deduce that:

$$\mathcal{K}(\Gamma^*) \cap \Gamma = \Gamma^* .$$

The cases where an endpoint of Γ^* is an endpoint of Γ can be treated in a similar way.

($\mathbf{2} \Rightarrow \mathbf{3}$)   Let Γ_1 and Γ_2 be two subarcs of Γ such that $\Gamma = \Gamma_1 \cup \Gamma_2$, $\Gamma_1 \cap \Gamma_2 = \{x\}$, and neither Γ_1 nor Γ_2 contains just one point. The intersection $\mathcal{K}(\Gamma_1) \cap \mathcal{K}(\Gamma_2)$ contains the point x. Assume that it contains another point z distinct from x. This point belongs to the chord xA_1 of Γ_1 and to a chord xA_2 of Γ_2. We distinguish two cases: A_1 belongs to the segment xA_2 (Fig. 14.9), or A_2 belongs to segment xA_1. In the first case, A_1 belongs to $\mathcal{K}(\Gamma_2)$. This means that $\mathcal{K}(\Gamma_2) \cap \Gamma$ contains a point, A_1 which does not belong to Γ_2. This contradicts the hypothesis. Similarly, the second case leads to a contradiction. Thus we conclude that the point z does not exist.

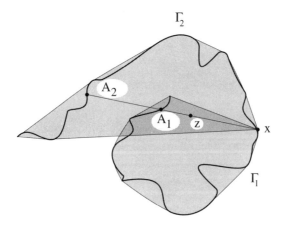

Fig. 14.9. *The curve* Γ *is divided into* $\Gamma_1 \cup \Gamma_2$ *where* $\Gamma_1 \cap \Gamma_2 = \{x\}$, *and* $\mathcal{K}(\Gamma_1) \cap \mathcal{K}(\Gamma_2)$ *contains a point* z *different from* x.

$(\mathbf{3} \Rightarrow \mathbf{1})$ Let x be a point of Γ. Assume first that $x = A$ is one of the endpoints of Γ. When a point y of Γ tends to x, the convex hull $\mathcal{K}(B^\frown y)$ tends to $\mathcal{K}(\Gamma)$. By hypothesis, x does not belong to $\mathcal{K}(B^\frown y)$. Therefore x belongs to the boundary of $\mathcal{K}(\Gamma)$, and there exists a line $\mathbf{D}$ passing through x that contains no other point of Γ. If x is not an endpoint of Γ, let $\Gamma_1 = A^\frown x$ and $\Gamma_2 = B^\frown x$. By hypothesis, $\mathcal{K}(\Gamma_1) \cap \mathcal{K}(\Gamma_2) = \{x\}$. Therefore, there exists a line $\mathbf{D}_x$ passing through x, such that the two convex hulls $\mathcal{K}(\Gamma_1)$ and $\mathcal{K}(\Gamma_2)$ are on opposite sides of this line. Moreover, we can choose $\mathbf{D}_x$ in such a manner that the boundary of $\mathcal{K}(\Gamma_1)$ does not intersect $\mathbf{D}_x$ at any point other than x. Similarly for $\mathcal{K}(\Gamma_2)$. Condition $\mathbf{1}$ is therefore satisfied.

Expansivity We are left to prove that if Γ satisfies conditions $\mathbf{1}$, $\mathbf{2}$, or $\mathbf{3}$, then it is expansive. Construct a cover of Γ by arcs of deviation ϵ as defined in §3. For $k = 1, \ldots, N$ these arcs are either disjoint or adjacent. Condition $\mathbf{3}$ applied to these N arcs implies that the convex hulls $\mathcal{K}(\Gamma_k^\epsilon)$ have at most one common point. We deduce that:

$$\mathcal{A}(\cup_1^N \mathcal{K}(\Gamma_k^\epsilon)) = \sum_1^N \mathcal{A}(\mathcal{K}(\Gamma_k^\epsilon)) .$$

However, if $A_N \neq B$, these arcs do not cover the entire curve Γ, and we must add an arc Γ_{N+1}^ϵ. This arc and Γ_N^ϵ are not disjoint. We then write

$$\sum_1^{N+1} \mathcal{A}(\mathcal{K}(\Gamma_k^\epsilon)) = \mathcal{A}(\cup_1^N \mathcal{K}(\Gamma_k^\epsilon)) + \mathcal{A}(\mathcal{K}(\Gamma_{N+1}^\epsilon))$$

$$\leq 2 \max \{ \mathcal{A}(\cup_1^N \mathcal{K}(\Gamma_k^\epsilon)), \, \mathcal{A}(\mathcal{K}(\Gamma_{N+1}^\epsilon)) \}$$

$$\leq 2 \mathcal{A}(\cup_1^{N+1} \mathcal{K}(\Gamma_k^\epsilon)) .$$

This family of arcs satisfies the two expansivity conditions of §4. ◀

Application Suppose that the curve Γ is a part of the boundary ∂K of a convex set K. Such a curve satisfies the criterion. In fact, a straight line intersects ∂K at 0, 1, or 2 points. If x is a point of Γ and y is a point of ∂K not in Γ, the line $\mathbf{D}_x$ passing through x and y does not have any point common with Γ other than x.

Another application Let Γ be the graph of a continuous function $z(t)$, drawn in a Cartesian coordinate system Ot, Oz. Every line that is parallel to Oz intersects Γ at 0 or 1 point. Thus Γ satisfies the criterion.

Attention!   The criterion gives a **sufficient** condition for a curve to be expansive. This condition is not a **necessary** one. Many expansive curves do not satisfy it. For example, a spiral does not satisfy this criterion, yet it is expansive. Another example: the Von Koch curve (Fig. 14.6) does not satisfy this criterion, yet it is expansive. The rest of this chapter is devoted to the study of more general criteria.

◇   Here is one:

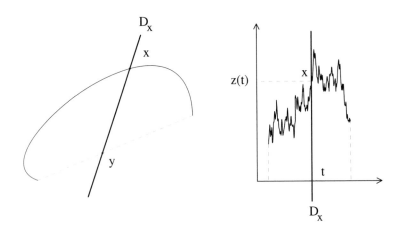

Fig. 14.10. *Two examples of curves satisfying the expansivity criterion of §5: a) Γ is included in the boundary of a convex set; b) Γ is the graph of a continuous function $z(t)$.*

Assume that for some constant c and all subarcs Γ^* of Γ there exists a convex set $W(\Gamma^*)$ included in $\mathcal{K}(\Gamma^*)$ such that

$$W(\Gamma^*) \cap \Gamma \subset \Gamma^*$$

and

$$\mathcal{A}(\mathcal{K}(\Gamma^*)) \leq c\,\mathcal{A}(W(\Gamma^*)) .$$

Then Γ is expansive.

We note that the hypothesis does not indicate whether or not the arc Γ^* intersects the convex set $W(\Gamma^*)$.

▶   Assume that Γ satisfies this condition for all $\Gamma^* \subset \Gamma$. Consider the family of arcs Γ_k^ϵ constructed in §3. For $k = 1, \ldots, N$, the convex sets $W(\Gamma_k^\epsilon)$ have disjoint interiors and $\mathcal{A}(\cup_1^N W(\Gamma_k^\epsilon)) = \sum_1^N \mathcal{A}(W(\Gamma_k^\epsilon))$. If there is a supplementary arc Γ_{N+1}^ϵ, we deduce:

$$\sum_1^{N+1} \mathcal{A}(\mathcal{K}(\Gamma_k^\epsilon)) \leq c \sum_1^{N+1} \mathcal{A}(W(\Gamma_k^\epsilon))$$
$$\leq c\left(\mathcal{A}(\cup_1^N W(\Gamma_k^\epsilon)) + \mathcal{A}(W(\Gamma_{N+1}^\epsilon))\right)$$
$$\leq 2c\,\mathcal{A}(\cup_1^{N+1} W(\Gamma_k^\epsilon))$$
$$\leq 2c\,\mathcal{A}(\cup_1^{N+1} \mathcal{K}(\Gamma_k^\epsilon)) ,$$

the curve Γ is therefore expansive.   ◀

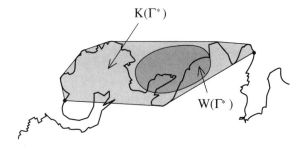

Fig. 14.11. *The convex hull $\mathcal{K}(\Gamma^*)$ of the subarc Γ^* of Γ can intersect with Γ outside of Γ^*; however, the convex set $W(\Gamma^*)$ does not contain any point of Γ that is not in Γ^*.*

14.6 Expansivity and self–similarity

Here is a condition for a self–similar curve to be expansive:

CONVEX SET CRITERION   *Given $N+1$ points A_1, ..., A_{N+1} and N similarities F_1, ..., F_N defining (as in Chap. 13, §2) a self–similar curve Γ. We assume that there exists a closed, bounded, and convex set D such that:*
   1) $F_i(D) \subset D$ for all $i = 1, \ldots, N$;
   2) $F_i(D) \cap F_{i+1}(D) = \{A_{i+1}\}$ for all $i = 1, \ldots, N-1$;
   3) $F_i(D)$ and $F_j(D)$ are disjoint whenever $|i - j| \geq 2$.
Then Γ is expansive.

We recognize the three conditions of the *closed-set criterion* (Chap. 13, §7). Therefore, these conditions imply that Γ is a simple curve. The only difference between these criteria consists of the supplementary hypothesis, namely, the convexity of D, which ensures the expansivity of Γ.

▶   We use the notation and some results from Chapter 13. Condition 1) implies that Γ is included in D. But D is convex; therefore it contains $\mathcal{K}(\Gamma)$, the convex hull of Γ. The closed domain $\mathcal{K}(\Gamma)$ also satisfies conditions 1), 2), and 3). Without loss of generality, we may assume that $D = \mathcal{K}(\Gamma)$.

   Let $\rho_i = \mathrm{dist}(A_i, A_{i+1})/\mathrm{dist}(A, B)$ be the ratio of the similarity F_i, and let $\rho_{\max} = \max\{\rho_i\}$. To each point x of Γ, there corresponds a sequence $(i_1, \ldots, i_k, \ldots)$ of integers such that

$$\{x\} = \bigcap_k F_{i_1}(\ldots (F_{i_k}(D)) \ldots) .$$

This sequence is unique unless x is a vertex of the polygonal curve $\mathbf{P}_k$ for some k. For each k, the point x then belongs to a unique convex domain

$F_{i_1}(\ldots(F_{i_k}(D))\ldots)$, the image of D by the similarities. The breadth of this convex set is equal to $\rho_{i_1}\ldots\rho_{i_k}\mathrm{breadth}(D)$.

Given a real number ϵ and any point x of Γ that is not a vertex of a polygonal curve, we can find an integer $k(x)$ that is the smallest integer k such that $\rho_{i_1}\ldots\rho_{i_k}\mathrm{breadth}(D) \leq \epsilon$. Let $D_\epsilon(x)$ be the corresponding domain $F_{i_1}(\ldots(F_{i_{k(x)}}(D))\ldots)$. Its breadth is between ϵ and $\epsilon\rho_{\max}$. If $x \neq x'$, then it is impossible to have $D_\epsilon(x)$ strictly included in $D_\epsilon(x')$, because of the choice of $k(x)$. If they are not identical, $D_\epsilon(x)$ and $D_\epsilon(x')$ are of disjoint interiors. The family $\{\,D_\epsilon(x)\,,\ x \in \Gamma\,\}$ is a finite family of domains that are convex hulls of arcs; we denote them by Γ_i^ϵ. These arcs form a cover of Γ. Moreover, by taking the deviation of an arc in the sense of the breadth of its convex hull, they satisfy the following two conditions:

(i) $\epsilon\rho_{\max} \leq \mathrm{dev}(\Gamma_i^\epsilon) \leq \epsilon$;

(ii) $\sum_k \mathcal{A}(\mathcal{K}(\Gamma_i^\epsilon)) = \mathcal{A}(\cup_k\mathcal{K}(\Gamma_i^\epsilon))$ (because the interiors of the convex hulls are disjoint). The curve Γ is therefore expansive.   ◀

14.7 How to construct an expansive curve

We are going to describe a model of a curve that does not use any similarity. This is a rich mine of examples, and it will be used in Chapter 15 to obtain new curves. We start with a convex domain D, and we construct successive chains E_k that constitute the approximations of the curve. This construction is done by iterating the following operation:

OPERATION OF TYPE $\mathcal{T}$ *Let $N \geq 2$ be an integer. Let D be a closed set and let*

$$A = A_1, A_2, \ldots, A_{N+1} = B$$

be $N + 1$ distinct points of D such that A and B are on the boundary ∂D. An **operation of type** $\mathcal{T}$ *consists of replacing D by N closed sets $D(1)$, ..., $D(N)$, such that the following hold:*

1) $D(i) \subset D$ for all $i = 1, \ldots, N$;

2) $D(i) \cap D(i+1) = \{A_{i+1}\}$ for all $i = 1, \ldots, N-1$;

3) $D(i)$ and $D(j)$ are disjoint whenever $|i - j| \geq 2$;

4) $D(1)$ contains $A(1)$, and $D(N)$ contains $A(N+1)$.

We see that many of the parameters are not fixed in this operation, for example, the position of the points $A(i)$, the form of the domains $D(i)$,.... Thus, there are many different ways to determine $\mathcal{T}$ exactly. In Chapter 13, we determined this operation by using similarities. We can also do it by using other plane transformations (affinities and various contractions) or direct geometrical constructions (Figs. 15.2, 15.3, 15.4).

To construct the curve Γ by induction, we need an infinite number of successive operations $\mathcal{T}$. First we fix an integer $N \geq 2$.

Step 1 Given a domain D and two points A and B on its boundary, we replace D with a chain $D(1)$, ..., $D(N)$ according to a $\mathcal{T}$ operation and we put:

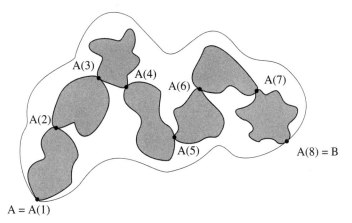

Fig. 14.12. *The operation* T *replaces the domain* D *by the disjoint or adjacent subdomains* $D(1)$, ..., $D(N)$ *(here* $N = 7$*) that join the two points* A *and* B *of the boundary* ∂D.

$$E_1 = \cup_{i=1}^{N} D(i) .$$

Step 2 For every i, replace D with $D(i)$, A with $A(i)$, and B with $A(i+1)$. With a new operation of type T, we replace $D(i)$ by a chain $D(i,1)$, ..., $D(i,N)$. For each j, the intersection $D(i,j) \cap D(i,j+1)$ is reduced to the unique point $A(i,j+1)$, and $A(i) = A(i,1) = A(i-1,N)$. We put

$$E_2 = \cup_{i=1}^{N} \cup_{j=1}^{N} D(i,j)$$

and so forth.

Step k Assume that we have constructed the chain E_{k-1}, formed by N^{k-1} closed sets $D(i_1,\ldots,i_{k-1})$, $1 \leq i_j \leq N$. These sets are either disjoint or they have a unique common point. We replace each of them by an operation of type T. The new domains, which are included in the previous ones, will be denoted by $D(i_1,\ldots,i_{k-1},i_k)$. The intersection $D(i_1,\ldots,i_{k-1},i_k) \cap D(i_1,\ldots,i_k+1)$ is reduced to one unique point $A(i_1,\ldots,i_k+1)$, and $A(i_1,\ldots,i_{k-1}) = A(i_1,\ldots,i_{k-1},1) = A(i_1,\ldots,i_{k-1}-1,N)$. We put

$$E_k = \bigcup_{\substack{1 \leq i_1 \leq N \\ \cdots \\ 1 \leq i_k \leq N}} D(i_1,\ldots,i_k) .$$

Having constructed the sets E_k, they form an embedded sequence of closed sets. The intersection

$$\Gamma = \cap_k E_k$$

is also a closed set. This set contains the points A and B; the points $A(i)$, $A(i,j)$; and so forth. Under what conditions will this set be a curve?

PROPOSITION 1 *Using the same notation, let $\Gamma = \cap E_k$. We assume that there exists a sequence of real numbers ϵ_k converging to 0 such that:*

$$\operatorname{diam}(D(i_1, \ldots, i_k)) \leq \epsilon_k$$

for all k and for every sequence i_1, ..., i_k of integers between 1 and N. Then Γ is a simple curve.

This condition is realized, for example, if there exists a constant $c < 1$ such that: $\operatorname{diam}(D(i_1, \ldots, i_{k-1}, i_k)) \leq c \operatorname{diam}(D(i_1, \ldots, i_{k-1}))$, for all k and all sequences i_1, ..., i_k.

▶   We shall only give the main ideas of the proof. It is a bit easier than the analogous proof of self–similarity (Chap. 13, §3 and §7), because similarities may contain symmetries that could change the role of $A(i)$ and $A(i+1)$. In the operations $\mathcal{T}$, we always go from E_k to E_{k+1} while preserving the direction from A to B. It is then easier to establish a direct correspondence between the points of Γ and those of the interval $[0, 1]$ written in base N.

Definition of the parameterization γ Every $t \in [0, 1]$ can be written as

$$t = 0, j_1 j_2 \ldots, j_k \ldots ,$$

where j_k is an integer between 0 and $N - 1$. We put

$$\gamma(t) = \bigcap_{k=1}^{\infty} D(j_1 + 1, \ldots, j_k + 1) .$$

This is an embedded sequence of closed sets whose diameters tend to 0 by hypothesis; their intersection is then reduced to a unique point. Moreover, we must verify that if t admits two different base N expansions, then the corresponding point $\gamma(t)$ remains the same. For example, if

$$t = 0, j_1 (N - 1) \ldots (N - 1) \ldots = 0, (j_1 + 1) 0 \ldots 0 \ldots ,$$

then the intersections $\cap D(j_1 + 1, N, \ldots, N)$ and $\cap D(j_1 + 2, 1, \ldots, 1)$ contain the same point $A(j_1 + 2)$ and similarly for all other points. Therefore the function $\gamma(t)$ is well defined. Its domain of definition is $[0, 1]$, and its range is Γ.

Continuity of γ If $0 < t - t' < N^{-k}$, there exists an integer i, $1 \leq i \leq N - 1$, such that

$$(i - 1) N^{-k} \leq t' \leq i N^{-k} \leq t \leq (i + 1) N^{-k} .$$

The points $\gamma(t)$ and $\gamma(t')$ belong to two domains of rank k that are touching at the point $\gamma(i N^{-k})$. We deduce that $\operatorname{dist}(\gamma(t), \gamma(t')) \leq 2 \epsilon_k$. For all $\epsilon > 0$, there exists an integer k such that $2 \epsilon_k < \epsilon$ and an $\eta = N^{-k}$ such that

$$|t - t'| \leq \eta \Rightarrow \operatorname{dist}(\gamma(t), \gamma(t')) \leq \epsilon .$$

Injectivity of γ If $t < t'$, we can always find i and k such that

$$t \leq i\,N^{-k} < (i+2)\,N^{-k} \leq t' \ .$$

Then the two points $\gamma(t)$ and $\gamma(t')$ belong to two disjoint domains of rank k. Therefore $\gamma(t) \neq \gamma(t')$.

    In conclusion, Γ is a simple curve parameterized by γ. Its endpoints are A and B.   ◀

    Now we give sufficient conditions for this curve to be expansive.

PROPOSITION 2   *To the hypothesis of Proposition 1, we add the assumption that all of the domains D, $D(i)$, $D(i_1, \ldots, i_k)$ are convex. Let $\Gamma(i_1, \ldots, i_k) = \Gamma \cap D(i_1, \ldots, i_k)$. We also suppose that for some constant $c > 0$,*

$$\mathrm{breadth}(\mathcal{K}(\Gamma(i_1, \ldots, i_{k-1}, i_k))) \geq c\,\mathrm{breadth}(\mathcal{K}(\Gamma(i_1, \ldots, i_{k-1})))$$

for all k and all sequences $i_1, \ldots, i_k$. Then Γ is expansive.

◇   The condition on the breadths may seem to be difficult to verify because we do not always know the convex hull of Γ nor those of the subarcs. Usually, we escape this difficulty by forming the convex hulls of all the points $A(i_1, \ldots, i_k, j)$ inside $D(i_1, \ldots, i_k)$, where $1 \leq j \leq N$. This is the set of all the vertices of step $k+1$ that belong to $D(i_1, \ldots, i_k)$. Then we use the following inequality:

$$\mathrm{breadth}(D(i_1, \ldots, i_k)) \geq \mathrm{breadth}(\mathcal{K}(\Gamma(i_1, \ldots, i_k)))$$
$$\geq \mathrm{breadth}(\mathcal{K}(\{\, A(i_1, \ldots, i_k, j)\,,\, 1 \leq j \leq N\})) \ .$$

▶   A similar argument to the one used in the proof of self–similarity (§6) can be used to prove this proposition. First we observe that without any loss of generality, we can replace the initial domain D with $\mathcal{K}(\Gamma)$ and similarly $D(i_1, \ldots, i_k)$ with $\mathcal{K}(\Gamma(i_1, \ldots, i_k))$. Given any ϵ with $0 < \epsilon < \mathrm{breadth}(\mathcal{K}(\Gamma))$, we can associate to each x that is not a vertex $A(i_1, \ldots, i_k)$ for some k, the smallest integer $k(x)$ such that $x \in \Gamma(i_1, \ldots, i_{k(x)})$ with $\mathrm{breadth}(\mathcal{K}(\Gamma(i_1, \ldots, i_{k(x)}))) \leq \epsilon$. Putting together all the arcs $\Gamma(i_1, \ldots, i_{k(x)})$ for all the points x, we obtain a cover of Γ. This family is finite: we denote its elements by Γ_i^ϵ. They are either disjoint or adjacent. Their deviation is between $\epsilon\,c$ and ϵ. Moreover, their convex hulls have disjoint interiors. This gives:

$$\sum_i \mathcal{A}(\mathcal{K}(\Gamma_i^\epsilon)) = \mathcal{A}(\cup\mathcal{K}(\Gamma_i^\epsilon)) \ .$$

Therefore the curve Γ is expansive.   ◀

Example   The image of a parallelogram by an affine transformation is a parallelogram. In Figure 14.13, seven affine functions $F_1, \ldots, F_7$ determine seven copies of the initial parallelogram D. Each of these applications can be noted by

$$F_i(x) = \mathcal{M}_i\,x + B_i \ ,$$

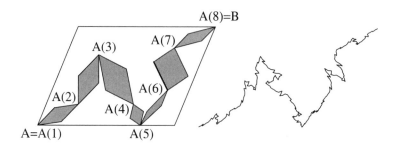

Fig. 14.13. *The operation T consists of replacing the initial parallelogram with seven copies that determine seven affine applications. By iterating this procedure, we obtain a self–affine expansive curve.*

where M_i is a 2×2-matrix, and B_i a translation vector. The operation T consists of replacing D with

$$T(D) = E_1 = \bigcup_{i=1}^{7} F_i(D) \ .$$

The domains of step 2 are also parallelograms, which can be written as $F_i(F_j(D))$, and

$$T(T(D)) = E_2 = \bigcup_{i=1}^{7} \bigcup_{j=1}^{7} F_i(F_j(D)) \ .$$

Iterating this procedure to rank k, we obtain a chain E_k formed by 7^k parallelograms, that are either disjoint or adjacent. At the limit stage we obtain (Proposition 1) a simple curve Γ. This curve is *self–affine* in the sense that it is the union of seven affine copies of itself.

Finally, Γ is expansive; in fact, since all the parallelograms $F_i(D)$ are of non-null areas, $\det \mathcal{M}_i \neq 0$. If we denote the transpose matrix of $\mathcal{M}_i$ by $\mathcal{M}_i^t$, the eigenvalues of the matrix $\mathcal{M}_i^t \mathcal{M}_i$ are strictly positive. Let λ_i be the smallest such eigenvalue; it can be shown that for all pairs of points (x, y) in the plane,

$$\mathrm{dist}(F_i(x), F_i(y)) \geq \sqrt{\lambda_i}\, \mathrm{dist}(x, y) \ .$$

Let $c = \min_i \sqrt{\lambda_i}$. We find that for all i,

$$\mathrm{breadth}(\mathcal{K}(F_i(\Gamma))) \geq c\,\mathrm{breadth}(\mathcal{K}(\Gamma)) \ ,$$

and in general, for any sequence $i_1, \ldots, i_k$,

$$\mathrm{breadth}(\mathcal{K}(\Gamma(i_1, \ldots, i_k))) \geq c\,\mathrm{breadth}(\mathcal{K}(\Gamma(i_1, \ldots, i_{k-1}))) \ .$$

The hypotheses of Proposition 2 are then verified, and Γ is expansive. Also, this result can be shown to hold even if some of the affine applications F_i are of null determinant.

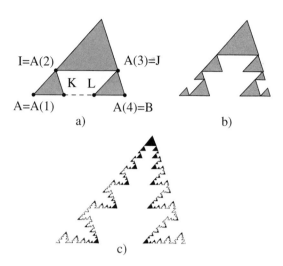

Fig. 14.14. *The operation* T *consists of replacing the initial triangle with three similar ones. By iterating this procedure, we obtain an expansive curve (which is not self–similar).*

Different example Here is another particular case of the general model of §7; we start with any triangle D with vertices ABC (Fig. 14.14). Let I and J be the points of AC and BC such that $\mathrm{dist}(A, I) = \mathrm{dist}(A, C)/3$ and $\mathrm{dist}(B, J) = \mathrm{dist}(B, C)/3$. Let K and L be the points dividing AB into three equal segments.

The operation T consists of replacing D with three triangles: $D(1) = AIK$, $D(2) = IJC$, $D(3) = JBL$. Thus, $A(1) = A$, $A(2) = I$, $A(3) = J$, $A(4) = B$. These three triangles are similar to D, with the ratios $1/3$, $2/3$, and $1/3$, respectively. But if D is not an isosceles triangle, it will be impossible to define T with similarities, because if F_1 is the similarity such that $F_1(D) = D(1)$, the point $A(2)$ is not the image of B by F_1.

The same operation T is iterated on the triangles $D(i)$ and so forth. We represent in a), b), c) the steps E_1, E_2, E_5 of this construction. The limit curve Γ is a simple curve (Proposition 1). It is self–affine, as in the previous example, but it is not self–similar in the strict sense, even though the set E_k is formed by 3^k triangles similar to the initial triangle D. All of the domains $D(i_1, \ldots, i_k)$ are triangles, therefore they are convex sets that satisfy the inequality:

$$\mathrm{breadth}(D(i_1, \ldots, i_k)) \geq \frac{1}{3}\,\mathrm{breadth}(D(i_1, \ldots, i_{k-1})) \,.$$

By Proposition 2, Γ is an expansive curve.

14.8 Bibliographical notes

Attractors of iterated function systems are such a rich mine of fractal images that one could be led to believe that all natural forms may be represented by an attractor. We find important works on this subject, for example, in [M. Barnsley].

A part of this chapter is summarized in [C. Tricot 7] (1990).

15 The Constant–Deviation Variable–Step Algorithm

15.1 A unified analysis of expansive curves

It is very rare to encounter a curve that cannot be classified as expansive. Theoretically, one must establish some tests of expansivity before using this notion, but this precaution will be useless in most cases.

What is in reality the interest of this notion? We shall cite two problems in the characterization of fractal curves, which, thanks to this notion and that of deviation, can be solved satisfactorily.

Problem 1 What is the meaning of "self–similarity" or "statistical self–similarity" in the case of experimental curves given with the necessary limited precision? We answer this in §6, explaining at the same time the drawback of the "compass" method, which is a consequence of Richardson's measure of geographical coastlines.

Problem 2 The universal methods for computing the fractal dimension, like the box-counting and Minkowski sausage methods, do not give very good results in general. Thus we need to find more efficient algorithms that take into account the particular geometry of each curve. These bear some relation to the function $\overline{T}_\tau$, the average local size, computed as a function of the parameterization. We already know that:

— For a strictly self–similar curve with equal similarity ratios, we use the **diameters method** (Chap. 13, §11):

$$\Delta(\Gamma) = \lim_{\tau \to 0} \frac{\log \tau}{\log \overline{T}_\tau} .$$

— For the graph of a continuous function $z(t)$ parameterized by the abscissa t, we have the **variation method** (Chap. 12, §4). Using the fact that Var_τ, the mean of the oscillations, is a function equivalent to $\overline{T}_\tau$, we can write it as

$$\Delta(\Gamma) = \lim_{\tau \to 0} \left(2 - \frac{\log \overline{T}_\tau}{\log \tau} \right) .$$

Since we cannot just stick to these two types of curves, what happens to a curve of any type? Mathematically, we can define infinitely many different types. Some are given as examples in this chapter, but some are not because they are too difficult to classify, not very homogeneous, or have a nonidentifiable structure. Yet they offer important irregularities at different scales; these irregularities justify a fractal analysis. Such are the experimental curves met in nearly all branches of science. We shall give universal algorithms to compute their dimensions (in §4 we shall discuss the discrete approach, and in §7 the continuous one). These algorithms constitute an interpolation between the above formulas, and they have the same efficiency. They allow us to analyze curves independently from their geometrical type or their parameterization. Theoretically, we only need one assumption: that these curves are expansive.

To summarize what will follow: we base our analysis on two distinct notions, the **size** and the **deviation**, as discussed in Chapters 11 and 14. The local convex hull will therefore play an important role.

15.2 The covering index

First, we must define an index associated to every family of intervals of the real line or of arcs of a curve.

> Let $\mathcal{F}$ be a family of closed intervals of the real line: the **covering index** $\omega(\mathcal{F})$ of $\mathcal{F}$ is the largest integer n such that there exist n intervals of $\mathcal{F}$ whose common intersection contains at least two points.

• We can also say: The largest integer n such that there exist n intervals of $\mathcal{F}$ with interiors that have a nonempty intersection.
• If the intersection contains two points, then it contains an interval.
• The intervals of $\mathcal{F}$ are *disjoint* if they have no common point. Two intervals are *adjacent* if they have one common endpoint. The intervals of $\mathcal{F}$ are disjoint or adjacent if any two of them have disjoint interiors and $\omega(\mathcal{F}) = 1$.
• The family of all the intervals $[n, n+2]$, n an integer, has an index equal to 2.
• The family of all the intervals $[n, 2n]$, n an integer, is of infinite index.

◇ The notion of covering index on the families of intervals can be directly generalized to the families of subarcs of a curve as follows:

> If $\{\mathcal{F}\}$ is a family of closed subarcs of a curve, its **covering index** *is the largest integer n such that there are n subarcs of $\mathcal{F}$ whose intersection contains at least two points.*

This index is 1 if the arcs are disjoint or adjacent. On a simple curve, the families of subarcs have the same covering index as the families of intervals whose images by the parameterization are the subarcs themselves.

15.3 Convex hulls and Minkowski sausages

Here is a fundamental result that establishes a relation between the Minkowski sausage and the local convex hulls.

LEMMA   *Let Γ be a simple expansive curve. Let $\mathcal{F} = \{\Gamma_i\}$ be a finite cover of Γ by arcs. Given a deviation function $\mathrm{dev}(\Gamma)$, let $\epsilon_{\max} = \max_i\{\mathrm{dev}(\Gamma_i)\}$ and $\epsilon_{\min} = \min_i\{\mathrm{dev}(\Gamma_i)\}$. Assume that $\epsilon_{\min} \neq 0$. Then we can find two non-null constants c_1 and c_2 such that*

$$c_2 \, \mathcal{A}(\Gamma(\epsilon_{\min})) \leq \sum_i \mathcal{A}(\mathcal{K}(\Gamma_i)) \leq c_1 \, \omega(\mathcal{F}) \, \frac{\epsilon_{\max}}{\epsilon_{\min}} \, \mathcal{A}(\Gamma(\epsilon_{\min})) \;.$$

▶   We shall use three constants. The first two, c_3 and c_4, are associated with the deviation function (Chap. 14, §1): for every curve $\mathbf{C}$,

$$c_3 \, \mathrm{breadth}(\mathcal{K}(\mathbf{C})) \leq \mathrm{dev}(\mathbf{C}) \leq c_4 \, \mathrm{breadth}(\mathcal{K}(\mathbf{C})) \;.$$

The constant c_4 can be assumed to be ≥ 1. The third constant c_5 is associated with the expansivity condition (Chap. 14, §4): for all $\epsilon > 0$, there exists a cover $\{\Gamma_n^\epsilon\}$ of Γ by arcs such that

$$(i) \qquad \frac{\epsilon}{c_5} \leq \mathrm{dev}(\Gamma_n^\epsilon) \leq \epsilon \;;$$

$$(ii) \qquad \sum \mathcal{A}(\mathcal{K}(\Gamma_n^\epsilon)) \leq c_5 \, \mathcal{A}(\cup \Gamma_n^\epsilon) \;.$$

Finally, to shorten the notation, let $K_i = \mathcal{K}(\Gamma_i)$.
a) We prove the first inequality. We observe that for all ϵ, $\Gamma(\epsilon)$ is included in $\cup \Gamma_i(\epsilon)$ and thus in $\cup K_i(\epsilon)$. Thus,

$$\mathcal{A}(\Gamma(\epsilon)) \leq \mathcal{A}(\cup_i K_i(\epsilon)) \;.$$

The dilated set $K_i(\epsilon)$ has a diameter $\mathrm{diam}\,(K_i) + 2\epsilon$ and breadth $\mathrm{breadth}(K_i) + 2\epsilon$. By taking an ϵ such that $\epsilon \leq \mathrm{breadth}(K_i) \leq \mathrm{diam}\,(K_i)$, we deduce that $\mathrm{diam}\,(K_i(\epsilon)) \leq 3\,\mathrm{diam}\,(K_i)$, and $\mathrm{breadth}(K_i(\epsilon)) \leq 3\,\mathrm{breadth}(K_i)$.

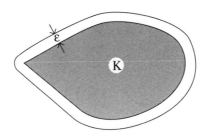

Fig. 15.1. *The diameter of the dilated convex set $K(\epsilon)$ is $\mathrm{diam}\,(K) + 2\epsilon$; its breadth is $\mathrm{breadth}(K) + 2\epsilon$.*

The general result

$$\frac{1}{2}\operatorname{diam}(K)\operatorname{breadth}(K) \le \mathcal{A}(K) \le \sqrt{2}\operatorname{diam}(K)\operatorname{breadth}(K) ,$$

which is proved for all convex sets K in Appendix C, §4, allows us to evaluate the area of $K_i(\epsilon)$:

$$\mathcal{A}(K_i(\epsilon)) \le 9\sqrt{2}\operatorname{diam}(K_i)\operatorname{breadth}(K_i) \le 18\sqrt{2}\,\mathcal{A}(K_i) .$$

We choose $\epsilon = \epsilon_{\min}/c_4$, which is smaller than $\operatorname{breadth}(K_i)$ for all i. We obtain

$$\mathcal{A}(\Gamma(\epsilon_{\min}/c_4)) \le 18\sqrt{2}\sum_i \mathcal{A}(K_i) .$$

Finally, we use the Minkowski sausage (lemma in Chap. 10, §6), to find

$$\mathcal{A}(\Gamma(\epsilon_{\min})) \le c_4^2\,\mathcal{A}(\Gamma(\epsilon_{\min}/c_4)) \le 18\sqrt{2}\,c_4^2\sum_i \mathcal{A}(K_i) ,$$

thus proving the desired inequality with $c_2 = 1/(18\sqrt{2}\,c_4^2)$.

b) For the other inequality, we shall use the cover $\{\Gamma_n^\epsilon\}$ of Γ, which proves the expansivity of the curve (see the definition of expansive curve, Chap. 14, §4). We take $\epsilon = \epsilon_{\min}$ and put $\mathbf{C}_n = \Gamma_n^{\epsilon_{\min}}$. We have $\operatorname{dev}(\mathbf{C}_n) \le \epsilon_{\min}$ for all n. Finally, let $\tilde{\Gamma}_i$ be the arc Γ_i without its endpoints. Since $\{\mathbf{C}_n\}$ is a cover of Γ, each $\tilde{\Gamma}_i$ is included in the union of the arcs $\mathbf{C}_n$. Using the subadditivity of the diameters (Chap. 14, §2), we get that

$$\operatorname{diam}(\Gamma_i) \le \sum_{\mathbf{C}_n \cap \tilde{\Gamma}_i \ne \emptyset} \operatorname{diam}(\mathbf{C}_n) .$$

For all integers i and n, we have $\operatorname{dev}(\mathbf{C}_n) \le \operatorname{dev}(\Gamma_i)$, and therefore, the fact that dev is an increasing function implies that none of the Γ_i can be included in one of the $\mathbf{C}_n$. Then every arc Γ_i that intersects $\mathbf{C}_n$ must contain at least one of the endpoints of $\mathbf{C}_n$. But each of these endpoints cannot belong to more than $\omega(\mathcal{F})$ arcs $\tilde{\Gamma}_i$. Summing the terms of the previous inequality over i, and knowing that every $\mathbf{C}_n$ cannot be counted more than $2\omega(\mathcal{F})$ times, we get:

$$\sum_i \operatorname{diam}(\Gamma_i) \le 2\omega(\mathcal{F})\sum_n \operatorname{diam}(\mathbf{C}_n) .$$

Thus, we may estimate the areas of the convex hulls as follows:

$$\sum_i \mathcal{A}(K_i) \le \sqrt{2}\sum_i \operatorname{diam}(K_i)\operatorname{breadth}(K_i)$$

$$\le \frac{\sqrt{2}}{c_3}\sum_i \operatorname{diam}(K_i)\operatorname{dev}(K_i)$$

$$\le \frac{2\sqrt{2}}{c_3}\epsilon_{\max}\,\omega(\mathcal{F})\sum_n \operatorname{diam}(\mathbf{C}_n) .$$

Since $\epsilon_{\min} \leq c_5 \operatorname{dev}(\mathbf{C}_n) \leq c_4 c_5 \operatorname{breadth}(\mathcal{K}(\mathbf{C}_n))$,

$$\sum_i A(K_i) \leq 2\sqrt{2}\, \frac{c_4 c_5}{c_3}\, \frac{\epsilon_{\max}}{\epsilon_{\min}}\, \omega(\mathcal{F}) \sum_n \operatorname{diam}(\mathbf{C}_n) \operatorname{breadth}(\mathcal{K}(\mathbf{C}_n))$$

$$\leq 4\sqrt{2}\, \frac{c_4 c_5}{c_3}\, \frac{\epsilon_{\max}}{\epsilon_{\min}}\, \omega(\mathcal{F}) \sum_i A(\mathcal{K}(\mathbf{C}_n))\, .$$

This is where we use the expansivity:

$$\sum_i A(\mathcal{K}(\mathbf{C}_n)) \leq c_6\, \frac{\epsilon_{\max}}{\epsilon_{\min}}\, \omega(\mathcal{F})\, A(\cup_n \mathcal{K}(\mathbf{C}_n))\, ,$$

with $c_6 = 4\sqrt{2}\, c_4 c_5^2 / c_3$.

Since the breadth of the convex sets $\mathcal{K}(\mathbf{C}_n)$ is $\leq \epsilon_{\min}/c_3$, each point of $\mathcal{K}(\mathbf{C}_n)$ is at a distance $\leq \epsilon_{\min}/c_3$ from Γ. Thus

$$A(\cup_n \mathcal{K}(\mathbf{C}_n)) \leq A(\Gamma(\frac{\epsilon_{\min}}{c_3})) \leq \frac{1}{c_3^2}\, A(\Gamma(\epsilon_{\min}))\, ,$$

which implies that:

$$\sum_i A(K_i) \leq c_1\, \frac{\epsilon_{\max}}{\epsilon_{\min}}\, \omega(\mathcal{F})\, A(\Gamma(\epsilon_{\min}))\, ,$$

where $c_1 = c_6/c_3^2$. ◄

15.4 A theorem on the dimension: the discrete form

Given ϵ, it is always possible to construct a cover of the curve Γ by arcs of deviation ϵ; the index of this cover is ≤ 2 (Chap. 14, §3). We fix ϵ, and we denote this cover by $\{\Gamma_i\}$. Let $\epsilon_{\max} = \epsilon_{\min} = \epsilon$ and $\omega(\mathcal{F}) = 2$. The lemma in §3 gives us

$$c_2\, A(\Gamma(\epsilon)) \leq \sum_i A(\mathcal{K}(\Gamma_i)) \leq 2\, c_1\, A(\Gamma(\epsilon))\, .$$

Consequently,

$$A(\Gamma(\epsilon)) \simeq \epsilon \sum_i \operatorname{diam}(\Gamma_i)\, .$$

It is then possible to evaluate the area of the ϵ–Minkowski sausage by taking steps of deviation ϵ along the curve and measuring the covered path. However, such a path is not unique, and the following theorem is stated in a general form.

THEOREM   Let Γ be a simple expansive curve. Let c be a constant $c \geq 1$, and let ω be an integer $\omega \geq 1$. Assume that for every ϵ with $0 < \epsilon < \operatorname{dev}(\Gamma)$ there exists a cover $\{\Gamma_i^\epsilon\}$ of Γ of index $\leq \omega$ such that

$$\frac{\epsilon}{c} \leq \operatorname{dev}(\Gamma_i^\epsilon) \leq \epsilon\, .$$

Then

$$\Delta(\Gamma) = \lim_{\epsilon \to 0} \left(1 + \frac{\log \sum_i \operatorname{diam}(\Gamma_i^\epsilon)}{|\log \epsilon|} \right) .$$

◇   If the dimension of Γ is not a limit, then we should replace $\lim_{\epsilon \to 0}$ with $\limsup_{\epsilon \to 0}$ in the formula.

◇   There is an important similarity between the formula and the formula of the "compass" method which will be discussed in §6. We shall see that the compass method does not compute the dimension, except for particular curves.

◇   Using this formula, one can devise an algorithm to compute the dimension. It will be called the "constant–deviation variable–step method" (see §9). This nomenclature covers many methods, because there are many different ways to define paths along Γ.

▶   The proof is a direct consequence of the lemma in §3. In fact, this gives

$$c_2 \, \mathcal{A}(\Gamma(\tfrac{\epsilon}{c})) \le \sum_i \mathcal{A}(\mathcal{K}(\Gamma_i^\epsilon)) \le c_1 \, c\omega \, \mathcal{A}(\Gamma(\tfrac{\epsilon}{c})) \,,$$

and therefore (Appendix B, §2):

$$\frac{c_2}{c^2} \, \mathcal{A}(\Gamma(\epsilon)) \le \sum_i \mathcal{A}(\mathcal{K}(\Gamma_i^\epsilon)) \le c_1 \, c\omega \, \mathcal{A}(\Gamma(\epsilon)) \,.$$

We deduce that:

$$\sum_i \mathcal{A}(\mathcal{K}(\Gamma_i^\epsilon)) \simeq \mathcal{A}(\Gamma(\epsilon)) \,.$$

Since $\mathcal{A}(\mathcal{K}(\Gamma_i^\epsilon)) \simeq \operatorname{diam}(\Gamma_i^\epsilon) \operatorname{dev}(\Gamma_i^\epsilon)$, we finally find that

$$\mathcal{A}(\Gamma(\epsilon)) \simeq \epsilon \sum_i \operatorname{diam}(\Gamma_i^\epsilon)$$

and

$$2 - \frac{\log \mathcal{A}(\Gamma(\epsilon))}{\log \epsilon} \simeq 1 - \frac{\log \sum_i \operatorname{diam}(\Gamma_i^\epsilon)}{\log \epsilon} \,. \qquad \blacktriangleleft$$

15.5 Applications

The constant–deviation variable–step algorithm has interesting applications in the computation of the dimension of experimental curves (§9). In this section we look at some of its consequences in some types of mathematical models.

1. Self–similar curves Self–similar curves are defined by $N+1$ distinct points $A = A_1, \ldots, A_{N+1} = B$ and by N similarities $F_1, \ldots, F_N$ such that $F_i(AB) = A_i A_{i+1}$. Moreover, we assume that there exists a closed, bounded, and convex set D such that

1) $F_i(D) \subset D$ for all $i = 1, \ldots, N$;
2) $F_i(D) \cap F_{i+1}(D) = \{A_{i+1}\}$ for all $i = 1, \ldots, N-1$;
3) $F_i(D)$ and $F_j(D)$ are disjoint whenever $|i - j| \geq 2$.

These conditions imply that Γ is a simple expansive curve (Chap. 14, §6). The similarity ratios are given by: $\rho_i = \text{dist}(A_i, A_{i+1})/\text{dist}(A, B)$. Let e be the real number satisfying the equation:

$$\sum_i \rho_i^e = 1 \ .$$

Using the theorem in §4, we will show that the following equality holds: $e = \Delta(\Gamma)$. For this we can construct a path on Γ formed by arcs of the same deviation.

▶ Assume that Γ is parameterized in the usual way (Chap. 13, §10): the measure μ of the arc $\Gamma^* = F_{i_1}(\ldots(F_{i_k}(\Gamma))\ldots)$, which is the image of Γ by similarities of ratios $\rho_{i_1} \ldots \rho_{i_k}$, is equal to $(\rho_{i_1} \ldots \rho_{i_k})^e$. This is the time spent in this arc by an object that runs through the whole curve in one unit of time. This measure may be written as: $\mu(\Gamma^*) = (\text{diam}\,(\Gamma^*)/\text{diam}\,(\Gamma))^e$. We have seen that for all ϵ, it is always possible to cover Γ by N_ϵ disjoint or adjacent arcs of this type whose deviation is between ϵ and $\epsilon\,\rho_{\max}$ (Chap. 14, §6). By symmetry, these arcs have the same size, which is equivalent to ϵ. If these arcs are denoted by Γ_i^ϵ, $i = 1, \ldots, N_\epsilon$, the total sum of time spent inside each of these arcs is

$$1 = \sum \mu(\Gamma_i^\epsilon) = (\text{diam}\,(\Gamma))^{-e} \sum_i (\text{diam}\,(\Gamma_i^\epsilon))^e \ .$$

We deduce that $N_\epsilon\,\epsilon^e \simeq 1$. Thus, the integer N_ϵ has order ϵ^{-e}, and $\sum_i \text{diam}\,(\Gamma_i^\epsilon) \simeq N_\epsilon\,\epsilon \simeq \epsilon^{1-e}$. A direct application of the theorem in §4 gives:

$$\Delta(\Gamma) = \lim_{\epsilon \to 0} \left(1 + \frac{\log N_\epsilon\,\epsilon}{|\log \epsilon|}\right) = 1 - (1 - e) = e \ . \quad ◀$$

Thus, we have proved a well-known result using a new approach (Chap. 13, §8). This is an interesting proof, because it opens the way to other dimension theorems, as we shall see later.

2. The generalized model of expansive curves We recall that this model was introduced in Chapter 14, §7. In what follows, we shall discuss some of its special cases. It is obtained by a sequence of infinite operations of type $\mathcal{T}$, starting with a closed convex set D that contains $N + 1$ distinct points:

$$A = A_1, A_2, \ldots, A_{N+1} = B ,$$

such that A and B are on the boundary ∂D. The operation $\mathcal{T}$ consists of replacing D with N closed convex sets $D(1)$, ..., $D(N)$, such that
   1) $D(i) \subset D$ for all $i = 1, \ldots, N$;
   2) $D(i) \cap D(i + 1) = \{A_{i+1}\}$ for all $i = 1, \ldots, N - 1$;
   3) $D(i)$ and $D(j)$ are disjoint whenever $|i - j| \geq 2$;
   4) $D(1)$ contains $A(1)$, and $D(N)$ contains $A(N + 1)$.
We repeat the operation $\mathcal{T}$ on each of the $D(i)$, and so forth.
   The domains of rank k are denoted by $D(i_1, \ldots, i_k)$. They form chains of sets E_k that constitute an increasingly precise approximation of Γ. Let $\Gamma(i_1, \ldots, i_k)$ be the part of Γ included in $D(i_1, \ldots, i_k)$. For this curve to be expansive we shall assume (Proposition 2, Chap. 14, §7) that for all k and every finite sequence i_1, ...,i_k of integers between 1 and N,

$$\mathrm{breadth}(\mathcal{K}(\Gamma(i_1, \ldots, i_{k-1}, i_k))) \geq c\,\mathrm{breadth}(\mathcal{K}(\Gamma(i_1, \ldots, i_{k-1}))) ,$$

where c is a constant.

3. A particular case of this model Suppose the sizes and deviations of the domains of rank k are obtained by a multiplicative process. That is, there exist $2\,N$ real numbers ρ_1, ..., ρ_N and b_1, ..., b_N such that

$$0 < b_i \leq \rho_i < 1 , \quad \sum_i \rho_i b_i \leq 1 , \quad \sum_i \rho_i \geq 1 ,$$
$$\mathrm{diam}\,(D(i)) = \rho_i\,\mathrm{diam}\,(D) , \quad \mathrm{breadth}(D(i)) = b_i\,\mathrm{breadth}(D) .$$

Similarly, at all the stages of the construction:

$$\mathrm{diam}\,(D(i_1, \ldots, i_k)) = \rho_{i_1} \ldots \rho_{i_k}\,\mathrm{diam}\,(D) ,$$
$$\mathrm{breadth}(D(i_1, \ldots, i_k)) = b_{i_1} \ldots b_{i_k}\,\mathrm{breadth}(D) .$$

The sought inequality on the breadths is satisfied with a constant c that equals $\min_i\{b_i\}$. We find examples of such curves in Figures 15.2, 15.3, and 15.4, not counting all the self–similar curves.
   Since $\sum_i \rho_i b_i^{x-1}$ is a decreasing continuous function that takes values ≥ 1 for $x = 1$ and ≤ 1 for $x = 2$, there exists a unique real number e that is a solution of the equation

$$\sum_i \rho_i b_i^{e-1} = 1 .$$

We shall show that:

This value e is precisely the dimension of Γ.

▶   a) First we define a measure on Γ by attributing to $\Gamma(i) = \Gamma \cap D(i)$ the measure $\rho_i\, b_i^{e-1}$ (this is the time needed to run through $\Gamma(i)$) and so forth by using the multiplicative process: the measure of $D(i_1, \ldots, i_k)$ is

$$(\rho_{i_1} \cdots \rho_{i_k})\,(b_{i_1} \cdots b_{i_k})^{e-1} \simeq \operatorname{diam}(D(i_1, \ldots, i_k))\,(\operatorname{breadth}(D(i_1, \ldots, i_k)))^{e-1} .$$

b) For every given ϵ, we select a family of N_ϵ arcs of Γ of type $\Gamma(i_1, \ldots, i_k)$, disjoint or adjacent, whose deviations are between ϵ and ϵ/c (such a construction was done in the proof of Proposition 2, Chap. 14, §7). We denote these arcs by Γ_i^ϵ. The sum of their measures is 1: this is the amount of time needed to run through the whole curve. This means that

$$\sum_i \operatorname{diam}(\Gamma_i^\epsilon)\,(\operatorname{breadth}(\Gamma_i^\epsilon))^{e-1} \simeq 1 ,$$

and therefore

$$\sum_i \operatorname{diam}(\Gamma_i^\epsilon) \simeq \epsilon^{1-e} .$$

The theorem in §4 allows us to conclude that:

$$\Delta(\Gamma) = 1 - (1 - e) = e . \quad ◀$$

◇   In the case of self–similarity, $b_i = \rho_i$ for all i, and the equation giving the dimension is reduced to the well-known formula: $\sum \rho_i^e = 1$.

4. Another particular case (Figs. 15.2 and 15.4): With the same hypothesis as in case 3, we assume that all the ρ_i are equal to ρ and that all the b_i are equal to b. The dimension is the solution of the equation

$$N\,\rho\, b^{e-1} = 1 .$$

That is,

$$\Delta(\Gamma) = 1 + \frac{\log N \rho}{|\log b|} .$$

Let α and β be two real numbers such that

$$\rho = N^{-\alpha} \qquad \text{and} \quad b = N^{-\beta} .$$

The natural parameterization associates the measure $\tau_k = N^{-k}$ to each arc of rank k. Such an arc is of size $\simeq \tau_k^\alpha$ and deviation $\simeq \tau_k^\beta$. Therefore, we can now interpret α and β as two parameters that govern the relation between the measure of local arcs and their geometrical shapes. And the dimension can be written as:

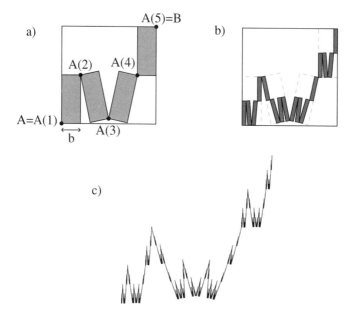

Fig. 15.2. *Let b be a parameter $< 1/4$. We start with the unit square. The operation T consists of constructing four rectangles of length $1/2$ and width b, as in a); in each of these rectangles, we construct four rectangles of length 2^{-2} and width b^2, as in b); and so forth. The final curve is drawn in c). The dimension is obtained from the equation*

$$\sum_{i=1}^{4} \rho_i\, b_i^{e-1} = 1 \,,$$

with $\rho_i = 1/2$ and $b_i = b$. That is,

$$e = 1 + \frac{\log 2}{|\log b|} \,.$$

There are two limit cases: (i) $b \to 0$, the case where Γ tends to a polygonal curve of dimension 1 with vertices $(0,0)$, $(0,1/2)$,$(1,1/2)$, $(1,1)$; (ii) $b \to 1/4$, case where Γ tends to the self-affine curve of dimension $3/2$ in Figure 12.10 ($H = 1/2$). When b runs through the interval $]0, 1/4[$, the dimension $\Delta(\Gamma)$ takes all values between 1 and $3/2$. Outside the limit case $b = 1/4$, this curve is not self-affine.

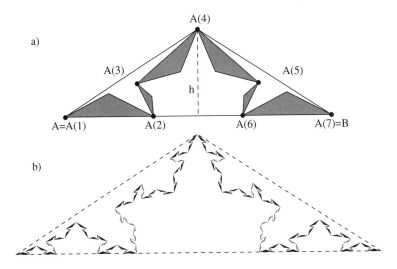

Fig. 15.3. *Given an isosceles triangle D with base 1 and height h, it is always possible to construct a polygonal curve inside it as in a). The curve is formed by six segments of lengths ρ_1, ρ_2, ρ_3, ρ_3, ρ_2, ρ_1, such that the following conditions hold:*

$$0 < \rho_1 < \frac{1}{2} \, , \, \rho_1 - \rho_2 + \rho_3 = \frac{1}{2} \, , \, \frac{1}{2} - \rho_1 \leq \rho_3 \leq \frac{1}{2(1 + 2\rho_1)} \, , \, h \leq 2\sqrt{\rho_2 \, \rho_3} \, .$$

Let $c > 1$. The operation $\mathcal{T}$ consists of replacing D with six isosceles triangles with base ρ_i and height $h \, \rho_i^c$. In each of them, we repeat an operation of the same type, and so forth: The kth step consists of 6^k isosceles triangles with bases $\rho_{i_1} \ldots \rho_{i_k}$ and heights $h(\rho_{i_1} \ldots \rho_{i_k})^c$. Their union tends to a limit curve Γ (in b) that is simple and expansive. The fact that c is larger than 1 implies that the structure of Γ looks more and more squashed when we increase the scale of observation. The dimension is the solution of the equation:

$$\sum_{i=1}^{6} \rho_i^{1+c(e-1)} = 1 \, .$$

In this figure, $\rho_1 = 1/3$, $\rho_2 = 2/15$, $\rho_3 = 3/10$, $h < 2/5$, and $c = 1.2$. Therefore $e = 1.2779....$

$$\Delta(\Gamma) = \frac{\beta + 1 - \alpha}{\beta} \, .$$

This is more than just another way of expressing the dimension. In §7 we shall see the full importance of the coefficients α and β in a general context.

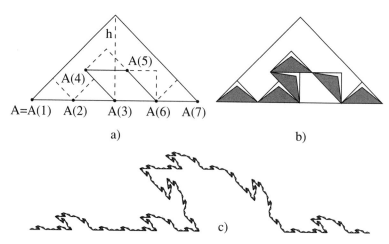

Fig. 15.4. *Let D be an isosceles triangle with base 1 and an acute angle $\phi < \pi/3$. We draw a polygonal curve* **G** *formed by six segments of length 1/4 with vertices $A(1), \ldots, A(7)$ in the triangle, as in a). Let $h = \tan(\phi)/2$ be the height of D, and let b be a parameter $< 1/4$. The operation $\mathcal{T}$ consists of replacing D with six isosceles triangles with base $A(i)A(i+1)$ and height $h\,b$, as in b). Each of these triangles is then replaced by six triangles whose base length is 4^{-2} and whose height is $h\,b^2$; and so forth. The limit curve Γ (drawn in c), with $b = 3/16$) is simple and expansive. Its fractal dimension is the solution of the equation: $N \rho\, b^{1-e} = 1$, with $N = 6$ and $\rho = 1/4$. We find*

$$\Delta(\Gamma) = 1 + \frac{\log 3/2}{|\log b|} \ .$$

The parameters α and β (§5) that govern the relation between the measure of a local arc, its size, and its deviation are equal to

$$\alpha = \frac{\log 4}{\log 6} \quad , \quad and \quad \beta = \frac{|\log b|}{\log 6} \ .$$

We can verify the relation $\Delta(\Gamma) = (\beta + 1 - \alpha)/\beta$. When $b \to 0$, $\beta \to \infty$ and $\Delta(\Gamma) \to 1$; the curve Γ tends to the curve **G**. *When $b \to 1/4$, $\beta \to \alpha$ and $\Delta(\Gamma) \to \log 6/\log 4$; the curve Γ tends to a self–similar set whose generator is* **G**.

15.6 Statistical self–similarity

How can we define the notion of "nonstrict" or "statistical" self–similarity of a curve? It all depends on the meaning of *similar arcs*. We assume that the notions of **size** and **deviation** can approximately characterize the geometry of subarcs of a curve; we shall say that "two arcs are similar if they have the same ratio size/deviation or if these ratios have the same order of growth." In the case of self–similarity, every subarc is similar to the entire curve. We naturally deduce the following definition:

A parameterized curve $\Gamma = \gamma([a,b])$ is **statistically self–similar** *if for all t and τ,*

$$\mathrm{dev}(\gamma([t-\tau, t+\tau])) \simeq \mathrm{size}(\gamma([t-\tau, t+\tau])) \ ;$$

that is, if there exists a constant c, independent from t and τ, such that

$$\mathrm{dev}(\gamma([t-\tau, t+\tau])) \geq c\,\mathrm{size}(\gamma([t-\tau, t+\tau])) \ .$$

Surely we can give a finer criterion, because our starting point is that size and deviation characterize the shape of an arc. In reality, other (even infinitely many) parameters are needed to completely characterize an arc in the same manner as the values of a function and its derivative at some given points are not sufficient to characterize the graph of an infinitely differentiable function. For this we need an infinite sequence of parameters, the coefficients of the Taylor series, for example. But our approach has the advantage of simplicity, and it can be used to find the dimension.

Computing the dimension Let Γ be an expansive curve and assume it is self–similar in the above sense. Let N_ϵ be the number of steps along Γ of size (or deviation) ϵ. In this case the theorem in §4 gives that:

$$\Delta(\Gamma) = \lim_{\epsilon \to 0}(1 + \frac{\log \epsilon\, N_\epsilon}{|\log \epsilon|}) \ ,$$

and therefore,

$$\Delta(\Gamma) = \lim_{\epsilon \to 0} \frac{\log N_\epsilon}{|\log \epsilon|} \ .$$

We can compare this formula to the one used in the "method of diameters" (Chap. 13, §10) for a parameterized curve; for every τ, we compute the average $\overline{T}_\tau$ of the sizes of the subarcs of the curve whose measure is τ, and we find the parameter:

$$\alpha = \lim_{\tau \to 0} \frac{\log \overline{T}_\tau}{\log \tau} \ .$$

Letting $\epsilon = \tau^\alpha$, the total measure of the curve is $\tau\,N_\epsilon \simeq 1$. Therefore, $N_\epsilon \simeq \tau^{-1}$, and we find the well-known formula:

$$\Delta(\Gamma) = \frac{1}{\alpha} \ .$$

Thus, the method of the diameters can be applied to all self–similar curves, where self-similarity is taken in the above general sense.

Comparison with the compass method   This method consists of running through the curve Γ with equal steps; that is, for all ϵ we construct a polygonal approximation $\mathbf{P}_\epsilon$ whose vertices belong to Γ and whose segments are all of length ϵ (Fig. 5.7). If $L(\mathbf{P}_\epsilon)$ is the length of $\mathbf{P}_\epsilon$, we calculate the following index:

$$\Delta'(\Gamma) = \lim_{\epsilon \to 0} \left(1 + \frac{\log L(\mathbf{P}_\epsilon)}{|\log \epsilon|} \right) .$$

When Γ is self–similar and the size of an arc is generally equivalent to the distance between its endpoints, the length $L(\mathbf{P}_\epsilon)$ has the same order as $\epsilon\, N_\epsilon$, and then

$$\Delta'(\Gamma) = \Delta(\Gamma) .$$

Let us keep in mind that:

   This result does not hold unless Γ is self–similar.

Example   Let $\epsilon = 4^{-k}$. If we run through the curve in Figure 15.4 with steps of length 4^{-k}, we need exactly 6^k steps. Thus, $L(\mathbf{P}_{4^{-k}}) = (3/2)^k$ and

$$\Delta'(\Gamma) = \frac{\log 6}{\log 4} .$$

We remark that this result does not depend on the value of the parameter b. The index $\Delta'(\Gamma)$ is never equal to the dimension $\Delta(\Gamma)$, except for the case where $b = 1/4$—in another words, the case where we have self–similarity. Generally, if the curve is not self–similar, the index $\Delta'(\Gamma)$ is strictly larger than the dimension; it does not have any precise metric meaning. To calculate the length of geographical coasts at an ϵ–scale, it is better to use the path of *constant deviation* (§4).

15.7 Curves of uniform deviation

By definition,

   *A curve $\Gamma = \gamma([a,b])$ is said to be of **uniform deviation** if arcs with the same measure have equivalent deviation. Thus, there exists a function $g(\tau)$ such that for $\tau > 0$,*

$$\mathrm{dev}(\gamma(t-\tau)\frown\gamma(t+\tau)) \simeq g(\tau) .$$

In the following theorem, this function $g(\tau)$ is a power function. Let us recall that the notation $T(t, \tau)$ indicates the size of local arcs (Chap. 11, §4), and

$$\overline{T}_\tau = \frac{1}{b-a} \int_a^b T(t, \tau)\, dt$$

is the mean of these sizes on $[a, b]$.

THEOREM *Let Γ be a simple expansive curve of uniform deviation. Assume that there exists a real number $\beta > 0$ such that for all $\tau > 0$*

$$\mathrm{dev}(\gamma(t-\tau)^\frown\gamma(t+\tau)) \simeq \tau^\beta .$$

Then

$$\Delta(\Gamma) = \lim_{\tau \to 0} \left(1 + \frac{1}{\beta} - \frac{1}{\beta}\frac{\log \overline{T}_\tau}{\log \tau} \right) .$$

▶ Suppose that $[0,1]$ is the interval of definition $[a,b]$. Let M be a fixed integer. We put $\tau = 1/2M$ and $\epsilon = \tau^\beta$. The sum

$$\frac{1}{M} \sum_{i=0}^{M-1} \mathrm{size}(\gamma([\frac{i}{M}, \frac{i+1}{M}]))$$

is a Riemannian sum for the integral $\overline{T}_\tau$. To find a good approximation of this integral we choose an integer N and compute the sum:

$$\frac{1}{MN} \sum_{i=0}^{MN-1} \mathrm{size}(\gamma([\frac{i}{MN}, \frac{i+1}{MN}])) .$$

This sum tends to $\overline{T}_\tau$ when N tends to infinity. Thus, we can choose N to be large enough so that the sum is between $\overline{T}_\tau/2$ and $2\overline{T}_\tau$.

The family of all the arcs $\gamma([i/MN, (i+1)/MN])$ can be distributed into N families $\mathcal{F}_1, \ldots, \mathcal{F}_N$ of index $\omega \leq 2$ so that each $\mathcal{F}_i$ covers Γ (some arcs may belong to different families at the same time). All these arcs are of deviation $\simeq \epsilon$. Using the lemma in §3, we deduce that for each $\mathcal{F}_i$,

$$\epsilon \sum_{\Gamma^* \in \mathcal{F}_i} \mathrm{size}(\Gamma^*) \simeq \sum_{\Gamma^* \in \mathcal{F}_i} \mathcal{A}(\mathcal{K}(\Gamma^*)) \simeq \mathcal{A}(\Gamma(\epsilon)) .$$

The previous Riemannian sum is, therefore, equivalent to both $\mathcal{A}(\Gamma(\epsilon))/M\epsilon$ and $\epsilon^{1/\beta-1}\mathcal{A}(\Gamma(\epsilon))$. Thus,

$$\overline{T}_\tau \simeq \epsilon^{\frac{1}{\beta}-1}\mathcal{A}(\Gamma(\epsilon)) .$$

A direct application of the formula

$$\Delta(\Gamma) = \lim_{\epsilon \to 0} \left(2 - \frac{\log \mathcal{A}(\Gamma(\epsilon))}{\log \epsilon} \right)$$

gives the desired result. ◀

◇ Recall that for all t, the image of the interval $[t, t+\tau_1(t,\epsilon)]$ by γ is an arc of deviation ϵ (Chap. 14, §3). Let $\Gamma(t,\epsilon)$ be this arc. Its measure (the time needed to run through it) is $\mu(\Gamma(t,\epsilon)) = \tau_1(t,\epsilon)$. This formula is a particular case of the following one:

$$\Delta(\Gamma) = \lim_{\epsilon \to 0} \left(1 + \frac{\log \int_a^b \frac{\text{size}(\Gamma(t,\epsilon))}{\mu(\Gamma(t,\epsilon))} \, dt}{|\log \epsilon|} \right) .$$

▶   In fact, it suffices to replace $\mu(\Gamma(t,\epsilon))$ with τ and ϵ with τ^β, to find the previous result.   ◀

We recognize in the ratio $\text{size}(\Gamma(t,\epsilon))/\mu(\Gamma(t,\epsilon))$ an evaluation of the *local velocity*. Its integral with respect to time is a length, the average of all paths of constant deviation that can be made along the curve.

This latter formula can be considered as the "continuous form" of the one proved in §4. It remains true in a more general setting than curves of uniform deviation.

15.8 Applications

Here is a direct application of the theorem:

If Γ is a simple expansive curve, such that

$$\text{dev}(\gamma(t-\tau) \frown \gamma(t+\tau)) \simeq \tau^\beta$$

and if

$$\overline{T}_\tau \simeq \tau^\alpha ,$$

then

$$\Delta(\Gamma) = \frac{\beta + 1 - \alpha}{\beta} .$$

We have already encountered this expression for the dimension in the discrete case (§5, Application 4). The two parameters α and β are two characteristics of Γ: α governs the relation between the measure of an arc and its size; and β governs the relation between the measure of an arc and its deviation.

Statistically self–similar curves   For such curves, sizes and deviation of local arcs are of the same order of growth (§6); therefore $\alpha = \beta$. By replacing them in the formula we find that:

$$\Delta(\Gamma) = \frac{1}{\alpha} ,$$

a well-known equation, corresponds to the "diameters method" to compute dimensions (Chap. 13, §10).

Graphs of continuous functions Let $z(t)$ be a continuous function defined on $[a, b]$ with graph Γ. In the typical case where Γ is a fractal curve, the breadth of the convex hull of an arc has the same order as its projection on Ot. Thus the arc that corresponds to the abscissas $[t - \tau, t + \tau]$ has a deviation with the same order as τ. Such a curve is of uniform deviation, with

$$\beta = 1 \ .$$

This is a particularly easy property to verify on the curve of Figure 12.10, for example. For this to hold in a general setting, it suffices to find, for all t and τ in the interval $[t - \tau, t + \tau]$, three real numbers t_1, t_2, t_3, such that $|t_1 - t_2| \simeq \tau$, $z(t_1) = z(t_2)$, and $|z(t_1) - z(t_3)| \simeq \tau$.

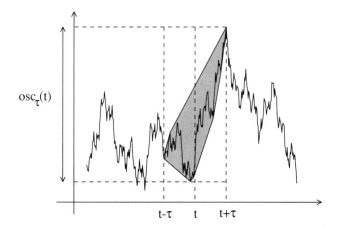

Fig. 15.5. *The arc of Γ that corresponds to the abscissas $[t - \tau, t + \tau]$ has a size with the same order as $\mathrm{osc}_\tau(t)$ and a deviation of order τ.*

For such graphs, the theorem in §7 gives

$$\Delta(\Gamma) = \lim_{\tau \to 0} \left(2 - \frac{\log \overline{T}_\tau}{\log \tau} \right) \ .$$

This formula corresponds to the method of "variation" (Chap. 12, §4), the variation of $z(t)$ is equivalent to $\overline{T}_\tau$ (Chap. 12, §3). When $\overline{T}_\tau \simeq \tau^\alpha$, we find

$$\Delta(\Gamma) = 2 - \alpha \ :$$

the parameter α is then interpreted as a Hölder exponent (Chap. 12, §5).

Values of the parameters α and β In the case where:

$$\Delta(\Gamma) = \frac{\beta + 1 - \alpha}{\beta} \ ,$$

what are the possible values of α and β? We can show that necessarily,

$$0 < \alpha \leq 1 \, , \, \alpha \leq \beta \, , \, \alpha + \beta \geq 1 \, .$$

▶   Assume that $z(t)$ is not constant on $[a, b]$, the only interesting case.
(i) Since $z(t)$ is continuous, $\overline{T}_\tau$ tends to 0 when τ tends to 0. Thus $\alpha > 0$.
(ii) We know that for some constant c, $\overline{T}_\tau \simeq \mathrm{Var}_\tau \geq c\tau$ (Chap. 12, §3).
Therefore $\alpha \leq 1$.
(iii) Since the breadth of a convex set is at most equal to its diameter, for all
size and deviation functions, there exists a constant c such that for all bounded
sets E,
$$\mathrm{dev}(E) \leq c \, \mathrm{size}(E) \, .$$
This implies that $\alpha \leq \beta$.
(iv) Finally, since Γ is expansive for all $\epsilon < \mathrm{dev}(\Gamma)$, there exists a cover of
$\Gamma = \cup \Gamma_i^\epsilon$ by arcs of deviation ϵ, such that

$$\sum \mathcal{A}(\mathcal{K}(\Gamma_i^\epsilon)) \simeq \mathcal{A}(\cup \mathcal{K}(\Gamma_i^\epsilon))$$

(Chap. 14, §4). For all ϵ, the right-hand-side term is bounded, while the left-
hand-side term is equivalent to

$$\epsilon \sum \mathrm{diam} \, (\Gamma_i^\epsilon) \simeq \tau^\beta \frac{1}{\tau} \overline{T}_\tau \simeq \tau^{\alpha + \beta - 1} \, .$$

Therefore, it is necessary to have: $\alpha + \beta - 1 \geq 0$ or $\alpha + \beta \geq 1$.
We remark that these inequalities evidently imply that:

$$1 \leq \Delta(\Gamma) \leq 2 \, . \quad ◀$$

◇   It is easy to obtain examples of curves such that $\beta > 1$ in the discrete case.
See, for example, Figure 15.4. When $\Delta > 1$, this gives locally angular curves.
When $\Delta = 1$, we must have $\alpha = 1$: this is the case of curves of finite length.
For example, an arc of a circle that is run through with a constant velocity is
characterized by $\Delta = 1$, $\alpha = 1$, $\beta = 2$.

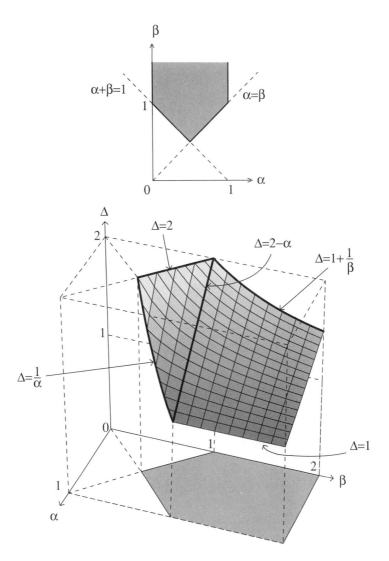

Fig. 15.6. *The function*

$$\Delta = \frac{\beta + 1 - \alpha}{\beta}$$

is a function of two variables defined on the domain represented in a) whose graph, represented in b), is a part of hyperboloid. We note the two limit cases: $\Delta = 1/\alpha$ *(*$\alpha = \beta$*, self–similarity) and* $\Delta = 2 - \alpha$ *(*$\beta = 1$*, graphs of continuous functions).*

15.9 The dimension of a curve

The notions developed in this chapter naturally lead us to new methods to compute the dimension. These methods have two advantages.

• By means of convex hulls, they are more adapted to the local geometry of the curve, which is not the case in other methods based on boxes and balls (Chap. 10). There exist two limit cases that lead to well-known and trustworthy methods: the **method of the diameters**, which can be applied to self–similar curves ($\Delta = 1/\alpha$); and the **variation method**, which is applied to graphs of continuous functions ($\Delta = 2 - \alpha$).

• In most cases, we need no preliminary hypothesis about the type of curve in question. We ask only that this curve be **simple** and **expansive**. This latter condition (Chap. 14) is a very loose one that is satisfied by experimental curves in the limit of data precision.

• The main drawback is the time needed to achieve the computation. This is mainly due to the search for the convex hulls.

1. The constant–deviation variable–step method.   This method follows from the theorem in §4:

$$\Delta(\Gamma) = \lim_{\epsilon \to 0} \left(1 + \frac{\log \sum_i \operatorname{diam}(\Gamma_i^\epsilon)}{|\log \epsilon|} \right) .$$

To evaluate $\sum_i \operatorname{diam}(\Gamma_i^\epsilon)$, we form steps of deviation ϵ along the curve, as was described in Chapter 14, §3. To avoid having a result that depends on the endpoints, we choose some points on the curve to be starting points. Then we take steps of deviation ϵ in both directions of the curve (Fig. 15.7). For each of these paths we compute the sum of diameters, then calculate the arithmetic mean of these sums. This (mean) procedure gives, as usual, a more precise result. Then we draw the logarithmic diagram

$$\left(|\log \epsilon| \, , \, \log \frac{1}{\epsilon} \sum_i \operatorname{diam}(\Gamma_i^\epsilon) \right) .$$

Its slope is the estimated value of the dimension. As we have said, this method also works for the Von Koch curve (Chap. 13), like the graphs in Chapter 12 or for geographic coastlines, etc.

2. The sausage of convex hulls   Starting with any point x of Γ, we find the point y such that $\operatorname{dev}(x{\frown}y) = \epsilon$ in the direction of the endpoint B, for example. If such a y does not exist, we put $y = B$. By taking the union of the convex hulls of all the arcs $x{\frown}y$, we obtain a sausage of a new type:

$$\mathcal{K}_\epsilon = \{ \, \mathcal{K}(x{\frown}y) \text{ such that } \operatorname{dev}(x{\frown}y) \leq \epsilon \, \} ,$$

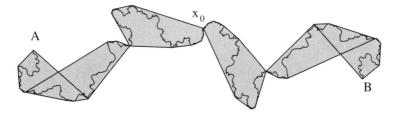

Fig. 15.7. *Starting with a point x_0 of the curve, we make steps of deviation ϵ in the direction of A and in the direction of B. The path thus created is an ϵ-approximation of the curve. The sum of the Γ_i^ϵ diameters serves to compute the dimension.*

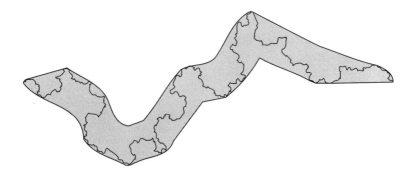

Fig. 15.8. *The union of all the convex hulls of local arcs of breadth ϵ constitute a sausage around Γ.*

which is perfectly adapted to the particular geometry of the curve.

Since Γ is a continuous curve, we can equivalently define $\mathcal{K}_\epsilon$ as the union of all the chords xy of Γ such that the arc $x\frown y$ is of deviation $\leq \epsilon$. Unlike the Minkowski sausage, the boundary of this sausage contains points of Γ; these are the points where the curve "changes direction" at an ϵ–precision. The evaluation of the area allows us to compute the dimension:

If Γ is a simple expansive curve, then

$$\Delta(\Gamma) = \lim_{\epsilon \to 0} \left(2 \frac{\log \mathcal{A}(\mathcal{K}_\epsilon)}{\log \epsilon} \right).$$

▶ If, for example, the deviation is taken in the sense of breadth, then $\mathcal{K}_\epsilon \subset \Gamma(\epsilon)$ and therefore $\mathcal{A}(\mathcal{K}_\epsilon) \leq \mathcal{A}(\Gamma(\epsilon))$. On the other hand, let (Γ_i^ϵ) be a path along Γ whose covering index is ≤ 2 with $\mathrm{dev}(\Gamma_i^\epsilon) = \epsilon$. We know (§4) that

$$\mathcal{A}(\cup \mathcal{K}(\Gamma_i^\epsilon)) \simeq \mathcal{A}(\Gamma(\epsilon)) \ .$$

Moreover, $\mathcal{K}(\Gamma_i^\epsilon) \subset \mathcal{K}_\epsilon$ implies that

$$\mathcal{A}(\cup \mathcal{K}(\Gamma_i^\epsilon)) \leq \mathcal{A}(\mathcal{K}_\epsilon) \ .$$

These results show that $\mathcal{A}(\mathcal{K}_\epsilon) \simeq \mathcal{A}(\Gamma(\epsilon))$. ◄

The corresponding logarithmic diagram is

$$\left(|\log \epsilon| \ , \ \log \frac{1}{\epsilon^2} \mathcal{A}(\mathcal{K}_\epsilon) \right) \ .$$

3. Integral of the local velocity   We can also use the formula in §7:

$$\Delta(\Gamma) = \lim_{\epsilon \to 0} \left(1 + \frac{\log \int_a^b \frac{\text{size}(\Gamma(t,\epsilon))}{\mu(\Gamma(t,\epsilon))} \, dt}{|\log \epsilon|} \right) \ .$$

It has the inconvenience of not being general. We have proved that it works
for curves of constant breadth. However, it gives a value that is at most equal
to $\Delta(\Gamma)$, and this is independent of the parameterization. This method is based
on principles that are similar to those of the constant–deviation variable–step
method, where we use an arithmetical mean of the obtained lengths by changing
the starting point. Its logarithmic diagram is

$$\left(|\log \epsilon| \ , \ \log \frac{1}{\epsilon} \int_a^b \frac{\text{size}(\Gamma(t, \epsilon))}{\mu(\Gamma(t, \epsilon))} \, dt \right) \ .$$

15.10 Bibliographical notes

For the compass method, the best reference is the book [B. Mandelbrot 2]. For
an application of the ideas of this chapter to cartography (automatic drawing of
geographical coastlines), see [F. Normant & C. Tricot].

16 Scanning a Curve with Straight Lines

16.1 Directional dimension

It is not always necessary to cover a set by balls or squares in order to compute its dimension. In the previous chapter we have introduced other methods that take into account the structure of the set. In particular, we have seen that some curves can be analyzed using some privileged directions. For a graph of a continuous function $z(t)$ drawn in a Cartesian coordinate system, we considered the direction of the axes: the Oz direction to measure the local Hölder exponent, and the Ot direction to cover the graph with horizontal segments. More generally, we will now discuss a *polarized analysis*.

We assume that Ox_1 and Ox_2 are the axes of a Cartesian coordinate system drawn in a given plane. This system allows us to define the polar coordinates (ρ, θ).

We recall that every straight line that does not pass through the origin can be determined by the orthogonal projection H of the origin on it (Chap. 8, §2). Therefore, a line can be determined by two numbers: the distance $\rho = OH$ and the angle $\theta = \angle(Ox, OH)$. This line is denoted by $\mathbf{D}(\rho, \theta)$.

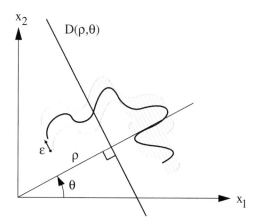

Fig. 16.1. *On the line $\mathbf{D}(\rho, \theta)$, we draw the segments of length 2ϵ centered on Γ. The angle θ being fixed, we let ρ vary: the union of the obtained segments forms the directional sausage $\Gamma(\theta, \epsilon)$.*

Let Γ be a curve. We recall that $\Gamma(\theta, \epsilon)$ denotes the union of all the segments of length 2ϵ centered on Γ that make an angle θ with Ox_2 (Chap. 9, §4). Also, this is the area scanned by Γ when it is displaced by a distance 2ϵ in the direction θ with respect to Ox_2. The surface $\Gamma(\theta, \epsilon)$ covers Γ and has the shape of a *directional sausage*.

Case of a segment The area of $\Gamma(\theta, \epsilon)$ is null if and only if Γ is a segment that makes an angle θ with Ox_2. If Γ is not a segment, $\mathcal{A}(\Gamma(\theta, \epsilon))$ cannot be null for any value of ρ and θ.

Polar dimension When Γ is of finite length, we have seen that $\mathcal{A}(\Gamma(\theta, \epsilon))$ can be related to the length of Γ. When Γ is of infinite length, we can associate to $\mathcal{A}(\Gamma(\theta, \epsilon))$ a dimensional coefficient, called *polar dimension* or *dimension with respect to θ*, in the following way:

$$\Delta_\theta(\Gamma) = \lim_{\epsilon \to 0} \left(2 \frac{\log \mathcal{A}(\Gamma(\theta, \epsilon))}{\log \epsilon} \right) .$$

If this limit does not exist, we replace lim by lim sup in the above.

◇ The directional dimension shares some of the properties of the classical dimension $\Delta(\Gamma)$. In particular, Δ_θ is **monotonous**: if Γ_1 is included in Γ_2, then

$$\Delta_\theta(\Gamma_1) \leq \Delta_\theta(\Gamma_2) .$$

◇ It is clear that the distance between each point of $\Gamma(\theta, \epsilon)$ and Γ is less than ϵ, therefore, it belongs to the Minkowski sausage $\Gamma(\epsilon)$. It follows that:

$$\Delta_\theta(\Gamma) \leq \Delta(\Gamma) ,$$

for any value of θ. The equality can be realized: for example, $\Gamma(\pi/2, \epsilon)$ is nothing but the sausage formed by horizontal segments of length 2ϵ, and we have proved that:

$$\mathcal{A}(\Gamma(\theta, \epsilon)) \simeq \mathcal{A}(\Gamma(\epsilon))$$

when Γ is a graph of a nonconstant continuous function (Chap. 12, §3). In this case

$$\Delta_{\pi/2}(\Gamma) = \Delta(\Gamma) .$$

We shall see that this equality holds whenever Γ is a simple curve and θ is any angle except a unique exceptional value θ_0.

◇ Consider a curve Γ, not included in a straight line. Fix the angle θ and let $[\rho_1, \rho_2]$ be the interval of values of ρ for which the line $\mathbf{D}(\rho, \theta)$ intersects Γ. Since

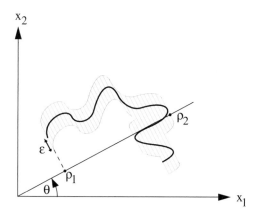

Fig. 16.2. *The area of the sausage* $\Gamma(\theta, \epsilon)$ *is at least equal to* $2\epsilon(\rho_2 - \rho_1)$.

$$\mathcal{A}(\Gamma(\theta, \epsilon)) \geq 2\,\epsilon\,(\rho_2 - \rho_1)\,,$$

where $\rho_1 < \rho_2$, we deduce that

$$\Delta_\theta(\Gamma) \geq 1\,.$$

16.2 Comparing the dimensions

We shall compare the values of the directional dimensions depending on the angle θ to the value of the classical dimension.

THEOREM   *Consider two values of the angle* $\theta_1 \neq \theta_2$ *that belong to the interval* $[0, \pi[$ *and a curve* Γ*; then*

$$\boxed{\Delta(\Gamma) = \max\{\Delta_{\theta_1}(\Gamma), \Delta_{\theta_2}(\Gamma)\}\,.}$$

In the case where there exists an exceptional value θ_0 for which $\Delta(\Gamma) \neq \Delta_{\theta_0}(\Gamma)$, this theorem implies that for any other value θ, we have $\Delta(\Gamma) = \Delta_\theta(\Gamma)$. Therefore, we can state the following corollary:

For every curve Γ*, there exists at most one exceptional value of the angle* θ_0 *(modulo* π*), such that:*

$$\Delta(\Gamma) \neq \Delta_{\theta_0}(\Gamma)\,.$$

▶   We already know that:

$$\Delta(\Gamma) \geq \max\{\Delta_{\theta_1}(\Gamma), \Delta_{\theta_2}(\Gamma)\} \ .$$

To obtain an inequality in the other direction, we construct, at every point x of Γ, the cross $X_\epsilon(x)$ formed by two segments of length 2ϵ, centered at x, which make angles θ_1 and θ_2 with Ox_2 (Chap. 10, §7). The union

$$X_\epsilon = \bigcup_{x \in \Gamma} X_\epsilon(x)$$

is a sausage whose area is equivalent to that of the Minkowski sausage, as was shown. But $X_\epsilon = \Gamma(\theta_1, \epsilon) \cup \Gamma(\theta_2, \epsilon)$. Therefore,

$$\epsilon^{\alpha-2}\mathcal{A}(\Gamma(\epsilon)) \preceq \epsilon^{\alpha-2}(\mathcal{A}(\Gamma(\theta_1, \epsilon)) + \mathcal{A}(\Gamma(\theta_2, \epsilon))) \ .$$

If we take $\alpha > \max\{\Delta_{\theta_1}(\Gamma), \Delta_{\theta_2}(\Gamma)\}$, we deduce that:

$$\epsilon^{\alpha-2}\mathcal{A}(\Gamma(\epsilon)) \to 0 \ .$$

This shows that $\alpha \geq \Delta(\Gamma)$.   ◄

16.3 Examples and applications

We reconsider our two classical "limit cases."

Graph of a continuous function If Γ is the graph of a function $z(t)$ (here, the axis Ox_1 becomes Ot, and Ox_2 becomes Oz) defined on the interval $[a, b]$, then the angle $\theta_0 = 0$ is exceptional once Γ is of dimension > 1 (Fig. 16.3). In fact, $\Gamma(0, \epsilon)$ is formed by vertical segments, so that integrating over $[a, b]$ we get:

$$\mathcal{A}(\Gamma(\epsilon, 0)) = 2\,\epsilon\,(b - a) \ ,$$

and it follows that

$$\Delta_0(\Gamma) = 1 \ .$$

We deduce from the preceding theorem that for all other values of the angle θ,

$$\Delta_\theta(\Gamma) = \Delta(\Gamma) \ .$$

We have already noticed that for $\theta = \pi/2$, the surface $\Gamma(\pi/2, \epsilon)$ has an area equivalent to the *variation* of $z(t)$ (Chap. 12, §3).

Strict self–similar curve Unlike the previous case, this curve is dimensionally isotropic. That is, it has no exceptional angle, unless it is reduced to a segment. This can be proved as follows.

►   Assume that

$$\Gamma = \cup \Gamma_i \ ,$$

where Γ_i is the image of Γ by the similarity F_i (Chap. 13, §2). This similarity is the product of a homothety of ratio ρ_i, a rotation of angle θ_i, an eventual symmetry with respect to Ox_1, and a translation. Assume that $\theta = 0$. Neither

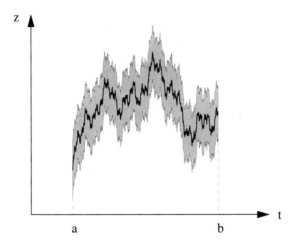

z

a   b

t

Fig. 16.3. *The angle 0 is an exceptional angle for* Γ *because* $\mathcal{A}(\Gamma(0,\epsilon)) = 2\,\epsilon\,(b-a)$ *and* $\Delta_0(\Gamma) = 1$. *For all angles* $\theta \neq 0$, *the area of the directional sausage* $\Gamma(\theta,\epsilon)$ *is equivalent to the area of the Minkowski sausage.*

the symmetry nor the translation will change the sausage $\Gamma(0,\epsilon)$. The rotation and the homothety change $\Gamma(0,\epsilon)$ into $\Gamma_i(\theta_i,\epsilon\,\rho_i)$. Since

$$\mathcal{A}(\Gamma_i(\theta_i,\epsilon\,\rho_i)) = \rho_i^2\,\mathcal{A}(\Gamma(0,\epsilon))\;,$$

we deduce that

$$\Delta_0(\Gamma) = \Delta_{\theta_i}(\Gamma_i)\;.$$

If 0 is an exceptional angle of Γ, we should have

$$\Delta_{\theta_i}(\Gamma_i) = \Delta_0(\Gamma) < \Delta(\Gamma) = \Delta(\Gamma_i)\;.$$

This shows that θ_i is an exceptional angle of Γ_i. But $\Gamma_i \subset \Gamma$, so that by monotony of the dimension:

$$\Delta_0(\Gamma_i) \leq \Delta_0(\Gamma) < \Delta(\Gamma) = \Delta(\Gamma_i)\;.$$

This proves that 0 is an exceptional angle of Γ_i. Since it cannot have two, such angles $\theta_i = 0(mod\pi)$ for all i. In this case, Γ is reduced to a segment. Therefore, if Γ is not a segment, the angle 0 cannot be exceptional, and by rotation, no other angle θ will be exceptional either. ◄

16.4 Coordinate systems

In a bounded domain $\mathcal{D}$ of the (Ox_1, Ox_2) plane, we define the new coordinates (u,v) by the equations

Fig. 16.4. *The area of the sausages $\Gamma(0, \epsilon)$, $\Gamma(-\pi/4, \epsilon)$, and $\Gamma(\pi/2, \epsilon)$, where Γ is the Von Koch curve, are all equivalent to the area of the Minkowski sausage $\Gamma(\epsilon)$.*

$$\begin{cases} x_1 = g(u, v) \\ x_2 = h(u, v) \end{cases},$$

where the two functions g and h have continuous partial derivatives. We assume that this transformation is bijective, and, moreover, that the Jacobian

$$\frac{\partial g}{\partial u}\frac{\partial h}{\partial v} - \frac{\partial g}{\partial v}\frac{\partial h}{\partial u}$$

is not null at any point of $\mathcal{D}$. Thus, the functions g and h define a coordinate transformation. The coordinate curves are $\mathbf{C}_u$ (u remains constant, v varies) and $\mathbf{C}_v$ (v remains constant, u varies), and they are differentiable. The family of all the curves $\mathbf{C}_u$ covers $\mathcal{D}$. However, no two of them intersect inside $\mathcal{D}$, and similarly for the $\mathbf{C}_v$. The fact that the Jacobian is not null indicates that at their point of intersection (u, v), the two curves $\mathbf{C}_u$ and $\mathbf{C}_v$ are not tangent. We can even say that since $\mathcal{D}$ is bounded, the acute angle between the tangents of $\mathbf{C}_u$ and $\mathbf{C}_v$ remains uniformly larger on $\mathcal{D}$ than a nonzero constant angle ϕ. In the case of Cartesian or polar coordinates, the angle ϕ is $\pi/2$ everywhere.

For every point x of $\mathcal{D}$ with coordinates (u, v) and for every ϵ, let $\mathbf{S}_\epsilon^{(u)}(x)$ be the arc of $\mathbf{C}_u$ whose points are situated between $(u, v - \epsilon)$ and $(u, v + \epsilon)$. Given a curve Γ included in the interior of $\mathcal{D}$, we can form the sausage $\Gamma^{(u)}(\epsilon)$ by taking the union of all these arcs:

$$\Gamma^{(u)}(\epsilon) = \bigcup_{x \in \Gamma} \mathbf{S}_\epsilon^{(u)}(x).$$

Similarly, we define the set of points $\mathbf{S}_\epsilon^{(v)}(x)$ on $\mathbf{C}_v$ between $(u - \epsilon, v)$ and $(u + \epsilon, v)$. The union of all these arcs, when x runs through Γ, constitutes the sausage $\Gamma^{(v)}(\epsilon)$. We deduce two definitions of the directional dimension that are dependent on the (u, v) coordinate system:

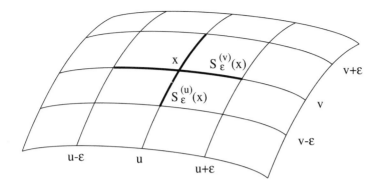

Fig. 16.5. *The coordinate curves* $\mathbf{C}_u$ *and* $\mathbf{C}_v$ *meeting at* $x = (u, v)$ *determine the arcs* $\mathbf{S}_\epsilon^{(u)}(x)$ *and* $\mathbf{S}_\epsilon^{(v)}(x)$.

$$\Delta_u(\Gamma) = \lim_{\epsilon \to 0} \left(2 - \frac{\log \mathcal{A}(\Gamma^{(u)}(\epsilon))}{\log \epsilon} \right)$$

$$\Delta_v(\Gamma) = \lim_{\epsilon \to 0} \left(2 - \frac{\log \mathcal{A}(\Gamma^{(v)}(\epsilon))}{\log \epsilon} \right) .$$

From the previous hypothesis on the derivatives, we conclude that the distances defined by these coordinate systems (x_1, x_2) and (u, v) are equivalent. The arcs $\mathbf{S}_\epsilon^{(u)}(x)$ and $\mathbf{S}_\epsilon^{(v)}(x)$ have lengths of order 2ϵ, and the union of the two sausages $\Gamma^{(u)}(\epsilon)$ and $\Gamma^{(v)}(\epsilon)$ has an area equivalent to the area of the Minkowski sausage $\Gamma(\epsilon)$, drawn with the Euclidian distance. The theorem of §2 becomes:

$$\Delta(\Gamma) = \max\{\Delta_u(\Gamma), \Delta_v(\Gamma)\} .$$

Example 1 Given an angle θ_1, we can, as we have seen in Section 1, cover the plane with parallel lines $\mathbf{D}(\rho, \theta_1)$. Taking $u = \rho$, these lines form the coordinate curves $\mathbf{C}_u$. For every curve Γ, the directional sausage $\Gamma(\theta_1, \epsilon)$ becomes $\Gamma^{(u)}(\epsilon)$, and $\Delta_{\theta_1}(\Gamma)$ becomes $\Delta_u(\Gamma)$. With another angle θ_2, we similarly create another network of lines, which are the curves of coordinates $\mathbf{C}_v$: $\Delta_{\theta_2}(\Gamma)$ becomes $\Delta_v(\Gamma)$. The acute angle ϕ between $\mathbf{C}_u$, and $\mathbf{C}_v$ is constant. The formulas for the change of variables $(x, y) \longrightarrow (u, v)$ are linear. The formula

$$\Delta(\Gamma) = \max\{\Delta_{\theta_1}(\Gamma), \Delta_{\theta_2}(\Gamma)\} ,$$

proved in §2, then, is a particular case of the previous one.

Example 2 In a bounded region of the plane (Ot, Oz) that does not contain the origin, we define the following change of variable:

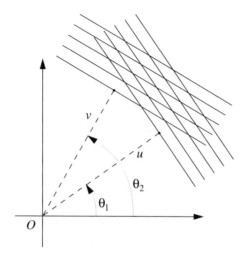

Fig. 16.6. *Distinct angles θ_1 and θ_2 determine a coordinate system.*

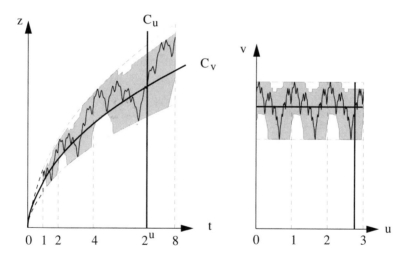

Fig. 16.7. *The change of variables $(t, z) \to (u, v)$ that transforms the graph Γ of $WM(t) = \sum_{-\infty}^{+\infty} 2^{-n/2}(1 - \cos 2^n t)$ to $WM^*(u) = \sum_{-\infty}^{+\infty} 2^{-(n+u)/2}(1 - \cos 2^{n+u})$ also transforms the sausage $\Gamma^{(v)}(\epsilon)$ to a sausage made with horizontal segments.*

$$\begin{cases} t = \omega^u \\ z = v\,\omega^{Hu} \end{cases},$$

where $0 < H < 1$, $\omega > 1$. The Jacobian $\omega^{(1+H)u} \log \omega$ cannot be null. The curves $\mathbf{C}_v$ follow the growth of the Weierstrass-Mandelbrot function

$$WM(t) = \sum_{-\infty}^{+\infty} \omega^{-nH}(1 - \cos\omega^n t) .$$

In the new coordinate system (u, v), this function can be written as

$$WM^*(u) = \sum_{-\infty}^{+\infty} \omega^{-(n+u)H}(1 - \cos\omega^{n+u}) ,$$

which is periodic of period 1 (Chap. 12, §11). Consider the graph Γ of $WM(t)$ and the directional sausage $\Gamma^{(v)}(\epsilon)$ (Fig. 16.7). Transformed by the change of variable, it becomes the sausage made by the horizontal segments of the graph of the function $WM^*(u)$.

Example 3 In a domain $\mathcal{D}$ that does not contain the origin, we define the transformation to polar coordinates:

$$\begin{cases} x = \rho \cos\theta \\ y = \rho \sin\theta \end{cases}$$

with Jacobian ρ. The arc $\mathbf{S}_\epsilon^{(\rho)}(x)$ is an arc of a circle of length $2\rho\epsilon$. The arc $\mathbf{S}_\epsilon^{(\theta)}(P)$ is a segment of length 2ϵ. In Figure 16.8 we represent the graph Γ of the periodic function

$$\rho(\theta) = \sum_0^\infty 2^{-n/2} \cos(2^n \theta) + c ,$$

where the value of the constant c $(c = \sqrt{2}/(\sqrt{2}-1))$ is taken to be large enough so that $\rho(\theta)$ can never be 0. Also, we represent the sausage $\Gamma^{(\rho)}(\epsilon)$. Its area can serve to compute the dimension of Γ. In fact, since every line originating at the origin intersects Γ at a unique point, we have

$$\Delta_\theta(\Gamma) = 1 ,$$

and therefore

$$\Delta_\rho(\Gamma) = \Delta(\Gamma) .$$

16.5 Intersections by straight lines

We study the way in which the line $\mathbf{D}(\rho, \theta)$ scans the curve Γ when ρ varies and θ is kept constant. The intersection is not empty when ρ runs through an interval $[\rho_1, \rho_2]$. We shall assume that Γ is not a line segment, therefore $\rho_1 \neq \rho_2$. When Γ has finite length, the points of the intersection $\mathbf{D}(\rho, \theta) \cap \Gamma$ are, for almost all values of θ, of finite number, and we have seen how we can use them to measure the length of Γ by computing their average (Chap. 8). When Γ is of infinite length, $\mathbf{D}(\rho, \theta) \cap \Gamma$ can be a Cantor set, that is, a nowhere dense subset of $\mathbf{D}(\rho, \theta)$, containing an infinite number of points. By taking all the points of

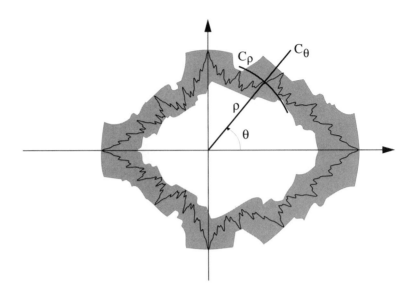

Fig. 16.8. *The curves* $\mathbf{C}_\rho$, $\mathbf{C}_\theta$ *are induced by polar coordinates. We represent the sausage* $\Gamma^{(\rho)}(\epsilon)$ *around the curve of equation* $\rho(\theta) = \sum_0^\infty 2^{-n/2} \cos(2^n \theta) + c$.

$\mathbf{D}(\rho,\theta)$ that are less than ϵ distance from this intersection, and by varying ρ, we construct the sausage $\Gamma(\theta, \epsilon)$ that allows the computation of the directional dimension $\Delta_\theta(\Gamma)$. In this section, we shall be interested in the relation between the dimensions $\Delta_\theta(\Gamma)$ or $\Delta(\Gamma)$ of the curve and the dimension of $\mathbf{D}(\rho,\theta) \cap \Gamma$.

◇   There exists a heuristic argument (that is, false but useful), which allows us to predict this relation in some cases:

We reconsider the notation $L_\Gamma(\rho,\theta,\epsilon)$ (Chap. 9, §4) for the length of the set of points of $\mathbf{D}(\rho,\theta)$ whose distance to a point of $\mathbf{D}(\rho,\theta) \cap \Gamma$ is $\leq \epsilon$. The scanning gives us:

$$\mathcal{A}(\Gamma(\epsilon,\theta)) = \int_{\rho_1}^{\rho_2} L_\Gamma(\rho,\theta,\epsilon)\, d\rho\ .$$

Now, if

$$L_\Gamma(\rho,\theta,\epsilon) \simeq \epsilon^{1-\Delta}$$

holds uniformly with respect to ρ for some value Δ, then Δ is the dimension of $\mathbf{D}(\rho,\theta) \cap \Gamma$, and by integration:

$$\mathcal{A}(\Gamma(\epsilon,\theta)) \simeq \epsilon^{1-\Delta}\ .$$

It follows that $2 - (\log \mathcal{A}(\Gamma(\epsilon,\theta))/\log \epsilon)$ converges to $1 + \Delta$. This is the value of $\Delta_\theta(\Gamma)$, which is, as we have already seen, the value of $\Delta(\Gamma)$, except when θ is the exceptional angle. Therefore, it seems reasonable to admit the following rule:

The dimension of Γ is obtained by adding 1 to the dimension of its intersection with a line for almost all lines.

Of course, this argument is false whenever Γ is constructed in such a way that every straight line intersects it in a finite number of points. That is what happens with the following curve.

Example Let Γ be the graph of the function

$$z(t) = \sqrt{t} + t^2 \cos t^{-5} , \, t \in [0, 1] .$$

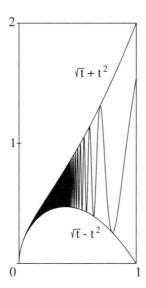

Fig. 16.9. *Graph of the function* $z(t) = \sqrt{t} + t^2 \cos t^{-5}$.

This curve is framed between the graphs of the two functions $\sqrt{t}+t^2$ and $\sqrt{t}-t^2$. Every line meets Γ at a finite number of points. The dimension of the intersection with any straight line is therefore null. However, the dimension of Γ is the same (Chap. 12, §4) as the dimension of the graph of $z_1(t) = t^2 \cos t^{-5}$; this is equal to $3/2$ (Chap. 10, §4). We conclude that

$$\Delta(\Gamma) = \frac{3}{2} .$$

In the next section, we shall give a general result that relates the different notions of dimension.

16.6 Essential upper bound

By definition,

*The **essential upper bound** of a real function $f(t)$ defined on a real interval is the number:*

$$\text{ess sup} f = \inf \{ \alpha \ \textit{such that the set} \{ t \ : \ f(t) > \alpha \}$$
$$\textit{has length zero} \}$$
$$= \sup \{ \alpha \ \textit{such that the set} \{ t \ : \ f(t) < \alpha \}$$
$$\textit{has a nonzero length} \} .$$

◇ This is the upper bound of the function $f(t)$ after taking away some "nonrepresentative values." The inequality

$$\text{ess sup} f \leq \sup f$$

always holds. Equality holds, for example, when $f(t)$ is continuous.

Here we consider the essential upper bound of $\Delta(\mathbf{D}(\rho, \theta) \cap \Gamma)$, considered as a function of ρ and defined on the interval $[\rho_1, \rho_2]$. In general, this is not a continuous function.

THEOREM   *For all angles θ,*

$$1 + \text{ess sup}_{\rho_1 \leq \rho \leq \rho_2} \Delta(\mathbf{D}(\rho, \theta) \cap \Gamma) \leq \Delta_\theta(\Gamma) .$$

Since $\Delta_\theta(\Gamma) \leq \Delta(\Gamma)$, we obtain a lower bound of the dimension.

▶   These dimensions can be written as:

$$\Delta_\theta(\Gamma) = \inf \{ \alpha \ : \ \epsilon^{\alpha-2} \mathcal{A}(\Gamma(\theta, \epsilon)) \to 0 \}$$
$$\Delta(\mathbf{D}(\rho, \theta) \cap \Gamma) = \inf \{ \alpha \ : \ \epsilon^{\alpha-1} L_\Gamma(\rho, \theta, \epsilon) \to 0 \} .$$

Take any number $\alpha > \Delta_\theta(\Gamma)$, the integral

$$\epsilon^{\alpha-2} \mathcal{A}(\Gamma(\theta, \epsilon)) = \epsilon^{\alpha-2} \int_{\rho_1}^{\rho_2} L_\Gamma(\rho, \theta, \epsilon) \, d\rho$$

tends to 0. A classical theorem from the theory of integrals allows us to deduce that: $\epsilon^{\alpha-2} L_\Gamma(\rho, \theta, \epsilon)$ tends to 0 for *almost all* values of ρ in $[\rho_1, \rho_2]$. Therefore $\alpha \geq 1 + \Delta(\mathbf{D}(\rho, \theta) \cap \Gamma)$ for all ρ except on a set of null measure. This proves that

$$\alpha \geq 1 + \text{ess sup}_{\rho_1 \leq \rho \leq \rho_2} \Delta(\mathbf{D}(\rho, \theta) \cap \Gamma) . \quad ◀$$

Fig. 16.10. *An estimate (by smaller values) of the dimension of Γ can be obtained by adding 1 to the dimension of the linear set $\mathbf{D}(\rho, \pi/2) \cap \Gamma$. In the case of the figure, $\mathbf{D}(\rho, \pi/2)$ is the horizontal line $z = \rho$. We must try many values of ρ so that we will not fall on exceptional values (which may exist, for example, when Γ contains horizontal regions).*

16.7 Uniform intersections

Following the "heuristic argument" of §5, we can give a theoretical condition for the equality

$$\Delta_\theta(\Gamma) = 1 + \text{ess sup}_{\rho_1 \le \rho \le \rho_2} \Delta(\mathbf{D}(\rho, \theta) \cap \Gamma)$$

to hold:

> *If there exists a value α and two non-null constants c_1 and c_2 such that for almost all values of ρ (that is, for all ρ in the interval $[\rho_1, \rho_2]$ less a subset of null measure) we have*
>
> $$c_1 \, \epsilon^{1-\alpha} \le L_\Gamma(\epsilon, \rho, \theta) \le c_2 \, \epsilon^{1-\alpha} ,$$
>
> *then*
>
> $$\Delta_\theta(\Gamma) = 1 + \alpha = 1 + \text{ess sup}_{\rho_1 \le \rho \le \rho_2} \Delta(\mathbf{D}(\rho, \theta) \cap \Gamma) .$$

▶ The hypothesis implies that for almost all ρ, $\Delta(\mathbf{D}(\rho, \theta) \cap \Gamma) = \alpha$. This is also the value of the essential upper bound of this dimension. On the other hand, the integral of a function does not change if we change the value of this function on a set of null measure. We deduce that

$$\mathcal{A}(\Gamma(\epsilon, \theta)) = \int_{\rho_1}^{\rho_2} L_\Gamma(\epsilon, \rho, \theta) \, d\rho \le c_2 \, (\rho_2 - \rho_1) \, \epsilon^{1-\alpha} .$$

Consequently, $\Delta_\theta(\Gamma) \leq 1 + \alpha$. The inequality in the other direction is always true, hence the equality. ◀

◇ There exists a method to compute the dimension of a curve. It consists of dividing the curve by a line and calculating the dimension of the intersection. It does not seem to be a very safe method; in fact in practice, the previous conditions are hard to satisfy. It is (with probability 1) true for curves resulting from some random process, and this finally justifies its use. Speaking of these procedures is outside the scope of this book. Thus we retain that this intersection method is very quick, and after some experiments (in order to avoid the exceptional values of ρ), we obtain a lower bound on the dimension. This method can be tested, with interesting results, on the Knopp or Weierstrass functions (Chap. 12) using their intersections with horizontal lines.

16.8 Intersection with an average curve

It is sometimes hard to measure the dimension of the intersection of a curve with a straight line when such an intersection does not offer enough information. Thus, it is natural to change the line to a rectifiable *average curve* that could better follow the original curve, thus increasing the probability that the intersection will contain a large number of points. This generalization of the intersection method is analogous to the generalization of the directional dimension by other systems of coordinates (§4).

1. If Γ is the **graph of a continuous function** $z(t)$, we construct an average function $z_1(t)$, and we look for the set of the zeros of the function $z(t) - z_1(t)$. This is a way to eliminate the low frequencies of the signal $z(t)$: the fractal dimension depends only on the high frequencies (in other words, the small oscillations of the curve).

A simple method to obtain an average curve consists of taking a window of width τ_0 and setting:

$$z_1(t) = \frac{1}{2\tau_0} \int_{t-\tau_0}^{t+\tau_0} z(s)\,ds\ .$$

This function is differentiable.

▶ In fact for all $h > 0$,

$$z_1(t+h) - z_1(t) = \frac{1}{2\tau_0} \left(\int_{t+\tau_0}^{t+\tau_0+h} z(s)\,ds - \int_{t-\tau_0}^{t-\tau_0+h} z(s)\,ds \right)\ .$$

The mean value theorem tells us that: $(1/h) \int_a^{a+h} z(s)\,ds = z(c)$, where $c \in [a, a+h]$. By the continuity of $z(t)$, this expression tends to $z(a)$ when h tends to 0. We deduce that

$$\lim_{h\to 0} \frac{z_1(t+h) - z_1(t)}{h} = \frac{z(t+\tau_0) - z(t-\tau_0)}{2\tau_0}\ :$$

this is the value of the derivative of $z_1(t)$.   ◀

The graph of z_1 is therefore rectifiable, and the transformation

$$(t, z) \longrightarrow (t, z - z_1(t))$$

is a regular transformation of a plane that does not change the value of the dimension. We deduce that the graphs of $z(t)$ and of $z(t) - z_1(t)$ have the same dimension. A simple computation of the dimension of the zeros of $z(t) - z_1(t)$ allows us to find an estimate of the dimension of the graph of $z(t)$ by adding 1.

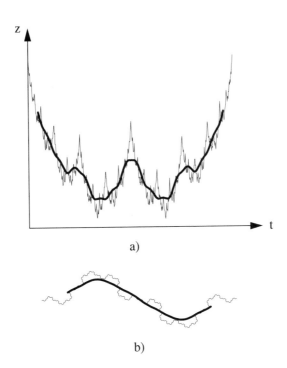

a)

b)

Fig. 16.11. *A drawing of the average curve for a) the graph of the Weierstrass function; b) a Gosper curve.*

2. If Γ is any **parameterized curve**, we can replace the average integral with the centroid. With a parameterization $\gamma(t)$ and a fixed window τ_0, we call $\gamma_1(t)$ the centroid of the arc

$$\gamma(t - \tau_0) \frown \gamma(t + \tau_0) .$$

This function γ_1 constitutes the parameterization of the average curve Γ_1, which is rectifiable with a tangent at each point. Assuming that the intersection $\Gamma \cap \Gamma_1$ is a Cantor set, we can directly compute its dimension in the plane (by the box method, or the Minkowski sausage method). Also, we can flatten the curve Γ_1 on a straight line to make $\Gamma \cap \Gamma_1$ a linear Cantor set; this is realized by calculating the parameterization of Γ_1 by the arc's length and then applying one of the methods of Chapter 3, §3.

◇   Let us recall that these intersection methods are less reliable than the direct sausage methods or steps along the curve.

16.9 Bibliographical notes

In theory, there exist more general and complete theorems that relate the dimension of a set to the dimension of the intersections. But then we would be talking about the *Hausdorff dimension*. This dimension is an excellent tool in pure mathematics, but it will not be discussed in this book. We are very much convinced that it has no practical utility. Thus, we prefer a partial and more complicated analysis, which seems to be more useful in experimental science. For this the reader may consult [C. Tricot, et al.].

17 Lateral Dimension of a Curve

17.1 Semisausages

If a boundary Γ is a rectifiable curve, then two observers sitting on both sides of this boundary (but not exactly over it), in general see the same thing: an approximate straight line, and the nearer they are to the curve, the straighter it appears. The Minkowski sausage

$$\Gamma(\epsilon) = \bigcup_{x \in \Gamma} B_\epsilon(x)$$

is then divided symmetrically by Γ; the areas on both sides are equivalent.

What happens when this boundary is of infinite length? It may happen that the areas of the two sides of the curve Γ are not equivalent. In this sense, there exist some symmetric curves and others that are not. It is then important to study with care what happens on each side. As an application, a "transition" through the interface Γ, whose rules depend on the fractal geometry, can have different properties depending on the direction in which Γ is crossed. We note here that some curves can be approached from one side only, for example, curves that are profiles of rough surfaces.

When Γ is a simple closed curve (a continuous image of a circle by a bijection), it is easy to define the two sides of the curve. The curve, in fact, divides the plane into two regions: the interior $\text{Int}(\Gamma)$ and the exterior $\text{Ext}(\Gamma)$, and thus, we denote the semisausages by:

$$\Gamma^{\text{Int}}(\epsilon) = \Gamma(\epsilon) \cap \text{Int}(\Gamma), \quad \Gamma^{\text{Ext}}(\epsilon) = \Gamma(\epsilon) \cap \text{Ext}(\Gamma).$$

If Γ is not closed, we can join its endpoints with another rectifiable curve Γ_1 such that $\tilde{\Gamma} = \Gamma \cup \Gamma_1$ is a simple closed curve. The right and left parts of the Minkowski sausage $\Gamma(\epsilon)$ can be defined as follows:

$$\Gamma^{\text{r}}(\epsilon) = \Gamma(\epsilon) \cap \text{Int}(\tilde{\Gamma}), \quad \Gamma^{\text{l}}(\epsilon) = \Gamma(\epsilon) \cap \text{Ext}(\tilde{\Gamma}).$$

Here, we use the notion of right and left with no prejudices to the observer's personal choice or to the way in which Γ_1 is drawn. They can be defined according to the direction of the parameterization of Γ. For the sake of simplicity we assume, in what follows, that the curve Γ is closed.

We call the index

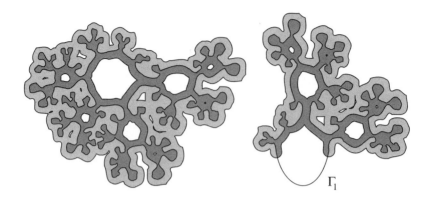

Fig. 17.1. *a) shows the two sausages, internal and external, of a simple closed curve Γ; in b), Γ is not closed, but it is completed by Γ_1 in such a way that we can define the two lateral sausages of Γ.*

$$\Delta^{\mathrm{Int}}(\Gamma) = \lim_{\epsilon \to 0}\left(2 - \frac{\log \mathcal{A}(\Gamma^{\mathrm{Int}}(\epsilon))}{\log \epsilon}\right)$$

the **interior dimension** *and the index*

$$\Delta^{\mathrm{Ext}}(\Gamma) = \lim_{\epsilon \to 0}\left(2 - \frac{\log \mathcal{A}(\Gamma^{\mathrm{Ext}}(\epsilon))}{\log \epsilon}\right)$$

the **exterior dimension**.

When we define the right and left of Γ, we may then speak of *left and right dimensions* and more generally of *lateral dimensions*.

17.2 Other expressions of the lateral dimensions

Maximum of disjoint balls   As we have done in the case of the fractal dimension $\Delta(\Gamma)$ itself, we can use disjoint balls to compute the lateral dimension of Γ. These are not centered on Γ, but rather, they are situated on one side or the other of Γ. Thus, we may define $M_\epsilon^{\mathrm{Int}}(\Gamma)$ to be the largest number of balls of diameter ϵ such that:

— *their interiors are disjoint and included in the interior of Γ; and*
— *they are adjacent to Γ (their boundaries touch Γ).*

Thus, Γ only intersects the boundaries of these balls. We shall show that:

$$\Delta^{\text{Int}}(\Gamma) = \lim_{\epsilon \to 0} \frac{\log M_\epsilon^{\text{Int}}(\Gamma)}{|\log \epsilon|} \, .$$

The only difficulty in the proof comes from the fact that $\text{Int}(\Gamma)$ could contain very thin *passages* in which we cannot squeeze any ball of diameter ϵ but whose area is not negligible compared to $\mathcal{A}(\Gamma^{\text{Int}}(\epsilon))$. That is why we should pack the interior of Γ with balls of different size in such a way that we can have a correct evaluation of the area.

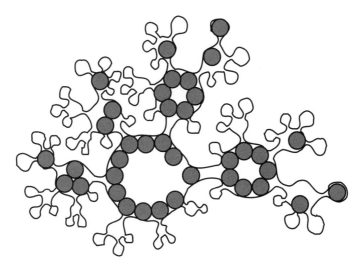

Fig. 17.2. *Disjoint balls of the same radius, adjacent to Γ and included in the interior of Γ. The curve may present thin "passages" where we cannot squeeze any ball.*

▶   All the chosen balls are situated in the internal sausage $\Gamma^{\text{Int}}(\epsilon)$; we deduce the inequality

$$\frac{\pi}{4}\, \epsilon^2 \, M_\epsilon^{\text{Int}}(\Gamma) \le \mathcal{A}(\Gamma^{\text{Int}}(\epsilon)) \, ,$$

and hence

$$\lim_{\epsilon \to 0} \frac{\log M_\epsilon^{\text{Int}}(\Gamma)}{|\log \epsilon|} \le \Delta^{\text{Int}}(\Gamma) \, .$$

For an inequality in the other direction, we form the family $\mathcal{F}_n$ of the $M_{2^{-n}\epsilon}^{\text{Int}}(\Gamma)$ balls adjacent to Γ whose diameter is $2^{-n}\epsilon$ and that have disjoint interiors. We consider the union of these families for all $n \ge 0$; we obtain a tiling, which is as precise as we wish, of the interior of Γ by balls (in fact, Γ is included in the closure of $\cup_n \mathcal{F}_n$). For each point x of the sausage $\Gamma^{\text{Int}}(\epsilon)$ not situated on Γ, let $n(x)$ be the smallest integer n such that there exists a ball of diameter $2^{-n}\epsilon$ containing x, adjacent to Γ, and included in $\Gamma^{\mathrm{Int}}(\epsilon)$: this integer somehow measures the *width* of $\Gamma^{\mathrm{Int}}(\epsilon)$ at the point x. If $n(x) = n$, this ball does not necessarily belong to $\mathcal{F}_n$. However, taking into account the fact that the number $M_{2^{-n}\epsilon}^{\mathrm{Int}}(\Gamma)$ is maximal, it intersects at least one ball of $\mathcal{F}_n$: we deduce that multiplying the diameter of the balls by three gives all the points x such that $n(x) = n$. Doing the same for all n, we cover all the semisausages $\Gamma^{\mathrm{Int}}(\epsilon)$. Thus we get:

$$\mathcal{A}(\Gamma^{\mathrm{Int}}(\epsilon)) \le \frac{9\pi}{4} \sum_{n=0}^{\infty} (\epsilon \, 2^{-n})^2 M_{2^{-n}\epsilon}^{\mathrm{Int}}(\Gamma) \ .$$

Therefore, for all real numbers α, we can write:

$$\epsilon^{\alpha-2} \mathcal{A}(\Gamma^{\mathrm{Int}}(\epsilon)) \le \frac{9\pi}{4} \sum_{n=0}^{\infty} 2^{-n(2-\alpha)} (\epsilon \, 2^{-n})^{\alpha} M_{2^{-n}\epsilon}^{\mathrm{Int}}(\Gamma) \ .$$

Assume that the limit of the ratio $\log M_{\epsilon}^{\mathrm{Int}}(\Gamma)/|\log \epsilon|$ is less than 2; this can be done without any loss of generality. And take α to be between this limit and 2. When ϵ is small enough,

$$\epsilon^{\alpha} M_{\epsilon}^{\mathrm{Int}}(\Gamma) \le 1 \ .$$

This is also true for $2^{-n}\epsilon$. Putting all this in the previous result we obtain

$$\epsilon^{\alpha-2} \mathcal{A}(\Gamma^{\mathrm{Int}}(\epsilon)) \le \frac{9\pi}{4} \sum_{n=0}^{\infty} 2^{-n(2-\alpha)} \ .$$

The right-hand-side term converges, so we deduce that $\Delta^{\mathrm{Int}}(\Gamma) \le \alpha$. And finally,

$$\Delta^{\mathrm{Int}}(\Gamma) \le \lim_{\epsilon \to 0} \frac{\log M_{\epsilon}^{\mathrm{Int}}(\Gamma)}{|\log \epsilon|} \ . \qquad \blacktriangleleft$$

$\diamond$ We can replace the continuous variable ϵ with a discrete one such as 2^{-k}, and let k go to infinity.

Adjacent boxes   We can also define $\Delta^{\mathrm{Int}}(\Gamma)$ with boxes; we draw a grid of parallel lines in the plane, determining squares of side ϵ. Let $\omega_{\epsilon}^{\mathrm{Int}}(\Gamma)$ be the total number of squares satisfying:

     — *the interiors of the squares are included in the interior of Γ; and*
     — *the squares have at least one common vertex with a box through which the curve passes.*

With an argument similar to the previous one, we show that:

$$\Delta^{\mathrm{Int}}(\Gamma) = \lim_{\epsilon \to 0} \frac{\log \omega_{\epsilon}^{\mathrm{Int}}(\Gamma)}{|\log \epsilon|} \ .$$

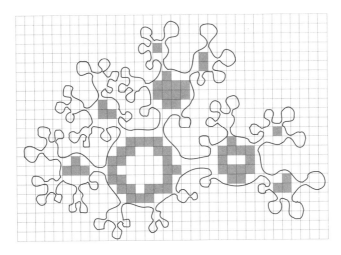

Fig. 17.3. *Boxes of a network situated in the interior of Γ and along Γ. Along Γ means that each of these boxes is adjacent to one of the boxes that cover Γ.*

17.3 Possible values of the lateral dimension

We recall that Γ is a simple closed curve. Its lateral dimension satisfies the following inequalities:

$$1 \leq \Delta^{\mathrm{Int}}(\Gamma) \leq \Delta(\Gamma) .$$

▶   Choose a point O in the interior of Γ. Let $R = \mathrm{dist}(O, \Gamma)$ and $\epsilon < R/2$. Each semisegment issued from O with slope θ meets $\Gamma^{\mathrm{Int}}(\epsilon)$ at a set of points with coordinates (ρ, θ), where ρ belongs to a set $E(\theta)$ whose length is $\geq 2\,\epsilon$. We have

$$\mathcal{A}(\Gamma^{\mathrm{Int}}(\epsilon)) = \int_0^{2\pi} \int_{E(\theta)} \rho \, d\rho \, d\theta .$$

Since all the values of $E(\theta)$ are larger than $R/2$,

$$\int_{E(\theta)} \rho \, d\rho \geq L(E(\theta)) \frac{R}{2} \geq \epsilon R .$$

We deduce that: $\mathcal{A}(\Gamma^{\mathrm{Int}}(\epsilon)) \geq 2\,\pi\,\epsilon\,R$. This shows that

$$\Delta^{\mathrm{Int}}(\Gamma) \geq 1 .$$

The other inequality $\Delta^{\mathrm{Int}}(\Gamma) \leq \Delta(\Gamma)$ follows from the fact that $\Gamma^{\mathrm{Int}}(\epsilon) \subset \Gamma(\epsilon)$.
◀

We can establish a more precise relation between the dimension of Γ and its lateral dimensions, that is,

$$\Delta(\Gamma) = \max\{\Delta^{\text{Int}}(\Gamma), \Delta^{\text{Ext}}(\Gamma)\} \ .$$

Thus,

the dimension of a curve is the maximum of its lateral dimensions.

▶   The area of the Minkowski sausage of Γ equals

$$\mathcal{A}(\Gamma(\epsilon)) = \mathcal{A}(\Gamma^{\text{Int}}(\epsilon)) + \mathcal{A}(\Gamma^{\text{Ext}}(\epsilon))$$

unless the curve itself has *non-null area*, a very particular case that will not be discussed, since the dimension of the curve is 2. Thus for all other curves, let α be between $\max\{\Delta^{\text{Int}}(\Gamma), \Delta^{\text{Ext}}(\Gamma)\}$ and 2. We have

$$\epsilon^{\alpha-2}\mathcal{A}(\Gamma^{\text{Int}}(\epsilon)) \longrightarrow 0 \ ,$$
$$\epsilon^{\alpha-2}\mathcal{A}(\Gamma^{\text{Ext}}(\epsilon)) \longrightarrow 0 \ ,$$

and hence

$$\epsilon^{\alpha-2}\mathcal{A}(\Gamma(\epsilon)) \longrightarrow 0 \ .$$

We deduce that $\Delta(\Gamma) \leq \alpha$.   ◀

◇ In the many cases where the lateral dimensions are equal, they are both equal to the dimension of the curve. In particular, we think of self–similar curves . But in general, curves may have different lateral dimensions.

17.4 Examples

All the examples of this section have a common feature: their interiors consist of *passages* or *peaks* that get thinner and thinner as the scale gets smaller and smaller in such a way that they reduce the value of the interior dimension.

Example 1 Consider two decreasing sequences (a_n) and (b_n) that converge to 0 and satisfy the following condition:

$$a_{n+1} + b_{n+1} < a_n - b_n \ .$$

In a Cartesian coordinate system, we place the points A_n with coordinates $(a_n + b_n, 0)$, B_n with coordinates (a_n, a_n), and C_n with coordinates $(a_n - b_n, 0)$. Let Γ be the curve formed by all the segments $A_n B_n$, $B_n C_n$, $C_n A_{n+1}$, $n \geq 1$, and with endpoints A_1 and O. By joining these endpoints with a rectifiable arc situated in the semiplane $x_2 < 0$, we then have that the *interior* of the curve means the region situated under the curve, while the *exterior* of the curve is the region above it. Now if (b_n) tends very quickly to 0, we can arrange the curve so that the interior of the peak $A_n B_n C_n$ is negligible; the interior dimension of Γ can then be made to equal 1. For the dimension, the internal side of the curve cannot be distinguished from the base segment $[0, 1]$. By contrast, (a_n) may converge very slowly to 0, so that when the curve is approached from above, it looks very bristly, and the neighborhoods of its peaks occupy an important place. The exterior dimension of Γ is then equal to 2. We show that these results can be obtained for $a_n \simeq (\log n)^{-1}$, and $b_n \simeq 2^{-n}$.

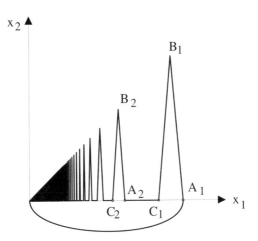

Fig. 17.4. *A curve that, seen from inside, has a dimension equal to 1 (very thin peaks); seen from outside, its dimension is 2 (many peaks that occupy an important place at all scales).*

▶   If $\epsilon \simeq b_n$,

$$\mathcal{A}(\Gamma^{\mathrm{Int}}(\epsilon)) \simeq \epsilon \sum_{i=1}^{n} a_i + \sum_{n+1}^{\infty} a_i b_i \; .$$

This gives $\mathcal{A}(\Gamma^{\mathrm{Int}}(\epsilon)) \preceq n\,\epsilon$, and hence

$$\Delta^{\mathrm{Int}}(\Gamma) = 1 \; .$$

If ϵ has the same order of growth as the segment $A_n C_{n+1}$, that is, $\epsilon \simeq a_n - a_{n+1} \simeq 1/n(\log /n)^2$, we obtain

$$\mathcal{A}(\Gamma^{\mathrm{Ext}}(\epsilon)) \succeq \epsilon \sum_{i=1}^{n} a_i \simeq (\log n)^{-3} \; .$$

Thus,

$$\Delta^{\mathrm{Ext}}(\Gamma) = 2 \; ,$$

which is the value of $\Delta(\Gamma)$.   ◀

Example 2 We construct a graph of a function in the same way as the graph of the Knopp function (Chap. 12, §6):

$$z(t) = \sum_{i=0}^{\infty} 2^{-n/2} g_n(2^n t) \, , \; t \in [0,1] \, .$$

Instead of always taking the same function, we assume that $g_n(t)$ depends on n. Given the sequence $b_n = 2^{-2n}$, we define

$$g_n(t) = \begin{cases} 0 & \text{if } t \in [0, \frac{1}{2} - b_n], \\ \frac{1}{b_n}(t - \frac{1}{2} + b_n) & \text{if } t \in [\frac{1}{2} - b_n, \frac{1}{2}], \\ \frac{1}{b_n}(-t + \frac{1}{2} + b_n) & \text{if } t \in [\frac{1}{2}, \frac{1}{2} + b_n], \\ 0 & \text{if } t \in [\frac{1}{2} + b_n, 1] \, . \end{cases}$$

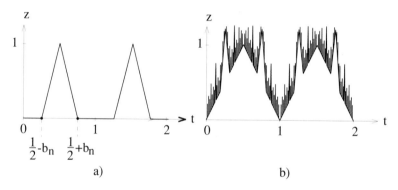

a)   b)

Fig. 17.5. *a) shows the graph of the periodic function $g_n(t)$; b) shows the graph of the function*

$$z(t) = \sum_{i=0}^{\infty} 2^{-n/2} g_n(2^n t) \, .$$

With arguments similar to those used in the construction of the Knopp function, we can show that $z(t)$, on average, has the same oscillations. Therefore, the dimension of its graph Γ is:

$$\Delta(\Gamma) = \frac{3}{2} \, .$$

This is independent of the values of b_n. As in the previous example, we consider the region under the curve to be the *interior*. The sequence (b_n) tends to 0 quickly enough so that the interior dimension will be strictly less than $3/2$.

▶   In fact, if $\epsilon \simeq b_n$,

$\mathcal{A}(\Gamma^{\mathrm{Int}}(\epsilon)) \simeq \epsilon \, (\text{length of the peaks up to rank } n)$

$\qquad + \, (\text{area of the interior of the peaks from } n \text{ to } \infty)$

$$\simeq \epsilon \sum_{i=0}^{n} 2^i \, 2^{-i/2} + \sum_{i=n}^{\infty} b_i \, 2^i \, 2^{-i/2}$$

$$\simeq 2^{-3n/2} \ .$$

We deduce that

$$\Delta^{\mathrm{Int}}(\Gamma) = \lim(2 \frac{\log \mathcal{A}(\Gamma^{\mathrm{Int}}(\epsilon))}{\log \epsilon}) = \frac{5}{4} \ . \quad \blacktriangleleft$$

Example 3 See Figure 17.6.

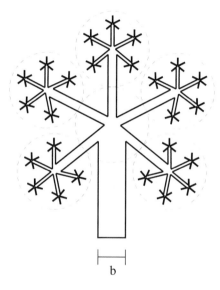

b

Fig. 17.6. *This figure shows how we can construct a simple closed tree-shaped curve whose dimension is strictly larger than its interior dimension. For all $k \geq 0$, there are 5^k branches of length $\simeq 3^{-k}$ and width b^{-k}, $b > 3$. The dimension is $\Delta(\Gamma) = \log 5/\log 3$. It does not depend on the value of b. The interior dimension is*

$$\Delta^{\mathrm{Int}}(\Gamma) = 1 + \frac{\log 5/3}{\log b} \ .$$

17.5 The inverse Minkowski operation

There is a method to draw a rectifiable curve in the internal part of a closed curve Γ that constitutes a good approximation of Γ. Here is how we proceed. Choose a point O in the interior of Γ and a value $\epsilon < \mathrm{dist}(O, \Gamma)$. First, we consider the set $\partial(\Gamma^{\mathrm{Int}}(\epsilon))$, the boundary of the internal semisausage. This set is formed by the curve Γ itself and a number of closed rectifiable curves that lie inside Γ. Among these curves, choose the one whose interior contains O and denote it by $\mathbf{C}_\epsilon$. Then we form the external semisausage of $\mathbf{C}_\epsilon$: its boundary $\partial(\mathbf{C}_\epsilon^{\mathrm{Ext}}(\epsilon))$ is formed by the curve $\mathbf{C}_\epsilon$ and another closed rectifiable curve $\mathbf{G}_\epsilon$. This new curve has points in common with Γ; the other points of $\mathbf{G}_\epsilon$ are inside Γ. This is our internal approximation of Γ at an ϵ–precision (Fig. 17.7), and $\mathbf{G}_\epsilon$ is the result of what we call "the inverse Minkowski operation."

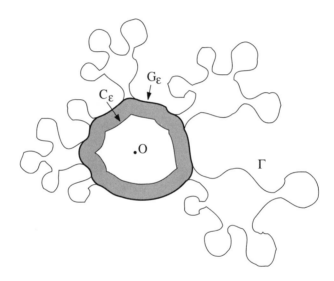

Fig. 17.7. *Take a closed curve Γ and a point O inside it. The curve $\mathbf{C}_\epsilon$ is formed by points circling O whose distance to Γ is ϵ. The curve $\mathbf{G}_\epsilon$ is formed by points circling $\mathbf{C}_\epsilon$ whose distance to $\mathbf{C}_\epsilon$ is ϵ. This is a rectifiable curve in the interior of Γ with many points in common with Γ.*

$\diamond$ The initial choice of O has no real importance, in the sense that if O' is another point in the interior and $\mathbf{G}'_\epsilon$ is the internal approximation of Γ that corresponds to O', then

$$\mathbf{G}_\epsilon = \mathbf{G}'_\epsilon$$

once ϵ is small.

$\diamond$ If Γ is itself rectifiable such that at each point it has an internal curvature radius larger than ϵ, then $\mathbf{G}_\epsilon$ itself is Γ.

We can describe $\mathbf{G}_\epsilon$ in another manner with help from a mechanical procedure:

Let us turn a wheel around O, inside the curve Γ, and along it (so that the wheel will always touch the curve). The surface scanned by the wheel during its motion has $\mathbf{G}_\epsilon$ as its external boundary.

Without going into details, let us indicate that the value $L(\mathbf{G}_\epsilon)/\epsilon$ is equivalent to the maximum number of disjoint balls of diameter ϵ that are included in Γ and touch $\mathbf{G}_\epsilon$. This latter number is at most equal to $M_\epsilon^{\text{Int}}(\Gamma)$ (§2). Thus, we deduce the following inequality:

$$\Delta(\Gamma) \geq \lim_{\epsilon \to 0} \left(1 + \frac{\log L(\mathbf{G}_\epsilon)}{\log \epsilon}\right) .$$

For some curves equality holds, but this does not happen (Fig. 17.9) when the interior of Γ is divided at every scale into "islands" related by "passages" of width $< \epsilon$, which reduces the length of $\mathbf{G}_\epsilon$.

◇ A notion of *accessibility* of the points of the interior of Γ, with respect to the central point O can be derived from the above discussion:

We say that the point x of Int(Γ) is ε–accessible if there exists an arc Γ* such that O and x belong to the sausage Γ*(ε), and Γ*(ε) ⊂ Int(Γ).

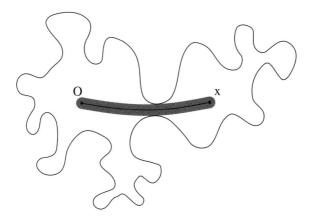

Fig. 17.8. *The point x is ε–accessible from the point O in the interior of Γ.*

The following example describes a curve whose set of accessible points is negligible at every scale. The fractal dimension of this curve cannot be equal to the limit of $1 + \log L(\mathbf{G}_\epsilon)/\log \epsilon$.

Example a) First, we define a self–similar set E_0 using the following five similarities (Chap. 13, §1):

$$F_i(x) = \mathcal{R}_{\theta_i}\left(\frac{1}{3}x + \begin{pmatrix} 2/3 \\ 0 \end{pmatrix}\right) , \quad i = 1, 2, 3, 4 ,$$

where the rotation angles θ_i have the following values:

$$\theta_1 = \frac{-\pi}{6} \, , \, \theta_2 = \frac{\pi}{6} \, , \, \theta_3 = \frac{\pi}{2} \, , \, \theta_4 = \frac{5\pi}{6} \, , \, \theta_5 = \frac{7\pi}{6} \, .$$

This set E_0, which satisfies $E_0 = \cup F_i(E_0)$, is nowhere dense in the plane.

b) Then we draw the disk $\mathbf{D}$ with center O and radius $1/3$ (rank 1). By the similarities F_i, this disk is transformed into five disks (rank 2) of radius 3^{-2}, which in turn are transformed into a total of twenty-five disks (rank 3) of radius 3^{-3}, and so forth.

c) To form this curve, we join all these disks or "islands" by "passages" (here, they are plain rectangles); $\mathbf{D}$ is related to the five disks of rank 2 by five passages of length $\simeq 1/3$ and width b^{-2}, where b is a parameter > 3. Similarly, each disk of rank k is related to its five images of rank $k+1$ by five passages of length $\simeq 3^{-k}$ and width b^{-k-1}.

d) Take the union of all these obtained sets: the set E_0, the disks, and the rectangles of all ranks. We obtain a closed set. The boundary of this closed set is a simple curve Γ (Fig. 17.9). As in the example in Figure 17.6, its dimension is

$$\Delta(\Gamma) = \frac{\log 5}{\log 3} \, .$$

But here the internal dimension is equal to $\log 5/\log 3$, because of the width of the islands. We may verify that the limit of $1 + \log L(\mathbf{G}_\epsilon)/\log \epsilon$ is strictly less than this value. This is due to the fact that the passages become thinner as the observation scale becomes smaller, which delays the expansion of $\mathbf{G}_\epsilon$ when ϵ tends to 0.

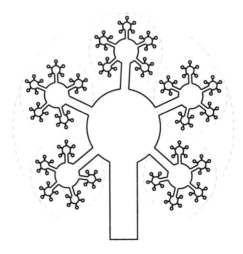

Fig. 17.9. *A curve whose interior is in its major part formed of points inaccessible from its center at all scales.*

Let $\epsilon = b^{-k}$. The length of the curve $\mathbf{G}_\epsilon$ is equivalent to the sum of the perimeters of the islands and the passages up to rank k, that is,

$$L(\mathbf{G}_\epsilon) \simeq \sum_{i=0}^{k} 5\,3^{-i} \simeq (5/3)^k \ .$$

We deduce that:
$$1 + \frac{\log L(\mathbf{G}_\epsilon)}{|\log \epsilon|} \simeq 1 + \frac{\log 5/3}{\log b} \ .$$

This is strictly less than $\log 5/\log 3$ whenever b is larger than 3.

17.6 Bibliographical notes

The idea of constructing an internal approximation of a curve Γ by turning a wheel inside it goes back to [J. Perkal] (1952) in a cartographic study. We note the important contributions made by geographers to the study of the geometry of irregular curves. Drawing a coastline can inspire many good and original ideas.

Some parts of this chapter, as well as the relation between the lateral geometry of curves and porous surfaces, can be found in [C. Tricot 5].

18 Dimensional Homogeneity

18.1 Local structures of some curves

We review some curves whose dimension is strictly larger than 1.

1. In Chapter 10, we found locally rectifiable curves that occupy a large part of the space (in the sense of the area of its Minkowski sausage) around a given point of accumulation. Thus, the spiral

$$\begin{cases} \rho(t) = t^\alpha \\ \theta(t) = \dfrac{2\pi}{t} \end{cases} \quad 0 < t \le 1,$$

where $0 < \alpha < 1$ to which we can add the origin 0 (corresponding to $t = 0$), has dimension:

$$\Delta(\Gamma) = \frac{2}{\alpha + 1} .$$

However, each arc Γ^* of Γ that does not contain the origin is of finite length and dimension 1. This curve is not "homogeneous" from the dimensional point of view. The same remark holds for the following:

2. The dimension of the graph of the function

$$z(t) = \begin{cases} t^\alpha \, \cos t^{-\beta} & \text{if } 0 < t \le 1; \\ 0 & \text{if } t = 0 , \end{cases}$$

where $0 < \alpha < \beta$, is

$$\Delta(\Gamma) = 2 - \frac{\alpha + 1}{\beta + 1} .$$

Like the previous example, every arc Γ^* of Γ whose endpoints are different from 0 is of dimension 1.

3. By contrast, the curves that are "fractal" in the restricted mathematical sense of Chapter 11, §1, can be written as

$$\Gamma = \cup F_i(\Gamma) ,$$

where the $F_i(\Gamma)$ are copies of Γ by contractions F_i, such as similarities, affinities, or others. These curves are dimensionally homogeneous; every subarc has the same dimension as the entire curve.

4. This also holds for the graphs of the Weierstrass or Weierstrass–Mandelbrot functions (Chap. 12, §7 and §11). The dimension of these curves is independent of the interval of definition of the function.

5. We can say the same about the particular examples of expansive curves given in Chapters 14 and 15. Their dimensional homogeneity is ultimately the natural result of iterative geometrical constructions. But it is easy to construct nonhomogeneous curves. For example, it suffices to concatenate two curves of different dimensions. We have also noted that experimental curves are often nonhomogeneous (Chap. 15, §1); these are curves for which the varying-steps algorithm may be used.

In light of these examples, we may now define the notion of "dimensional homogeneity":

A curve Γ is **dimensionally homogeneous** *if for every subarc Γ^* (not reduced to one point) of Γ,*

$$\Delta(\Gamma^*) = \Delta(\Gamma) .$$

A subarc of a simple curve Γ that is not reduced to one point will always contain a set of the form $\Gamma \cap B_\epsilon(x)$, where x is a point of the curve and $B_\epsilon(x)$ is a ball with center x and radius ϵ. Therefore, we can generalize the previous definition to any bounded set:

A set E is **dimensionally homogeneous** *if for every point x of E and all $\epsilon > 0$:*

$$\Delta(E \cap B_\epsilon(x)) = \Delta(E) .$$

18.2 Local dimension

Given a point x in a set E, we remark that the function

$$f(\epsilon) = \Delta(E \cap B_\epsilon(x))$$

of the variable ϵ is increasing: when ϵ decreases to 0, $f(\epsilon)$ decreases to a limit value, the **local dimension of E at the point** x. We will denote it by

$$\Delta(E, x) = \lim_{\epsilon \to 0} \Delta(E \cap B_\epsilon(x)) .$$

For some sets E, the local dimension could be independent of ϵ and of the point x. In this case, $\Delta(E \cap B_\epsilon(x))$ is always equal to $\Delta(E)$.

A compact set E is dimensionally homogeneous if and only if its local dimension is $\Delta(E)$ at every point of E.

Point of maximal dimension   If E is a compact set that is not dimensionally homogeneous, then there exists a point x_0 of E where the local dimension is $\Delta(E)$.

▶   This results from the stability of the dimension Δ (Chap. 10, §3):

$$\Delta(E_1 \cup E_2) = \max\{\Delta(E_1), \Delta(E_2)\} .$$

In fact, suppose we cover the plane with closed boxes of side 1. The set E is covered by a finite number of them so that by the stability of the dimension, there exists one box, say C_1, such that:

$$\Delta(E \cap C_1) = \Delta(E) .$$

We cover $E \cap C_1$ with boxes of side $1/2$: one of them, say C_2, is such that

$$\Delta(E \cap C_1 \cap C_2) = \Delta(E \cap C_1) = \Delta(E) ;$$

and so forth. We construct an infinite sequence of boxes (C_n) such that:

$$\Delta(E \cap (\cap_{i=1}^{n} C_i)) = \Delta(E) .$$

The sequence $(E \cap (\cap_{i=1}^{n} C_i))$ is a sequence of embedded closed sets whose intersection contains a unique point x_0 of E. For all $\epsilon > 0$, there exists an integer n such that $\cap_{i=1}^{n} C_i \subset B_\epsilon(x_0)$. We deduce that:

$$\Delta(E \cap B_\epsilon(x_0)) = \Delta(E) .   ◀$$

Example   We can easily construct a curve whose local dimension varies depending on the considered point. For example, we can transform the Weierstrass series (Chap. 12, §7) in the following manner:

$$z(t) = \begin{cases} t^\alpha \sum_{i=0}^{\infty} \omega^{-nt} \cos(\omega^n t) & \text{if } 0 < t \le 1; \\ 0 & \text{if } t = 0, \end{cases}$$

where $\alpha > 1$ and $\omega > 1$. The factor t^α is designed to ensure that $z(t)$ is defined and continuous at 0.

▶   Indeed the sum of the series $\sum \omega^{-nt}$ is equivalent to $1/(t \log \omega)$ in neighborhoods of 0. Therefore, $\lim_{t \to 0} z(t) = 0$ when $\alpha > 1$.   ◀

The amplitude of the nth term of the series is ω^{-nt}. By comparing it to the ω^{-nH} term of the Weierstrass series, we can predict that the local Hölder exponent is equal to t. Thus the local dimension of the graph at the point $x = (t, z(t))$ is equal to

$$\boxed{\Delta(\Gamma, x) = 2 - t .}$$

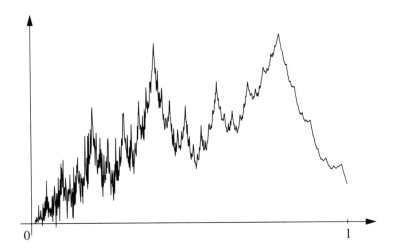

Fig. 18.1. *A graph of the function $t^{3/2} \sum_{i=0}^{\infty} 2^{-nt} \cos(2^n t)$ on the interval $[0,1]$. The local dimension at a point of abscissa t is $2 - t$; it varies continuously along the curve between the values 2 and 1.*

▶ a) First, let

$$u(t) = \sum_{i=0}^{\infty} \omega^{-nt} \cos(\omega^n t) \ .$$

This is a well-defined continuous function on the interval $]0, 1]$. Let $\mathbf{G}(a, b)$ be the graph of $u(t)$ when t belongs to the interval $[a, b]$, $0 < a < b \le 1$. We shall prove that

$$\Delta(\mathbf{G}([a, 1])) = 2 - a \ .$$

The proof uses arguments similar to those used in Chapter 12, §7.

a1) Let $u_n(t) = \sum_{i=0}^{n} \omega^{-it} \cos(\omega^i t)$ and calculate its derivative. We show that

$$|u_n'(t)| = |\sum_{i=0}^{n} \omega^{-it} \left(-i \log \omega \cos(\omega^i t) - \omega^i \sin(\omega^i t)\right)| \preceq \omega^{n(1-a)} \ .$$

Therefore, if $\tau = \omega^{-n}$, we obtain that $\mathrm{osc}_\tau(t)(u_n) \preceq \tau^a$. On the other hand,

$$\mathrm{osc}_\tau(t)(u - u_n) \le \sum_{i=n+1}^{\infty} \omega^{-it} \preceq \tau^a \ .$$

We deduce that for all $t \in [a, 1]$, $\mathrm{osc}_\tau(t)(u) \preceq \tau^a$, and therefore

$$\Delta(\mathbf{G}([a, 1])) \le 2 - a \ .$$

a2) We take any real b in $]a, 1[$, and we show that the variation of $u(t)$ on $[a, b]$ satisfies the inequality

$$|\mathrm{Var}_\tau| \geq \left| \int_a^b \mathrm{osc}_\tau(t)(u)\, dt \right| \succeq \tau^b .$$

We deduce

$$\Delta(\mathbf{G}([a,1])) \leq \Delta(\mathbf{G}([a,b])) \leq 2 - b .$$

When b converges to a, we finally obtain the equality

$$\Delta(\mathbf{G}([a,1])) = 2 - a .$$

b) Now, we consider the function $z(t) = t^\alpha u(t)$.

b1) Let $\Gamma(a,b)$ be the graph of $z(t)$ when t belongs to the interval $[a,b]$. If $a > 0$, the transformation that changes $\mathbf{G}([a,b])$ to $\Gamma([a,b])$ is a diffeomorphism of the plane (multiplication of the coordinates by t^α) that does not change the dimension. Thus,

$$\Delta(\Gamma([a,b])) = 2 - a .$$

We deduce that when $0 < t - \epsilon < t \leq 1$,

$$\Delta(\Gamma([t - \epsilon, t + \epsilon])) = 2 - t + \epsilon .$$

When ϵ tends to 0, the local dimension at any point $t \in]0,1]$ is:

$$\Delta(\Gamma, z(t)) = 2 - t .$$

b2) Finally, $\Delta(\Gamma([0, 2\epsilon])) \geq \Delta(\Gamma([\epsilon, 2\epsilon])) = 2 - \epsilon$. By letting ϵ tend to 0, we can deduce that $\Delta(\Gamma, 0) \geq 2$. Therefore,

$$\Delta(\Gamma, 0) = 2 .$$

At each $t \in]0,1]$, the local Hölder exponent is the same for the functions $u(t)$ and $z(t)$. Its value is equal to t. Figure 18.1 shows that the complexity of the graph of $z(t)$ grows when t goes from 1 to 0. ◀

18.3 The packing dimension

Now we introduce a new concept of dimension, the packing dimension, which allows a finer analysis of a set. In particular, it attributes the value 1 to the locally rectifiable curves, and the value 0 to finite or countable sets; this is not, in general, the case of $\Delta(E)$. However, it directly follows from the Bouligand dimension. By definition:

> Let E be a set. We divide it into a finite or countable union of bounded sets E_n and compute $\sup \Delta(E_n)$. The smallest among these values, over all possible partitions, is called the **packing dimension** $\hat{\Delta}(E)$.

In a more symbolic language:

$$\hat{\Delta}(E) = \inf\{\, \sup \Delta(E_n) \,, \text{where the } E_n \text{ are bounded, and } E \subset \cup E_n \,\} \ .$$

Deep down, this new dimension "regularizes" the Bouligand dimension. We often call $\hat{\Delta}$ the **packing dimension**, because it is associated to some particular measures known as "packing measures," as opposed to "covering measures," which are associated with the Hausdorff dimension. We do not intend to discuss the notion of measure, whose study is outside the scope of this book. Here are some properties of $\hat{\Delta}$:

1. $\hat{\Delta}$ is **monotonous**: *if E_1 is included in E_2, then*

$$\hat{\Delta}(E_1) \leq \hat{\Delta}(E_2) \ .$$

2. $\hat{\Delta}$ is σ–**stable**: *if (E_n) is a finite or infinite sequence of bounded sets,*

$$\hat{\Delta}(\cup_n E_n) = \sup_n \hat{\Delta}(E_n) \ .$$

3.   *For any bounded set E,*

$$\hat{\Delta}(E) \leq \Delta(E) \ .$$

4. $\hat{\Delta}$ is **invariant** *under all plane transformations T such that for all x, the ratio*

$$\frac{\log \operatorname{dist}(T(y), T(z))}{\log \operatorname{dist}(y, z)}$$

tends to 1 when y and z both tend to x, and $y \neq z$. For such transformations,

$$\hat{\Delta}(T(E)) = \hat{\Delta}(E) \ .$$

5.   *If Γ is a simple curve:*

$$\hat{\Delta}(\Gamma) \geq 1 \ .$$

▶   **1.** and **4.** follow from the corresponding properties of Δ (Chap. 10, §3).
    3. follows directly from the definition, with $E_n = E$.
    2. Let $E = \cup E_n$. For all ϵ and all n, we can find a sequence $(E_{n,k})$ such that $E_n \subset \cup_k E_{n,k}$, and

$$\sup_k \Delta(E_{n,k}) \leq \hat{\Delta}(E_n) + \epsilon \ .$$

Therefore $E \subset \cup_{n,k} E_{n,k}$, and

$$\sup_{n,k} E_{n,k} \leq \sup_n \hat{\Delta}(E_n) + \epsilon \ .$$

Letting ϵ go to 0 proves the inequality

$$\hat{\Delta}(E) \leq \sup_n \hat{\Delta}(E_n) .$$

The inequality in the other direction follows from Property **1**.

5. A curve is a compact set. If $\Gamma \subset \cup E_n$, then at least one of the E_n has the following property: it is *somewhere dense* in Γ, that is, its closure $\overline{E}_n$ contains a subarc of Γ (this results from Baire's theorem, which will be stated in §4). We deduce that:

$$\Delta(E_n) = \Delta(\overline{E}_n) \geq 1 .$$

Therefore, $\sup_n \Delta(E_n) \geq 1.$ ◀

◇   The dimension $\hat{\Delta}$ is also defined on unbounded sets. In the definition we specify that the sets E_n are bounded, so that we can measure their dimension Δ.

◇   The most characteristic property of $\hat{\Delta}$ is the σ–stability. The dimension Δ is stable but not σ–stable. In revenge, Δ has the property $\Delta(\overline{E}) = \Delta(E)$, which means that this dimension cannot be associated with the topological properties of a set.

18.4 Possible values of the packing dimension

• If E is a finite or countable set of points, then $\hat{\Delta}(E) = 0$.

▶ E can be covered by a finite or countable union of one point sets whose dimension Δ is 0. ◀

• For every curve of finite length,

$$\hat{\Delta}(\Gamma) = \Delta(\Gamma) = 1 .$$

• The spiral

$$\begin{cases} \rho(t) = t \\ \theta(t) = \dfrac{2\pi}{t} \end{cases} \quad 0 < t \leq 1,$$

to which we add the origin 0, has packing dimension 1.

▶   In fact, it can be considered as $\cup E_n$, where $E_0 = \{0\}$, and the E_n are all arcs of finite length. Since $\Delta(E_0) = 0$ and $\Delta(E_n) = 1$, we obtain $\hat{\Delta}(\Gamma) \leq 1.$ ◀

• More generally, every set E that can be written as

$$E = E_0 \cup (\cup_{n \geq 1} E_n) ,$$

where E_n, $n \geq 1$, is an arc of finite length (eventually reduced to one point) and where E_0 is a set of dimension $\Delta(E_0) \leq 1$, has a packing dimension 1.

• For a large class of sets, Δ and $\hat{\Delta}$ are identical:

If E is a compact set that is dimensionally homogeneous with respect to Δ, then

$$\Delta(E) = \hat{\Delta}(E) .$$

This follows immediately from the following.

THEOREM   *If E is a closed set, then*

$$\inf_{x \in E} \Delta(E, x) \leq \hat{\Delta}(E) \leq \sup_{x \in E} \Delta(E, x) .$$

▶   a) We prove the left inequality. Assume that $\inf_{x \in E} \Delta(E, x) > 0$, and let α satisfy $0 < \alpha < \inf_{x \in E} \Delta(E, x)$. We shall show that

$$\alpha \leq \hat{\Delta}(E) .$$

Without any loss of generality, we can assume that E is bounded. We use the following theorem due to Baire:

> Let E be a compact set. Every subset of E that contains a set of the form $E \cap B_\epsilon(x)$, where $\epsilon > 0$ and $x \in E$, is called a **portion** of E. Let (E_n) be a cover, finite or countable, of E by closed sets. Then one of the E_n contains a portion of E.

This is a characteristic property of compact sets. It can be said in an equivalent manner as follows: *a compact set E cannot be covered by a union of finite or countable nowhere dense subsets of E.* For example, an interval of the real line cannot be covered by closed nowhere dense Cantor sets, even if their lengths are not null.

Assume that the compact set E is included in $\cup E_n$, where the E_n are bounded. Since the $\overline{E}_n$ are compact, there exists an integer n_0 such that $\overline{E}_{n_0}$ contains a portion of E. So there is a point x of E and an $\epsilon > 0$ such that

$$E \cap B_\epsilon(x) \subset \overline{E}_{n_0} .$$

We deduce:

$$\alpha \leq \Delta(E, x_0) \leq \Delta(E \cap B_\epsilon(x_0)) \leq \Delta(\overline{E}_{n_0}) = \Delta(E_{n_0}) .$$

Therefore, $\alpha \leq \sup \Delta(E_n)$. Since this inequality is true for any cover (E_n), we obtain $\alpha \leq \hat{\Delta}(E)$.

b) Now we show the right inequality. When E is a bounded set, we know that the right term is $\Delta(E)$ (§2). The inequality is then reduced to $\hat{\Delta}(E) \leq \Delta(E)$. Therefore, assume that E is unbounded.

We take a real number $\beta > \sup_{x \in E} \Delta(E, x)$, and we prove that

$$\beta \geq \hat{\Delta}(E) .$$

Let (E_n) be a cover of E by bounded sets. For all $\epsilon > 0$, there exists an E_{n_0} such that $\Delta(E_{n_0}) \geq \sup \Delta(E_n) - \epsilon$. From §2 we know that there exists an $x_0 \in \overline{E}_{n_0}$ such that

$$\varDelta(\overline{E}_{n_0}, x_0) = \varDelta(\overline{E}_{n_0}) = \varDelta(E_{n_0}) \ .$$

Moreover,

$$\varDelta(\overline{E}_{n_0}, x_0) \leq \varDelta(E, x_0) \ .$$

From these inequalities we deduce that:

$$\sup_n \varDelta(E_n) \leq \varDelta(E_{n_0}) + \epsilon \leq \varDelta(E, x_0) + \epsilon \leq \beta + \epsilon \ .$$

Letting ϵ tend to 0, we obtain the sought inequality.   ◄

• If E is a set that is not dimensionally homogeneous with respect to $\varDelta$, we can sometimes find a direct relation between $\hat{\varDelta}(E)$ and the local dimension. For example:

Assume that $\varDelta(E, x)$ is a continuous function of x. Then

$$\hat{\varDelta}(E) = \sup_{x \in E} \varDelta(E, x) \ .$$

► Let α be the term on the right. The inequality $\hat{\varDelta}(E) \leq \alpha$ is always true. In the other direction: assume that $\alpha > 0$ (if not, we would have $\hat{\varDelta}(E) = 0$). Let $\epsilon > 0$. There exists a point x_0 of E such that

$$\varDelta(E, x_0) \geq \alpha - \epsilon \ .$$

Such a point is not isolated in E, otherwise $\varDelta(E, x_0)$ would be null. Since the function $\varDelta(E, x)$ is continuous with respect to x, we can find a real η such that

$$x \in E \ , \text{ and } \operatorname{dist}(x, x_0) \leq \eta \Longrightarrow \varDelta(E, x) \geq \varDelta(E, x_0) - \epsilon \ .$$

We deduce, using the previous theorem, that:

$$\hat{\varDelta}(E \cap B_\eta(x_0)) \geq \varDelta(E, x_0) - \epsilon \ .$$

Therefore $\hat{\varDelta}(E) \geq \alpha - 2\epsilon$. Then we make ϵ tend to 0.   ◄

• This latter result can be applied to the curve Γ in Figure 18.1, where $\varDelta(\Gamma, z(t)) = 2 - t$ is a continuous function. If we define the local packing dimension by:

$$\hat{\varDelta}(E, x) = \lim_{\epsilon \to 0} \hat{\varDelta}(E \cap B_\epsilon(x)) \ ,$$

we obtain for this particular curve

$$\varDelta(\Gamma, (t, z(t))) = \hat{\varDelta}(\Gamma, (t, z(t))) = 2 - t \ .$$

18.5 The σ–stabilization

The operation that consists of changing Δ to $\hat{\Delta}$ is in fact general. It can be applied to any function of bounded sets $\alpha(E)$ by putting

$$\hat{\alpha}(E) = \inf\{\,\sup \alpha(E_n)\,,\,\text{where the } E_n \text{ are bounded, and } E \subset \cup E_n\,\}\,.$$

It has the following properties:

1. $\hat{\alpha}$ *is* **monotonous**.

$$E_1 \subset E_2 \Longrightarrow \hat{\alpha}(E_1) \le \hat{\alpha}(E_2)\,.$$

2. $\hat{\alpha}$ *is* σ–**stable**:

$$\hat{\alpha}(\cup E_n) = \sup_n \hat{\alpha}(E_n)\,.$$

3. *For all* E,

$$\hat{\alpha}(E) \le \alpha(E)\,.$$

4. *If* α *is* **invariant** *by a transformation* T, *then so is* $\hat{\alpha}$:

$$\hat{\alpha}(T(E)) = \hat{\alpha}(E)\,.$$

▶   **1.** If $E_1 \subset E_2$, then every cover of E_2 is a cover of E_1. Therefore, if (F_n) is a cover of E_2, then $\hat{\alpha}(E_1) \le \sup \alpha(F_n)$.
   2. If $E = \cup E_n$, $\hat{\alpha}(E_n) \le \hat{\alpha}(E)$ by monotonicity, and we have $\sup_n \hat{\alpha}(E_n) \le \hat{\alpha}(E)$. In the other direction: fix $\epsilon > 0$. For all n, there exists a cover $(E_{n,k})$ of E_n such that

$$\sup_k \alpha(E_{n,k}) \le \hat{\alpha}(E_n) + \epsilon\,.$$

The family $(E_{n,k})$ covers E, and

$$\hat{\alpha}(E) \le \sup_{n,k} \alpha(E_{n,k}) = \sup_n(\sup_k \alpha(E_{n,k})) \le \sup_n \hat{\alpha}(E_n) + \epsilon\,.$$

We make ϵ tend to 0.
   3. and **4.** are immediate.   ◀

Example 1 The function

$$\alpha(E) = \begin{cases} 0 & \text{if } E \text{ is a finite set;} \\ 1 & \text{otherwise} \end{cases}$$

is stable, but not σ-stable; it can be σ-stabilized as follows:

$$\hat{\alpha}(E) = \begin{cases} 0 & \text{if } E \text{ is a finite or countable set;} \\ 1 & \text{if it is uncountable.} \end{cases}$$

Example 2 The functions $\Delta(E)$ and $\delta(E)$ both give two σ–stable indices $\hat{\Delta}(E)$ and $\hat{\delta}(E)$, which satisfy the following inequalities:

$$\hat{\delta}(E) \le \hat{\Delta}(E) \le \Delta(E),$$

$$\hat{\delta}(E) \le \delta(E) \le \Delta(E) .$$

We remark that $\hat{\delta}$ is σ–stable, even if δ is not stable (Chap. 2, §9 and Chap. 10, §4). The index $\hat{\delta}$ has many properties in common with the Hausdorff dimension, however they are not identical.

Localization of α When $\alpha(E)$ is a monotonous set function, we can always define a local α at each point of E. We put

$$\alpha(E,x) = \lim_{\epsilon \to 0} E \cap B_\epsilon(x) .$$

This index α is said to be *uniform* on the compact set E if $\alpha(E,x)$ remains constant at every point of E. We denote this value by $\alpha_{\mathrm{loc}}(E)$. It is not always equal to $\alpha(E)$. For example, in the example of Figure 2.2, $\delta(E) = 1/2$, while $\delta_{\mathrm{loc}}(E) = \hat{\delta}(E) = 1/3$. In general, α_{loc} and $\hat{\alpha}$ are related by the following theorem:

If E is a closed set for which α is uniform,

$$\alpha_{\mathrm{loc}}(E) = \hat{\alpha}(E) .$$

The index $\hat{\alpha}$ can be interpreted as a "regularization" of the index α.

18.6 Bibliographical notes

We find the Baire theorem on compact sets in all the good textbooks; this is one of the most interesting results in topology.

It is difficult to find a first clear reference concerning the notion of local dimension. It often appears in an implicit manner. We find one in [C. Tricot], in addition to the search for the point of maximal dimension. The first approach to the packing dimension, and generally to the regularized index $\hat{\alpha}$, can be found in [C. Tricot 1] (1979). This regularization allows the correction of what look like "anomalies" in the Bouligand dimension. At the same time, we are led into the family of indices defined in a nonconstructive manner, and therefore, of no practical utility. The Hausdorff dimension belongs to this family. The Hausdorff dimensions, the packing dimension, and the $\hat{\delta}$-dimension give rise to interesting comparisons [C. Tricot 3].

A. Upper Limit and Lower Limit

A.1 Convergence

There are two ways to look at convergence: The convergence of a function $f(x)$ of a real *continuous* variable x in the neighborhood of a point or at infinity; or the convergence of a sequence $a(n)$ when the *discrete* variable n tends to infinity. If a sequence is considered to be a function of integer variables, then the second approach can be seen as a particular case of the first one. However, the limit of sequences has undoubtedly a pedagogical advantage: the theorems look more intuitive, and sequences are more like experimental data, which are usually related to discrete variables. For this reason, we discuss both points of view.

We start by recalling some results about sequences.

- A sequence $a(n)$ of real numbers is said to be **bounded** if there exists a real number K such that for all n,

$$|a(n)| \leq K .$$

- The sequence $a(n)$ converges to a limit a if the values of $a(n)$ get as near to a as we wish when n is sufficiently large. In other words, for all $\epsilon > 0$, there exists an integer $N(\epsilon)$ such that

$$n \geq N(\epsilon) \implies |a(n) - a| \leq \epsilon .$$

We write that:

$$a(n) \longrightarrow a ,$$

and $a = \lim_{n \to +\infty} a(n)$.

- By extension, $a(n)$ tends to infinity if, when n is sufficiently large, $a(n)$ is positive and non-null, and $1/a(n)$ converges to 0. This is the same as saying for all real numbers K, there exists an integer $N(K)$ such that

$$n \geq N(K) \implies a(n) \geq K .$$

We write:

$$a(n) \longrightarrow +\infty .$$

- The sequence is said to be *increasing* if $a(n) \leq a(n+1)$ for all n. If such a sequence is bounded, then it is necessarily convergent. The sequence is *strictly increasing* if $a(n) < a(n+1)$ for all n.

• Symmetrically, the sequence is *decreasing* if $a(n) \geq a(n+1)$ for all n. If such a sequence is bounded, then it is convergent.

• A *subsequence* of $a(n)$ consists of values $a(k_n)$, where k_n is a strictly increasing sequence of integers. Every bounded sequence contains a convergent subsequence.

Example The sequence $a(n) = (-1)^n$ does not converge, yet it is bounded (we can take $K = 1$), and the subsequence $a(2n)$ converges to the limit 1. Here, $k_n = 2n$.

Now, we consider a function $f(x)$, defined on the set of real numbers.

• $f(x)$ converges to a limit y_0 in the neighborhood of x_0 if for any $\epsilon > 0$, there exists a real value $\eta(\epsilon)$ such that

$$x \neq x_0 , \ |x - x_0| \leq \eta(\epsilon) \Longrightarrow |f(x) - y_0| \leq \epsilon .$$

We write:

$$y_0 = \lim_{\substack{x \to x_0 \\ x \neq x_0}} f(x) .$$

• $f(x)$ is continuous at x_0 if $f(x)$ converges to the limit $f(x_0)$ when x tends to x_0.

• By extension, $f(x)$ tends to $+\infty$ in the neighborhood of x_0 if for all real K there exists a real $\eta(K)$ such that

$$x \neq x_0 , \ |x - x_0| \leq \eta(K) \Longrightarrow f(x) \geq K .$$

Sometimes we write:

$$\lim_{\substack{x \to x_0 \\ x \neq x_0}} f(x) = +\infty .$$

• Other conventions of the same type are used when the variable x tends to infinity: the behavior of $f(x)$ is characterized in the same manner as that of a sequence (a_n). For example, we write:

$$\lim_{x \to +\infty} f(x) = y_0$$

when, for all $\epsilon > 0$, there exists a real A such that

$$x > A \Longrightarrow |f(x) - y_0| \leq \epsilon .$$

• Here is an important result linking sequences and continuous functions: if $a(n)$ converges to a finite limit a and if $f(x)$ is continuous at a, then

$$f(a) = \lim_{n \to \infty} f(a(n)) .$$

A.2 Nonconvergent sequences

1. The least upper bound of a set Let E be a bounded subset of the real line. An **upper bound** of E is a real number larger than or equal to every element of E. The **least upper bound** of E is the smallest among its upper bounds. This is also the right endpoint of the smallest interval containing E. If E is closed, then it contains its least upper bound, which will be denoted by $\sup E$.

2. Upper limit of a sequence Let $a(n)$ be a bounded sequence, and for every n, let $E(n)$ be the set of all the $a(k)$ such that $k \geq n$ ($E(n)$ is sometimes called the "tail" of the sequence starting at rank n). The sequence $(E(n))$ is a decreasing sequence of sets in the sense that

$$E(n+1) \subset E(n) \, .$$

The sequence $\sup E(n)$ of least upper bounds is therefore a decreasing sequence and so it converges to a limit that is the "upper limit" of $a(n)$, denoted:

$$\limsup_{n \to \infty} a(n) = \lim_{n \to \infty} (\sup E(n)) \, .$$

An equivalent definition can be given by using the limit values of a sequence. In fact, we can extract a convergent subsequence of $a(n)$, and the limit of this subsequence is called a **limit value** of $a(n)$. It may happen that all the convergent subsequences of $a(n)$ have the same limit: in this case $a(n)$ is convergent. If not there are at least two limit values (sometimes infinitely many). In any case the set Ω of these limit values is closed.

▶   Consider a sequence $x(k)$ of limit points of the sequence $a(n)$. For each k, $x(k)$ is a limit of a subsequence $a(n,k)$ of $a(n)$. Let x^* be the limit of $x(k)$. For all $\epsilon > 0$, we can find an integer k and an integer n_k such that $|x(k) - x^*| \leq \epsilon/2$ and $|x(k) - a(k, n_k)| \leq \epsilon/2$. Since $|a(k, n_k) - x^*| \leq \epsilon$, x^* is a limit of a subsequence of $a(n)$. Thus, every limit of a convergent sequence of Ω belongs to Ω. That is, Ω is closed. ◀

Therefore the set Ω contains its least upper bound, which is exactly the lim sup of $a(n)$:

$$\limsup a(n) = \sup \Omega \, .$$

Finally, $\limsup a(n)$ can be axiomatically characterized as follows:

The upper limit of the sequence $a(n)$ is the value α such that for all $\epsilon > 0$ there exists an integer $N(\epsilon)$ for which the following two propositions hold:
(i) for all $n \geq N(\epsilon)$, $a(n) \leq \alpha + \epsilon$;

(ii) there exists $n_0 \geq N(\epsilon)$ such that $a(n_0) \geq \alpha - \epsilon$.

Examples

If $a(n) = (-1)^n$, $\limsup a(n) = 1$.

If $a(n) = n\pi - $ (integer part of $n\pi$), then the sequence $a(n)$ "fills" the interval $[0, 1]$. The sequence is dense in the interval. Thus, every point of $[0, 1]$ is a limit value of the sequence: $\limsup a(n) = 1$.

If $a(n)$ converges, $\limsup a(n) = \lim a(n)$.

3. Lower limit of a sequence This is a the notion symmetrical to the previous one:

$$\liminf_{n \to \infty} a(n) = \lim_{n \to \infty} (\inf E(n)) .$$

The lower limit is also the smallest limit value of $a(n)$.

◇   We observe that if $f(x)$ is continuous and increasing,

$$f(\limsup a(n)) = \limsup f(a(n))$$

and

$$f(\liminf a(n)) = \liminf f(a(n)) ,$$

while if $f(x)$ is decreasing,

$$f(\limsup a(n)) = \liminf f(a(n))$$

and

$$f(\liminf a(n)) = \limsup f(a(n)) .$$

For example, $\liminf a(n) = -\limsup(-a(n))$; and if $a(n) \neq 0$ for all n,

$$\liminf a(n) = \frac{1}{\limsup \dfrac{1}{a(n)}} .$$

A.3 Nonconvergent functions

We extend the notions of Section 2 to a real function of a real variable. Thus

$$\limsup_{x \to x_0} f(x) = \lim_{\epsilon \to 0} (\sup f([x_0 - \epsilon, x_0 + \epsilon])) .$$

This definition is possible because the images $f([x_0 - \epsilon, x_0 + \epsilon])$ are embedded sets, and therefore $\sup f([x_0 - \epsilon, x_0 + \epsilon])$ is a decreasing function of ϵ with a limit when ϵ tends to 0.

But we can also define the upper limit of a function at x_0 by considering the **limit values** of $f(x)$. A limit value is a real number α satisfying

$$\alpha = \lim_{n \to \infty} f(x_n) ,$$

where x_n is a sequence that converges to x_0. The set of all limit values is a closed set that is reduced to a unique value $f(x_0)$ when f is continuous at x_0. Otherwise, the set of limit values contains more than one point, and the lim sup of f is the largest of these values.

We can axiomatically characterize the lim sup as follows:

The upper limit of $f(x)$ at x_0 is the value $\alpha = \limsup f(x)$ such that, for all $\epsilon > 0$, there exists $\eta(\epsilon)$ for which the following two propositions hold:
(i) $x \neq x_0$, $|x - x_0| \leq \eta(\epsilon) \Longrightarrow f(x) \leq \alpha + \epsilon$;
(ii) there exists an x_ϵ such that $|x_\epsilon - x_0| \leq \eta(\epsilon)$ and $f(x_\epsilon) \geq \alpha - \epsilon$.

Examples

Let
$$f(x) = \begin{cases} 1 & \text{if } x \text{ is rational;} \\ 0 & \text{if } x \text{ is irrational.} \end{cases}$$

At each point x_0, we have $\limsup_{x \to x_0} f(x) = 1$.

Let $f(x) = \sin(1/x)$. The set of limit values of this function at $x_0 = 0$ is the interval $[-1, 1]$, and $\limsup_{x \to 0} f(x) = 1$.

If $f(x)$ is continuous at x_0, $\limsup_{x \to x_0} f(x) = f(x_0)$.

$\diamond$ We can symmetrically define the lower limit (lim inf) of a function when x tends to infinity.

A.4 Limits of the ratio $\log f(\epsilon)/\log g(\epsilon)$

The logarithmic ratios are often used to determine the dimension (see Chap. 2, §5). The technical results of this section allow an easier manipulation of the formulas defining $\Delta(E)$ and $\delta(E)$.

Once and for all, we assume that the two functions $f(\epsilon)$ and $g(\epsilon)$ are defined in a neighborhood of 0 and that they are both strictly positive. Moreover, we assume that
$$\lim_{\epsilon \to 0} f(\epsilon) = \lim_{\epsilon \to 0} g(\epsilon) = 0 .$$

- If there exist two constants $0 < c_1 \leq c_2$ such that

$$c_1 \leq \frac{f(\epsilon)}{g(\epsilon)} \leq c_2 ,$$

then
$$\lim_{\epsilon \to 0} \frac{\log f(\epsilon)}{\log g(\epsilon)} = 1 .$$

▶   Because then

$$\log c_1 + \log g(\epsilon) \leq \log f(\epsilon) \leq \log c_2 + \log g(\epsilon) .$$

Without loss of generality, we can assume that $|g(\epsilon)| < 1$. Then

$$1 + \frac{\log c_2}{\log g(\epsilon)} \leq \frac{\log f(\epsilon)}{\log g(\epsilon)} \leq 1 + \frac{\log c_1}{\log g(\epsilon)} \ .$$

The left and right terms tend to 1 when ϵ tends to 0.   ◀

◇   The converse is not always true. For example, the functions $f(x) = x^\alpha |\log x|$ and $g(x) = x^\alpha$ do not satisfy the hypothesis, yet the ratio $\log f(\epsilon) / \log g(\epsilon)$ also tends to 1.

• Here are two representations of the same number:

$$\limsup_{\epsilon \to 0} \frac{\log f(\epsilon)}{\log g(\epsilon)} = \inf \{\, \alpha \text{ such that } g(\epsilon)^\alpha / f(\epsilon) \to 0 \,\}$$
$$= \inf \{\, \alpha \text{ such that } g(\epsilon)^\alpha / f(\epsilon)$$
$$\text{is bounded in the neighborhood of } 0 \,\} \ .$$

▶   Let α_1, α_2, α_3 be these numbers in the given order.
a) First, we remark that

$$\alpha > \alpha_2 \implies g(\epsilon)^\alpha / f(\epsilon) \to 0 \ .$$

Indeed, by definition of the greatest lower bound, there exists a real number β such that $\alpha > \beta > \alpha_2$, and $g(\epsilon)^\beta / f(\epsilon) \to 0$. But $g(\epsilon)^\alpha / f(\epsilon) = g(\epsilon)^{\alpha-\beta} g(\epsilon)^\beta / f(\epsilon)$ is the product of two functions tending to 0.
b) Similarly,

$$\alpha > \alpha_3 \implies g(\epsilon)^\alpha / f(\epsilon) \text{ is bounded in the neighborhood of } 0 \ .$$

Indeed, there exists a real β, $\alpha > \beta > \alpha_3$, such that for some K, $g(\epsilon)^\beta / f(\epsilon) \leq K$. But $g(\epsilon)^\alpha / f(\epsilon)$ is less than $K g(\epsilon)^{\alpha-\beta}$, which tends to 0, and therefore it is bounded.
c) We show that

$$\alpha_1 \leq \alpha_2 \ .$$

Let $\alpha > \alpha_2$. It suffices to show that $\alpha \geq \alpha_1$. Since according to a), $g(\epsilon)^\alpha / f(\epsilon)$ tends to 0, there exists an ϵ_0 such that for all $\epsilon \leq \epsilon_0$, $g(\epsilon)^\alpha / f(\epsilon) < 1$ and $g(\epsilon) < 1$. We deduce that $\log f(\epsilon) / \log g(\epsilon) < \alpha$. Therefore $\alpha_1 \leq \alpha$.
d) We show that

$$\alpha_2 \leq \alpha_3 \ .$$

Let $\alpha > \alpha_3$. It suffices to show that $\alpha \geq \alpha_2$. For all pairs α, β such that $\alpha_3 < \beta < \alpha$, there exists a K such that $g(\epsilon)^\beta / f(\epsilon) \leq K$. Therefore, $g(\epsilon)^\alpha / f(\epsilon) \leq K g(\epsilon)^{\alpha-\beta}$ which tends to 0. Therefore $\alpha \geq \alpha_2$.
e) We show that

$$\alpha_3 \le \alpha_1 \, .$$

Let $\alpha > \alpha_1$. It suffices to show that $\alpha \ge \alpha_3$. There exists an ϵ_0 such that for all $\epsilon \le \epsilon_0$, $\log f(\epsilon)/\log g(\epsilon) < \alpha$ and $g(\epsilon) < 1$. This implies the inequality $g(\epsilon)^\alpha/f(\epsilon) < 1$. Therefore $\alpha \ge \alpha_3$.   ◀

• The same formulas remain true if we replace $f(\epsilon)$ and $g(\epsilon)$ with two positive sequences $a(n)$ and $b(n)$ that tend to 0.

• Finally we obtain the symmetrical formulas for the lower limit

$$\liminf_{\epsilon \to 0} \frac{\log f(\epsilon)}{\log g(\epsilon)} = \sup\{\, \alpha \text{ such that } g(\epsilon)^\alpha/f(\epsilon) \to +\infty \,\}$$

$$= \sup\{\, \alpha \text{ for which there exists a } h > 0 \text{ such that}$$
$$g(\epsilon)^\alpha/f(\epsilon) > h \text{ in the neighborhood of } 0 \,\} \, .$$

A.5 Some applications

1. The dimension on the real line (Chap. 2) can be written as:

$$\Delta(E) = \limsup(1 - \frac{\log L(E(\epsilon))}{\log \epsilon}) = \frac{\log(\epsilon/L(E(\epsilon)))}{\log \epsilon} \, .$$

With $f(\epsilon) = \epsilon/L(E(\epsilon))$ and $g(\epsilon) = \epsilon$, we obtain

$$\Delta(E) = \inf\{\, \alpha \text{ such that } \epsilon^{\alpha-1}L(E(\epsilon)) \to 0 \,\} \, .$$

Similarly,

$$\delta(E) = \sup\{\, \alpha \text{ such that } \epsilon^{\alpha-1}L(E(\epsilon)) \to +\infty \,\} \, .$$

2. The dimension in the plane (Chap. 10) can be written as:

$$\Delta(E) = \limsup(2 - \frac{\log \mathcal{A}(E(\epsilon))}{\log \epsilon}) \, .$$

With $f(\epsilon) = \epsilon^2/L(E(\epsilon))$ and $g(\epsilon) = \epsilon$, we obtain

$$\Delta(E) = \inf\{\, \alpha \text{ such that } \epsilon^{\alpha-2}\mathcal{A}(E(\epsilon)) \to 0 \,\} \, .$$

3. The index e_B is defined (Chap. 2, §3) by

$$e_B = \limsup \frac{\log n}{|\log \frac{1}{n} \sum_{i=n}^{\infty} c_i|} = \limsup \frac{\log(1/n)}{\log \frac{1}{n} \sum_{i=n}^{\infty} c_i} \, .$$

This gives, with $a(n) = 1/n$ and $b(n) = \sum_{i=n}^{\infty} c_i/n$:

$$e_B = \inf \{ \alpha \text{ such that } n^{1-\alpha} (\sum_{i=n}^{n} c_i)^\alpha \to 0 \} .$$

Similarly, we find:

$$e = \limsup \frac{\log n}{|\log c_n|} = \inf \{ \alpha \text{ such that } n \, c_n^\alpha \to 0 \}$$

and

$$e_{BM} = \limsup(1 - \frac{\log \sum_{i=n}^{\infty} c_i}{\log c_n}) = \inf \{ \alpha \text{ such that } c_n^{\alpha-1} \sum_{i=n}^{\infty} c_i \to 0 \} .$$

B. Two Covering Lemmas

B.1 Vitali's lemma

Vitali's lemma is, without any doubt, a milestone of the geometrical theory of measure. It has many variants in a more or less general setting. Here, we discuss it from a particular point of view, which is related to its use in the present book (Chap. 2, §1 and Chap. 7, §4). We consider covers of bounded subsets of the real line by intervals (or similarly, the covers of a subset of a curve by arcs). The measure used is the length. We start by defining these covers.

Let E be a subset of the real line. Let $\mathcal{F}$ be a family of closed intervals such that for all $\epsilon > 0$ we can extract a cover of E with intervals of length less than ϵ. Then $\mathcal{F}$ is a **Vitali cover** *of E.*

This is a very rich family, because for all x in E, we can find a sequence $(u_n(x))$ of intervals of $\mathcal{F}$ containing x with lengths tending to 0.

Example 1 The simplest example could be the family of *dyadic* intervals of type $[k/2^n, (k+1)/2^n]$ for all integers n, k. This is a Vitali cover of the real line and, therefore, of each subset of it.

Example 2 The family of all closed intervals containing a point of E is a Vitali cover of E and the family of all the intervals $[x - \epsilon, x + \epsilon]$, for all $x \in E$ and $0 < \epsilon \leq 1$, is a Vitali cover of E.

The Vitali lemma shows that we can extract a cover on almost all of E by disjoint intervals from such a family.

LEMMA *Let E be a bounded subset of the real line and $\mathcal{F}$ be a Vitali cover of E. We can extract from $\mathcal{F}$ a subfamily of disjoint intervals $J_1, J_2, \ldots, J_n, \ldots$, such that*

$$L(E - \cup_n J_n) = 0 .$$

As usual, L here designates the *length*, in other words, the Borel measure on the real line.

▶ a) We construct the family of intervals J_n. Without loss of generality, we may assume that the length of all the intervals of $\mathcal{F}$ is less than a given number. That is,

$$s_1 = \sup_{u \in \mathcal{F}} L(u) ,$$

the least upper bound of these lengths. We can always find an interval J_1 of $\mathcal{F}$ such that

$$L(J_1) > \frac{s_1}{2} .$$

Let $\mathcal{F}_1 = \mathcal{F}$ and let $\mathcal{F}_2$ be the family of all the intervals of $\mathcal{F}_1$ that are disjoint from J_1. We denote

$$s_2 = \sup_{u \in \mathcal{F}_2} L(u) ,$$

and find an interval J_2 of $\mathcal{F}_2$ such that

$$L(J_2) > \frac{s_2}{2} .$$

The proof proceeds by induction. Assume that the disjoint intervals $J_1, J_2, \ldots, J_{n-1}$ have been constructed. Let $\mathcal{F}_n$ be the family of all the intervals of $\mathcal{F}$ that do not intersect any of the previous $(n-1)$ intervals. We denote

$$s_n = \sup_{u \in \mathcal{F}_n} L(u) ,$$

and we find an interval J_n in $\mathcal{F}_n$ such that

$$L(J_n) > \frac{s_n}{2} .$$

If there is no such interval, the procedure ends here. In this case, the distance between any point of E and the union of $J_1, \ldots, J_{n-1}$ is 0. But this is a union of a finite number of closed intervals, so it is closed. Therefore,

$$E \subset \bigcup_{i=1}^{n-1} J_i ,$$

and the lemma is proved (the set $E - \cup J_i$ is empty). This is where we make use of the fact that the elements of $\mathcal{F}$ are closed intervals.

If the procedure does not end at any value of n, then we obtain an infinite family of (J_n).

b) All that is left to be proved in this case is that

$$L(E - \cup_n J_n) = 0 .$$

For this, we use the intervals J_n^* that have the same center as J_n but with length $L(J_n^*) = 5\,L(J_n)$. We fix an integer n and consider the interval J of $\mathcal{F}_n$. There exists an integer $k \geq n$ such that

$$s_{k+1} < L(J) \leq s_k .$$

Since J does not belong to $\mathcal{F}_{k+1}$, J necessarily intersects one of the intervals $J_1, \ldots, J_k$. Since J belongs to $\mathcal{F}_n$, it is disjoint from $J_1, \ldots, J_{n-1}$. Therefore J intersects one of the intervals $J_n, \ldots, J_k$. Let J_p be this interval. We have

$$L(J) \leq s_k \leq s_p \leq 2L(J_p) ,$$

and therefore
$$J \subset J_p^* \, .$$

Every point of E that is not in $\bigcup_{i=1}^{n-1} J_i$ must be in an interval of $\mathcal{F}_n$. We deduce that it belongs to the union of the intervals $J_n^*, J_{n+1}^*, \dots$. In other words,

$$E - \bigcup_{i=1}^{n-1} J_i \subset \bigcup_{p=n}^{\infty} J_p^* \, .$$

But this implies that:

$$L(E - \bigcup_{i=1}^{\infty} J_i) \le L(E - \bigcup_{i=1}^{n-1} J_i)$$
$$\le L(\bigcup_{p=n}^{\infty} J_p^*)$$
$$\le \sum_{p=n}^{\infty} L(J_p^*)$$
$$\le 5 \sum_{p=n}^{\infty} L(J_p) \, .$$

Since E is bounded and since the J_n are disjoint, the series $\sum L(J_n)$ converges. Therefore the last of the previous inequalities tends to 0 when n tends to infinity. We deduce that the first inequality, which is independent from n, must be null. This proves the lemma. ◀

We usually use the following partial result rather than the whole lemma:

COROLLARY *Let E be a bounded subset of the real line and let $\mathcal{F}$ be a Vitali cover of E. For all $\epsilon > 0$, we can extract from $\mathcal{F}$ a finite family of disjoint intervals $J_1, J_2, \dots, J_n$ such that*

$$L(E) \le \sum_{i=1}^{n} L(J_i) + \epsilon \, ,$$

that is, the part of E not covered by the J_i has a maximum length of at most ϵ. This corollary follows directly from the above proof.

▶   Because if n is such that $L(\cup_{p=n+1}^{\infty} J_p^*) \le \epsilon$, then

$$L(E) - \sum_{1}^{n} L(J_i) \le L(E - \cup_1^n J_i) \le \epsilon \, .$$ ◀

◇   Using the notion of parameterization by lengths of arcs, the above results can be directly transferred to curves of finite length (Chap. 7): the arcs will replace the intervals in the definition of a Vitali cover, and the above result remains true.

◇   The Vitali lemma can equally be enunciated in the case of the plane, where intervals are replaced by geometrical figures like balls or squares of non-null areas. The length is replaced by the area. In particular, it is not possible to cover a square with disjoint disks that are included in it. But the Vitali lemma asserts that we can pack the square with disks (which are included in it) so that the uncovered part of the square is of null area.

A similar type of generalization can be done to n–dimensional spaces. The proof of the Vitali lemma in these cases is similar to the above.

B.2 Covers by homothetic convex sets

We cover a set E with convex domains of the same size, and we enlarge these domains by a homothety. How can we evaluate the area of this new cover? Here is a result in this direction.

THEOREM   *In the (Ox_1, Ox_2) plane we consider a nondegenerate convex set K (a set that is not reduced to a segment) containing the origin. We denote by*

$$K_\epsilon(x) = \epsilon K + x$$

the image of K by a homothety of center O and ratio ϵ followed by a translation of vector $\overrightarrow{Ox}$. Finally, let E be a bounded set of the plane. For all values $0 < \epsilon \leq \eta$,

$$\mathcal{A}(\bigcup_{x \in E} K_\eta(x)) \leq \frac{\eta^2}{\epsilon^2} \mathcal{A}(\bigcup_{x \in E} K_\epsilon(x)) \ .$$

◇   Here, we note that the translation of vector $\overrightarrow{xy}$ transforms $K_\epsilon(x)$ to $K_\epsilon(y)$, while the homothety of center x and ratio η/ϵ transforms $K_\epsilon(x)$ to $K_\eta(x)$. Consequently,

$$\mathcal{A}(K_\eta(x)) = \frac{\eta^2}{\epsilon^2} \mathcal{A}(K_\epsilon(x)) \ .$$

The above inequality becomes an equality whenever the $K_\eta(x)$ are disjoint, that is, whenever E is finite and the distance between its points is $\geq 2\eta$.

▶   The proof is divided into three parts.

a) Let x, y be two points of E, and let h be the homothety with center x and ratio η/ϵ. We shall prove that if the convex sets $K_\epsilon(x)$ and $K_\epsilon(y)$ are touching, then:

$$h(K_\epsilon(x) \cap K_\epsilon(y)) \subset K_\eta(x) \cap K_\eta(y) \ .$$

Let z be a point of $K_\epsilon(x) \cap K_\epsilon(y)$, $w = h(z)$, g be the homothety with center y and ratio η/ϵ, z' be such that $g(z') = w$, and finally, let z'' be the point such that $\overrightarrow{xz''} = \overrightarrow{xy} + \overrightarrow{xz}$. The point w belongs to $K_\eta(x)$, and the point z'' belongs to $K_\epsilon(y)$. Moreover, since $\overrightarrow{zz'} = a\,\overrightarrow{xy}$, where $a = 1 - \epsilon/\eta < 1$ (Fig. B1), z' belongs to the segment whose endpoints are z and z''. By convexity, we deduce that $z' \in K_\epsilon(y)$.

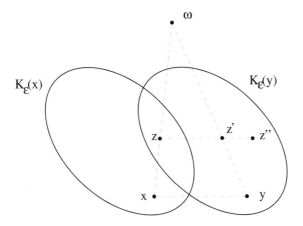

Fig. B.1. *If the point z belongs to $K_\epsilon(x) \cap K_\epsilon(y)$, then the point z' will belong to $K_\epsilon(y)$.*

Since $g(z') = \omega$, the point ω belongs to $K_\eta(y)$. Therefore, $\omega \in K_\eta(x) \cap K_\eta(y)$. This proves the desired inclusion.

  b) Assume that E is countable, and enumerate its points to form a sequence $(x_n)_{n\geq 0}$. We create new covers as follows:

$$C_0 = K_\epsilon(x_0), \ldots, C_n = K_\epsilon(x_n) - \cup_{i=0}^{n-1} K_\epsilon(x_i) ,$$
$$C_0' = K_\eta(x_0), \ldots, C_n' = K_\eta(x_n) - \cup_{i=0}^{n-1} K_\eta(x_i) .$$

The sets (C_n) are no longer convex, but they are disjoint; similarly for the (C_n'). Moreover, if h_n denotes the homothety with center x_n and ratio η/ϵ, then

$$C_n' \subset h_n(C_n) .$$

In fact, we deduce from a) that for all $i < n$,

$$h_n(K_\epsilon(x_n) \cap K_\epsilon(x_i)) \subset K_\eta(x_n) \cap K_\eta(x_i) ,$$

and therefore:

$$K_\eta(x_n) \cap h_n(K_\epsilon(x_i)) \subset K_\eta(x_n) \cap K_\eta(x_i) .$$

Using the fact that h_n is a bijection,

$$K_\eta(x_n) \cap h_n(\cup_{i=0}^{n-1} K_\epsilon(x_i)) = \cup_{i=0}^{n-1}(K_\eta(x_n) \cap h_n(K_\epsilon(x_i)))$$
$$\subset \cup_{i=0}^{n-1}(K_\eta(x_n) \cap K_\eta(x_i)) = K_\eta(x_n) \cap (\cup_{i=0}^{n-1} K_\eta(x_i)) .$$

Therefore,

$$K_\eta(x_n) - \cup_{i=0}^{n-1} K_\eta(x_i)) \subset K_\eta(x_n) - h_n(\cup_{i=0}^{n-1} K_\epsilon(x_i))$$
$$= h_n(K_\epsilon(x_n) - \cup_{i=0}^{n-1} K_\epsilon(x_i)) .$$

This proves the inclusion $C_n' \subset h_n(C_n)$. We immediately deduce that

$$A(C'_n) \leq \frac{\eta^2}{\epsilon^2} A(C_n) \, .$$

Taking the sum over n of the two members of this inequality, we find

$$A(\cup K_\eta(x_n)) \leq \frac{\eta^2}{\epsilon^2} A(\cup K_\epsilon(x_n)) \, .$$

c) Assume that E is any bounded set. We can always find a countable *dense* subset F of E such that $F \subset E \subset \overline{F}$. Let $a > 1$, then

$$\cup_{x \in E} K_\eta(x) \subset \cup_{x \in F} K_{a\eta}(x) \, .$$

We deduce from b) that

$$A(\cup_{x \in F} K_{a\eta}(x)) \leq \frac{a^2 \, \eta^2}{\epsilon^2} A(\cup_{x \in F} K_\epsilon(x))$$

and therefore:

$$A(\cup_{x \in E} K_\eta(x)) \leq \frac{a^2 \, \eta^2}{\epsilon^2} A(\cup_{x \in E} K_\epsilon(x)) \, .$$

By making a go to 1, we get the desired result. ◀

Application to the norms Assume that the convex set K of the above theorem has the origin O as its center of symmetry. Then it is associated, in a unique manner, to a real positive function $N(x)$, called **norm**, defined by

$$N(x) = \inf\{\, a > 0 \text{ such that } x \in K_a(0) \,\} \, .$$

This function satisfies the following three properties:

(i) $N(x) = 0 \Leftrightarrow x = 0$;

(ii) $N(\lambda x) = |\lambda| \, N(x)$ for all real λ ;

(iii) $N(x + y) \leq N(x) + N(y)$.

It induces a *distance* (Chap. 4, §2) on the plane by the relation

$$\mathrm{dist}(x, y) = N(x - y) \, .$$

The convex set K is nothing but its *unit ball*

$$K = K_1(O) = \{\, x \text{ such that } N(x) \leq 1 \,\} \, ,$$

and more generally,

$$K_\epsilon(x) = \{\, y \text{ such that } N(x - y) \leq \epsilon \,\} \, .$$

The *Euclidian norm* is one of many examples. It is defined by $N(x) = \sqrt{x_1^2 + x_2^2}$, and it is associated with the convex set K, which is the disk centered at O with radius 1 (Fig. B2). In turn, these norms define new *Hausdorff distances* between compact sets that correspond to *Minkowski sausages* of different forms. We have already encountered (Chap. 10) the sausages formed by centered balls $B_\epsilon(x)$

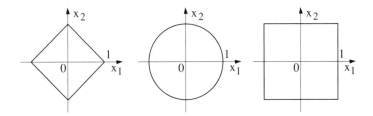

Fig. B.2. *Three cases of norms in the plane, determined by their unit balls* $K_1(O)$: *Manhattan norm* $N_1(x) = |x_1| + |x_2|$; *Euclidian norm* $N_2(x) = \sqrt{x_1^2 + x_2^2}$; *norm of the maximum* $N_\infty(x) = \max\{\,|x_1|, |x_2|\,\}$.

(Euclidian norm) and those formed by centered squares $C_\epsilon(x)$ (norm of the maximum). The theorem of this section allows us to establish a relation between these two sausages $E(\epsilon) = \cup_{x \in E} K_\epsilon(x)$ and $E(\eta) = \cup_{x \in E} K_\eta(x)$ (used in Chap. 10, §6). But let us note that there exist other types of applications; in the theorem, the origin O is not necessarily the center of symmetry of K.

Application to the length of the boundary of a sausage Let E be a set and ϵ be a positive real number. We are interested in the Minkowski sausage $E(\epsilon)$ (constructed by centered balls; but we can adopt a more general approach as suggested by the above theorem), more precisely, in its boundary F:

$$F = \partial(E(\epsilon)) = \{\, x \text{ such that dist}(x, E) = \epsilon \,\} \;.$$

This boundary is formed by one or more closed curves. The set F is of *finite length*: this fact, which looks evident, can be deduced from the geometry of F, which has at each point, and at least from one side of F, a curvature of radius $\geq \epsilon$. But the theorem allows us to give a simpler proof; in fact, we can show that

$$\limsup_{u \to 0} \frac{\mathcal{A}(F(u))}{2\,u} \leq \frac{2}{\epsilon} \mathcal{A}(E(\epsilon)) \;.$$

We immediately deduce (Chap. 9) that the length of F is bounded by the right term. Therefore, this length is finite.

▶   We verify the above formula by using the inclusion

$$F(u) \subset E(\epsilon + u) - E(\epsilon - v) \;,$$

for all $0 < u < v < \epsilon$. We deduce, while making v tend to u, that

$$\mathcal{A}(F(u)) \leq \mathcal{A}(E(\epsilon + u)) - \mathcal{A}(E(\epsilon - u))$$

$$\leq \left((\frac{\epsilon + u}{\epsilon - u})^2 - 1 \right) \mathcal{A}(E(\epsilon - u)) \;.$$

Therefore

$$\frac{\mathcal{A}(F(u))}{2\,u} \leq \frac{2\,\epsilon}{(\epsilon - u)^2}\mathcal{A}(E(\epsilon))\ .$$

When u tends to 0, the right term tends to $2\,\mathcal{A}(E(\epsilon))/\epsilon$.   ◀

C. Convex Sets in the Plane

C.1 Convexity

A convex set K is determined by the following property:

If any two points A, B belong to K, then the segment AB is included in K.

In addition:

*If a straight line **D** intersects K, then the intersection is a segment (possibly reduced to a point).*

In particular, it follows that through each point outside of K there passes a straight line that does not meet K. In the limit case, through each point of the boundary ∂K of the convex set, there passes a straight line **D** such that K is entirely situated on one side of **D**. Such a line is called a *support line* of K. Thus, we have the following characteristic property of convex sets:

Through each point of the boundary ∂K, there passes a support line of K.

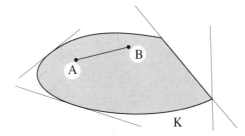

Fig. C.1. *The set K is convex if for all the points A and B of K, the segment AB is included in K. A convex set is always on one side of each of its support lines.*

Our interest in the geometry of convex sets stems from the fact that they are invariant under affine transformations. In fact, an affine transformation F, defined by a matrix and a translation vector (Chap. 12, §9), transforms a segment to a segment, therefore it transforms a convex set to a convex set.

Examples A disk is a convex set. Transformed by an affine transformation, it becomes a surface bounded by an ellipse, which is also a convex set. A square is a convex set. Its image by an affine transformation is a parallelogram. A triangle and its trianglular images by affine transformations are convex sets. A regular polygon is a convex set; its images are convex *polytops*. A segment is a convex set.

$\diamond$ If V is a convex open set, then its closure $\overline{V}$ is also a convex set: $\overline{V}$ is called a *convex body*. Thus, every convex closed set is a convex body; the segments are the only exception. All the convex sets K that will be considered are closed. They contain their boundaries ∂K. We assume that they are *bounded*.

$\diamond$ Another interesting property of the family of convex sets is the following:

The intersection of two convex sets is a convex set, provided that it is not empty.

▶   In fact, if the two points A and B belong to K_1 and K_2 at the same time, then the segment AB belongs to $K_1 \cap K_2$.   ◀

$\diamond$ Finally, if a sequence of convex sets (K_n), converges in the sense of Hausdorff distance (Chap. 5, §2), *its limit K is also a convex set.*

▶   This can be shown easily, by remarking that every segment AB whose endpoints belong to K is a limit of a sequence of segments $A_n B_n$ included in K_n.   ◀

C.2 Size of a convex set

What we call the **size** of a convex set K can be determined by different parameters (Chap. 11, §3), which are, in fact, equivalent (in the sense that the ratio of any two of these parameters is between two non-null constants).

• The **diameter** of K is defined as

$$\mathrm{diam}\,(K) = \max \{\, \mathrm{dist}(A, B) \; : \; A \in K \,, \, B \in K \,\} \,.$$

Since K is assumed to be closed, there exists a pair C and D in ∂K such that $\mathrm{dist}(C, D) = \mathrm{diam}\,(K)$. However, we note that the pair (C, D) is not unique (for example, in the case of a disk, any two diametrically opposed points of the boundary will do). Taking two lines perpendicular to CD that pass through the points C and D respectively, we obtain two support lines of K; otherwise $\mathrm{dist}(C, D)$ would not be the largest distance between the points of K. Thus, two support lines that are perpendicular pass through the endpoints of a diameter of the set K. Therefore, we deduce the following equivalent definition:

The diameter of a set K is the largest distance between any two parallel lines meeting K.

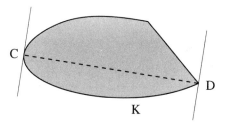

Fig. C.2. *If CD is a diameter of the convex set K, the lines perpendicular to this diameter at C and at D are support lines.*

● The **perimeter** of K is the length of the boundary ∂K. In the plane in which we have drawn a reference axis Ox, there is an integral formula giving the exact value of the perimeter of K, that is,

$$L(\partial K) = \int_0^\pi p(\theta)\, d\theta \ .$$

This was proved in a more general setting in Chapter 8. The function $p(\theta)$ designates the length of the projection of K on the line making an angle θ with Ox. The segment, a limit case, is considered to be a convex set whose perimeter is twice its length.

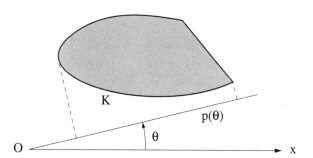

Fig. C.3. *The perimeter of K is proportional to the average over θ of the orthogonal projection $p(\theta)$ of K on a line of angle θ.*

The perimeter is a parameter equivalent to the diameter. We can show that

$$2\operatorname{diam}(K) \le L(\partial K) \le \pi \operatorname{diam}(K)\ .$$

▶ The first inequality follows from the fact that if CD is a diameter of K, the length of CD is smaller than the length of the two arcs $C\frown D$ of ∂K, situated on both sides of CD. The second can be verified by using the integral formula and the fact that for all θ, $p(\theta) \leq \operatorname{diam}(K)$. ◄

We note that these limits can be reached: $L(\partial K) = 2\operatorname{diam}(K)$ for a segment, and $L(\partial K) = \pi\operatorname{diam}(K)$ for a disk.

- The **diameter of the circumscribed circle** of K.

We denote this circle by $C(K)$. This diameter is also equivalent to the diameter of K. In fact,

$$\operatorname{diam}(K) \leq \operatorname{diam}(C(K)) \leq \frac{2}{\sqrt{3}}\operatorname{diam}(K) .$$

▶   The first inequality follows from the fact that $K \subset C(K)$. The second is associated with the properties of the triangle. It is shown first for convex sets whose boundaries are made of finite segments (polytops). Then it is shown to hold for any convex set using an approximation argument. ◄

The limits of these inequalities are reached: $\operatorname{diam}(C(K)) = \operatorname{diam}(K)$ for a disk, and $\operatorname{diam}(C(K)) = 2/\sqrt{3}\operatorname{diam}(K)$ for an equilateral triangle.

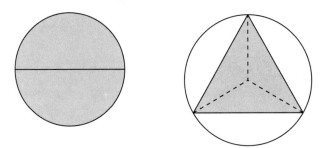

Fig. C.4. *We draw the circumscribed circle of K; its diameter equals $\operatorname{diam}(K)$ for a disk and $2\operatorname{diam}(K)/\sqrt{3}$ for an equilateral triangle.*

These parameters satisfy the following conditions, which make them good tools for analyzing sets:

If $F(K)$ denotes any of the functions $\operatorname{diam}(K)$, $L(\partial K)$, or $\operatorname{diam}(C(K))$, then $F(K)$ is increasing:

$$K_1 \subset K_2 \Longrightarrow F(K_1) \leq F(K_2)$$

and continuous:

$$\text{If } K = \lim K_n, \ F(K) = \lim F(K_n) .$$

▶   These two properties are easy to establish for diam (K) and diam $(C(K))$. Regarding $L(\partial K)$, no doubt the best method uses the integral property. If $L(\partial K_1) = \int_0^\pi p_1(\theta)\, d\theta$ and $L(\partial K_2) = \int_0^\pi p_2(\theta)\, d\theta$, then the inclusion $K_1 \subset K_2$ implies that $p_1(\theta) \leq p_2(\theta)$, and therefore $L(\partial K_1) \leq L(\partial K_2)$. And if (K_n) tends to K in the Hausdorff distance, then for all θ, $(p_n(\theta))$ tends to $p(\theta)$. We deduce that $\int_0^\pi p_n(\theta)\, d\theta$ tends to $\int_0^\pi p(\theta)\, d\theta$ by applying the Lebesgue-dominated convergence theorem.   ◀

C.3 Breadth of a convex set

Now we shall discuss the parameters that characterize the "thickness" of a convex set:

- The **breadth** of K, denoted by

$$\text{breadth}(K)\,,$$

is by definition the smallest distance between any two parallel lines enclosing K. We immediately deduce that

$$\text{breadth}(K) \leq \text{diam}\,(K)\,.$$

This notion generalizes the width of a rectangle. We can always find a pair of points (not necessarily unique) in the boundary ∂K through which two support lines of K pass and such that the distance between them is exactly the breadth of K. Thus, every convex set K can be inscribed in a parallelogram satisfying:

—*the distance between its longest sides is* breadth(K)*;*
—*the distance between its shortest sides is* diam (K)*.*

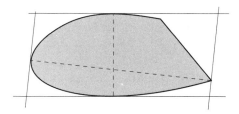

Fig. C.5. *The convex set K is inscribed in a parallelogram with the property that the distance between two of its sides is* diam (K) *and the distance between the other sides is* breadth(K)*.*

◇ In a rectangle, the breadth is equal to the length of the shortest side, while the diameter is the length of a diagonal. The angle between the shortest side and the diagonal is not a right angle. In general,

If we call ϕ, $0 \leq \phi \leq \pi/2$, the angle between the direction of the diameter and the breadth: we have

$$\cos \phi \leq \frac{\text{breadth}(K)}{\text{diam}\,(K)} .$$

▶   See Figure C6: we can find three points P_1, P_2, P_3 on the parallelogram that circumscribes K, forming a right–angled triangle with angle ϕ and with $\text{dist}(P_1, P_2) = \text{diam}\,(K)$ and $\text{dist}(P_2, P_3) \leq \text{breadth}(K)$.   ◀

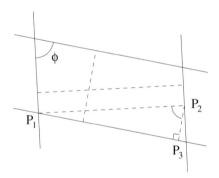

Fig. C.6. *A proof of the inequality* $\text{diam}\,(K)\cos\phi \leq \text{breadth}(K)$.

• Another parameter that can be compared with the breadth is the **interior diameter** of K, which is the diameter of the largest circle included in K. It will be denoted by

$$\text{diam}\,\text{int}(K) .$$

We always have the following inequalities:

$$\frac{2}{3}\,\text{breadth}(K) \leq \text{diam}\,\text{int}(K) \leq \text{breadth}(K) .$$

The proof of the first inequality is too technical, we shall not give it here. The second is immediate. We remark that the limits of these inequalities can be reached: $\text{diam}\,\text{int}(K) = (2/3)\,\text{breadth}(K)$ in the case of an equilateral triangle,

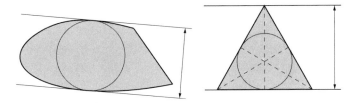

Fig. C.7. *The interior diameter is not always equal to the breadth. For an equilateral triangle, it equals* $(2/3)$ *breadth*(K).

and diam int(K) = breadth(K) in the case of a disk, whose breadth is also equal to the diameter.

• Finally, it is sometime convenient to use the notion of the **breadth perpendicular to the diameter**: given a diameter of K, we shall denote by the smallest distance between any two lines parallel to the diameter and enclosing K

$$L^\perp(K) .$$

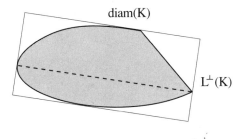

Fig. C.8. *The convex set K can be inscribed inside a rectangle. One of its sides is of length* diam (K), *and the length of the other is* $L^\perp(K)$.

If for some convex set K there is more than one segment whose length is the diameter, the above definition will be ambiguous. We can then define $L^\perp(K)$ as the minimum of all the values obtained when we vary the diameter. It is clear that $L^\perp(K)$ = breadth(K) in the case of a disk or an ellipse, while $L^\perp(K) >$ breadth(K) for a rectangle. In general:

$$\boxed{\text{breadth}(K) \leq L^\perp(K) \leq \sqrt{2}\,\text{breadth}(K) .}$$

▶ Only the second equality may pose some difficulties. To prove it, we first note that $L^\perp(K) \leq$ breadth$(K)/\sin\phi$, where ϕ is, as previously, the angle between the considered diameter and breadth. On the other hand, we clearly have $L^\perp(K) \leq \operatorname{diam}(K)$. If we replace $\operatorname{diam}(K)$ by h_1, $\operatorname{breadth}(K)$ by h_2, and $\sin\phi$ by $\sqrt{1 - \cos^2\phi}$, where $\cos\phi \leq h_2/h_1$, we obtain

$$L^\perp(K) \leq h_2 \min \left\{ \frac{h_1}{h_2}, \frac{1}{\sqrt{1 - \left(\frac{h_2}{h_1}\right)^2}} \right\}.$$

This minimum is obtained when the two terms are equal. Their common value is then equal to $\sqrt{2}$. ◀

Here also, the bounds are reached (Fig. C9).

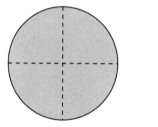

Fig. C.9. *For a disk,* $L^\perp(K) = \operatorname{breadth}(K)$. $L^\perp(K) = \sqrt{2}\operatorname{breadth}(K)$ *for a square.*

◇   These parameters satisfy the same properties as those of §2. The functions $\operatorname{breadth}(K)$, $\operatorname{diam int}(K)$, $L^\perp(K)$ and are increasing and continuous.

◇   A convex set K is said to be of **constant breadth** if

$$\operatorname{diam}(K) = \operatorname{breadth}(K).$$

A set of constant breadth is not necessarily a disk. In Figure C10 we give an example known as the *Reuleaux polygon*, constructed with the help of arcs of circle of the same radius.

◇   If we take into account the breadth of a set K, we can find a relation between the perimeter of K and its diameter that is finer than the inequality $L(\partial K) \geq 2\operatorname{diam}(K)$ of §2. We shall prove that

$$L(\partial K) \geq 2\sqrt{\operatorname{diam}(K)^2 + L^\perp(K)^2}.$$

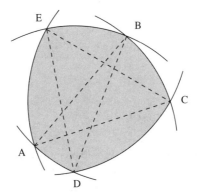

Fig. C.10. *Construction of a Reuleaux polygon with 5 sides. Its diameter is equal to its breadth. Its orthogonal projection on any line has a constant value: its perimeter is then equal to* $\pi \operatorname{diam}(K)$.

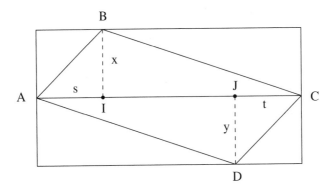

Fig. C.11. *The smallest perimeter of a quadrilateral $ABCD$ is obtained when the four sides are equal.*

▶ The convex set K is inscribed in a rectangle $h_1 \times h_2$, where $h_1 = \operatorname{diam}(K)$ and $h_2 = L^\perp(K)$. Let A, B, C, D be the four points of contact of the four sides with K (Figure C11; in the limit case, some of these points may be identical). Let I and J be the projections of B and D on AC. We put $x = \operatorname{dist}(B, I)$, $y = \operatorname{dist}(D, J)$, $s = \operatorname{dist}(A, I)$, and $t = \operatorname{dist}(C, J)$. Since K contains the quadrilateral $ABCD$, its perimeter is at least the sum of the sides of this quadrilateral, that is

$$L(\partial K) \geq \sqrt{x^2 + s^2} + \sqrt{x^2 + (h_1 - s)^2} + \sqrt{y^2 + t^2} + \sqrt{y^2 + (h_1 - t)^2}\,.$$

We need to find the minimum of this function with respect to the 4 variables s, t, x and y, subject to the restriction $x + y = h_2$. We obtain it when $s = t = h_1/2$ and $x = y = h_2/2$ (we can use the fact that the minimum of the function $f(z) = \sqrt{z^2 + a^2} + \sqrt{(z - b)^2 + a^2}$ on $[a, b]$ is obtained for $z = b/2$). We deduce:

$$L(\partial K) \geq 2\sqrt{h_1^2 + h_2^2} \ .$$

This inequality becomes an equality when K is a quadrilateral with four equal sides. ◀

C.4 Area of a convex set

The two parameters: diameter and breadth, suffice for getting an order of the size of the area $\mathcal{A}(K)$ of a convex set K, but they are not sufficient for giving the exact value of this area. We can show that:

$$\frac{1}{2}\operatorname{diam}(K)\operatorname{breadth}(K) \leq \mathcal{A}(K) \leq \sqrt{2}\operatorname{diam}(K)\operatorname{breadth}(K) \ .$$

The first inequality becomes an equality if K is a triangle. While the constant $\sqrt{2}$ of the right hand side term is not the best possible, it has the advantage of making the proof easier.

▶   It suffices to show the following inequalities:

$$\frac{1}{2}\operatorname{diam}(K)\,L^{\perp}(K) \leq \mathcal{A}(K) \leq \operatorname{diam}(K)\,L^{\perp}(K) \ .$$

Using the notation of Figure C11, the area of the polygon of vertices $ABCD$ is $\operatorname{diam}(K)\,L^{\perp}(K)/2$, and it is included in K. On the other hand, K is included in a rectangle whose area is $\operatorname{diam}(K)\,L^{\perp}(K)$. ◀

◇   We should not conclude that the maximum area of a convex set is given by $\sqrt{2}\operatorname{diam}(K)^2$. In fact, *the isoperimetric inequality* gives a better result:

$$\mathcal{A}(K) \leq \frac{1}{4\pi}L(\partial K)^2 \ ,$$

with equality in the case of a disk. Since $L(\partial K) \leq \pi \operatorname{diam}(K)$, we deduce that:

$$\mathcal{A}(K) \leq \frac{\pi}{4}\operatorname{diam}(K)^2 \ .$$

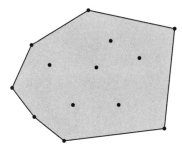

Fig. C.12. *Convex hull of a finite number of points.*

C.5 Convex hull

Given a set E in the plane, we call the smallest convex set containing E the **convex hull** of E, denoted

$$\mathcal{K}(E) \, .$$

Such a set does exist. Recall that the intersection of a family of convex sets, if it is not empty, is also a convex set. The convex hull of E may then be defined as the intersection of all the convex sets containing E.

$\diamondsuit$ If E itself is convex, then $\mathcal{K}(E) = E$.

$\diamondsuit$ If E is a finite set of points, then the boundary $\partial\mathcal{K}(E)$ of the convex hull consists a finite number of segments. In fact, $\mathcal{K}(E)$ is the union of all the triangles having as vertices any three chosen points of E.

$\diamondsuit$ If $E = \Gamma$ is a simple curve, then

$\mathcal{K}(\Gamma)$ *is the union of all the segments whose endpoints belong to the curve.*

▶   Let $\mathcal{U}$ be the union of the segments. It must be included in every convex set containing Γ. Therefore $\mathcal{U} \subset \mathcal{K}(\Gamma)$. Now it suffices to show that $\mathcal{U}$ is a convex set.

First, we show that if A, B, C are three points of Γ, then the triangle ABC is included in $\mathcal{U}$. We assume that the order of the points in question is the same as the order of Γ. Let P be a point inside the triangle and $\mathbf{D}_1$, $\mathbf{D}_2$ be the two semi–lines passing through P with origin at A and C, respectively. $\mathbf{D}_1$ intersects the arc $B\frown C$ in at least one point, let Q_1 be the furthest points from A. Similarly, $\mathbf{D}_2$ meets the arc $A\frown B$ is at least one point. Let Q_2 be the point furthest from C. Because Γ is simple, there are two possible cases: either P, which belongs to $\mathbf{D}_1$, is situated between A and Q_1, or P, which belongs to $\mathbf{D}_2$, is situated between C and Q_2. In both cases, P belongs to $\mathcal{U}$.

Similarly, every convex quadrilateral whose vertices belong to Γ is included in $\mathcal{U}$ because this quadrilateral can be divided into two triangles whose vertices belong to Γ. Now, take any two points A and B in $\mathcal{U}$. The point A belongs to a segment $A_1 A_2$ whose endpoints belong to Γ, and the point B belongs to the

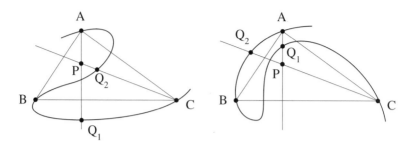

Fig. C.13. *The point P belongs either to the segment AQ_1 or to the segment CQ_2 (or to both of them).*

segment $B_1 B_2$ with endpoints in Γ. The convex hull of the 4 points A_1, A_2, B_1, B_2 could be either a quadrilateral, a triangle, or a segment, with vertices on Γ. This convex hull, which contains the segment AB, is included in $\mathcal{U}$. Therefore $AB \subset \mathcal{U}$. ◀

C.6 Perimeter of the convex hull

With the same notation as in the previous sections, $L(\partial\mathcal{K}(E))$ designates the perimeter of the convex hull of the set E. This length has a geometrical meaning:

The perimeter of $\mathcal{K}(E)$ is the smallest length of a closed curve circumscribing E.

▶   To prove this, we must first of all show that the length of any closed curve Γ is always larger than the perimeter of its convex hull:

$$L(\partial\mathcal{K}(\Gamma)) \leq L(\Gamma) \, .$$

Without going into details, let us say that $\partial\mathcal{K}(\Gamma)$ is formed of points of Γ and of segments whose length is smaller than those of the arcs of Γ above them.

Now we take a set E and any curve Γ circumscribing E. Since $\mathcal{K}(E) \subset \mathcal{K}(\Gamma)$, we know that

$$L(\partial\mathcal{K}(E)) \leq L(\partial\mathcal{K}(\Gamma)) \, .$$

Consequently, $L(\partial\mathcal{K}(E)) \leq L(\Gamma)$. ◀

From the inequality

$$L(\partial\mathcal{K}(\Gamma)) \leq L(\Gamma) \, ,$$

which holds if Γ is a closed curve, we immediately deduce that if Γ is any curve with endpoints A and B, then

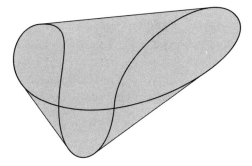

Fig. C.14. *The length of a closed curve is at least equal to the perimeter of its convex hull.*

$$L(\partial \mathcal{K}(\Gamma)) \leq L(\Gamma) + \mathrm{dist}(A, B) \ .$$

▶   In fact, $\Gamma \cup AB$ is a closed curve, and $\mathcal{K}(\Gamma \cup AB) = \mathcal{K}(\Gamma)$.   ◀

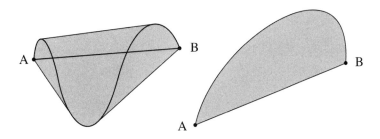

Fig. C.15. *The perimeter of $\mathcal{K}(\Gamma)$ is at most equal to $L(\Gamma) + \mathrm{dist}(A, B)$.*

This lemma was used in Chapter 8, §4. We obtain equality when Γ is a subarc of the boundary of a convex set because then, AB is the chord of this arc and $\partial \mathcal{K}(\Gamma \cup AB) = \Gamma \cup AB$.

C.7 Area of the convex hull of a curve

What is the relation between the length of a curve, the distance between its endpoints, and the area of its convex hull? Here is a result in this direction (used in Chapter 7, §2), For a curve Γ with endpoints A and B:

$$\mathcal{A}(\mathcal{K}(\Gamma)) \leq L(\Gamma)^{3/2}\sqrt{L(\Gamma) - \operatorname{dist}(A, B)} \,.$$

▶   We know (§4) that

$$\mathcal{A}(\mathcal{K}(\Gamma)) \leq \operatorname{diam}(\Gamma)\,L^{\perp}(\mathcal{K}(\Gamma)) \,.$$

Using the inequalities

$$2\sqrt{\operatorname{diam}(\Gamma)^2 + L^{\perp}(\mathcal{K}(\Gamma))^2} \leq L(\partial\mathcal{K}(\Gamma))$$

in §3, and

$$L(\partial\mathcal{K}(\Gamma)) \leq L(\Gamma) + \operatorname{dist}(A, B)$$

in §6, we find that:

$$L^{\perp}(\mathcal{K}(\Gamma))^2 \leq \frac{1}{4}(L(\Gamma) + \operatorname{dist}(A, B))^2 - \operatorname{diam}(\Gamma)^2 \,.$$

But $\operatorname{dist}(A, B) \leq \operatorname{diam}(\Gamma) \leq L(\Gamma)$, and so we obtain

$$L^{\perp}(\mathcal{K}(\Gamma)) \leq \sqrt{L(\Gamma)}\sqrt{L(\Gamma) - \operatorname{dist}(A, B)} \,,$$

thus proving the desired inequality.   ◀

References

M. Barnsley, *Fractals everywhere* (1988), Academic Press.

A.S. Besicovitch, "On the existence of tangent to rectifiable curves," *J. London Math. Soc.* **19** (1944), 205–207.

A.S. Besicovitch & S.J. Taylor, "On the complementary intervals of a linear closed set of zero Lebesgue measure," *J. Lond. Math. Soc.* **29** (1954), 449–459.

W. Blaschkee, *Vorlesungen über Integralgeometrie* (1936), Leipzig.

P. du Bois–Reymond, "Sur la grandeur relative des infinis de fonctions," *Annali di Mathematica* **4** (1871), 338–353.

E. Borel 1, *Euvres* (1972), éditions du C.N.R.S., Paris.

E. Borel 2, *Leçons sur la théorie de la croissance* (1910), Gauthiers–Villars.

E. Borel 3, *Eléments de la théorie des ensembles* (1949), Albin Michel.

G. Bouligand 1, "Dimension, Etendue, Densité," *C. R. Acad. Sc. Paris* **180** (1925), 246–248.

G. Bouligand 2, "Sur le potentiel et quelques théories connexes," *C.R. Acad. Sc. Paris* **184** (1927), 430–431.

G. Bouligand 3, "Ensembles impropres et nombre dimensionnel," *Bull. Sc. Math.* **52** (1928), 320–334 and 361–376.

G. Bouligand 4, "Sur la notion d'ordre de mesure d'un ensemble plan," *Bull. Sc. Math.* **53** (1929), 185–192.

G. Bouligand 5, *Les définitions modernes de la dimension* (1935), Hermann.

N. Bourbaki, *Fonctions d'une variable réelle* **4** (1961), Hermann.

J.C. & H. Burkill, *A second course in mathematical analysis* (1970), Cambridge University Press.

G. Cantor, Traductions parues dans *Acta Mathematica*, t. 2 and after (1883).

A. Cauchy, "Note sur divers théorèmes relatifs à la rectification des courbes, et à la quadrature des surfaces," *C. R. Acad. Sc. Paris* **13** (1841). Also, see *Œuvres de Cauchy*, t. VI, 369–379.

M.W. Crofton, "On the theory of local probability, applied to straight lines at random in a plane," *Phil. Trans. Royal Soc.* **158** (1868), 181–199.

E. Czuber, *Geometrische Wahrscheinlichkeiten und Mittelwerte* (1984), Leipzig. Under the title *Probabilités et moyennes géométriques* (1902), Hermann.

R. Deltheil, *Probabilités géométriques* (1926), Gauthiers–Villars.

A. Denjoy 1, "Sur une classe d'ensembles parfaits discontinus," *Atti Re. Ac. Naz. Lincei* **29** (1920), 291–294.

A. Denjoy 2, "Les degrés de nullité dans la mesure des ensembles parfaits linéaires," *C. R. Acad. Sc. Paris* **259** (1964), 4449–4451.

P. Dugac, *Sur les fondements de l'Analyse de Cauchy à Baire* (1978), doctoral thesis, U. de Paris VI.

K. Falconer, *The geometry of fractal sets* (1985), Cambridge University Press.

J. Favard, "Une définition de la longueur et de l'aire," *C. R. Acad. Sc. Paris* **194** (1932), 344–346.

H. Federer, *Geometric Measure Theory* (1969), Springer–Verlag.

M. Fréchet, "Une généralisation de la raréfaction d'un ensemble de mesure nulle," *C. R. Acad. Sc. Paris* **252** (1961), 1245–1250.

C. Grebogi, S. McDonald, E. Ott & J.A. Yorke, "Exterior dimension of fat fractals," *Phys. Lett. A* **110** (1985), 1–4.

G.H. Hardy 1, *Orders of infinity* (1910), Cambridge University Press.

G.H. Hardy 2, "Weierstrass non–differentiable function," *Trans. Amer. Math. Soc.* **17** (1916), 301–325.

M. Hata & M. Yamaguti, "The Takagi function and its generalization," *Japan J. Applied Math.* **1** (1984), 183–199.

M. Hata, "Fractals in Mathematics," *Stud. appl. Math.* **18** (1986), 259–278.

F. Hausdorff, "Dimension und äusseres Mass," *Math. Ann.* **79** (1919), 157–179.

J. Hawkes, "Hausdorff measure, entropy and the independance of small sets," *Proc. London Math. Soc.* **28** (1974), 700–724.

E. Hille & J.D. Tamarkin, "Remarks on a known example of a monotone function," *Amer. Monthly* **36** (1929), 255–264.

E.W. Hobson, *The theory of functions of a real variable*, third edition (1927), Cambridge University Press.

J.E. Hutchinson, "Fractal and self–similarity," *Ind. U. Math. J.* **30** (1981), 713–747.

C. Jordan, *Cours d'Analyse*, second edition (1893).

J.–P. Kahane & R. Salem, *Ensembles Parfaits et Séries Trigonométriques* (1963), Hermann.

J. Kaplan, J. Mallet–Paret, & J.A. Yorke, "The Lyapounov dimension of a nowhere differentiable attracting torus," *Ergod. Th. and Dyn. Syst.* **4** (1984), 261–281.

K. Knopp, "Ein Einfaches Verfarhen zur Bildung stetiger nirgends differenzier-barer Funktionen," *Math. Zeits.* **2** (1918), 1–26. Cited in [E.W. Hobson].

A.N. Kolmogorov & V.M. Tihomirov, "Epsilon–entropy and epsilon–capacity of sets in functional spaces," *Amer. Math. Soc. Transl.* **17** (1961), 277–364.

N. Kôno, "On self–affine functions I and II," *Japan J. Appl. Math.* **3** (1986), 271–280 et **5** (1988), 441–454.

K. Kuratowski, *Topologie* (1958), edited by the Polish Acad. of Science, Warsaw.

H. Lebesgue 1, *Œuvres scientifiques* (1972), Enseignement mathématique, Genève.

H. Lebesgue 2, *La mesure des grandeurs* (1956), Monographie de l'Enseignement Mathématique, reprinted by the Librairie Scientifique et Technique (1975), A. Blanchard.

B. Mandelbrot 1, *Fractals: Form, Chance and Dimension* (1977), Freeman.

B. Mandelbrot 2, *The Fractal Geometry of Nature* (1982), Freeman.

B. Mandelbrot 3, "Self–affine fractal sets and fractal dimension," *Phys. Scr.* **32** (1986), 257–260.

M. Mendès–France, Y. Dupain & C. Tricot, "Dimension de spirales," *Bull. Soc. Math. France* **111** (1983), 193–201.

H. Minkowski, "Ueber die Begriffe Länge, Oberfläche und Volumen," *Jahr. Deut. Math.* **9** (1901), 115–121.

P.A.P. Moran, "Measuring the length of a curve," *Biometrika* **53** (1966), 359–364.

Z. Moszner, "Sur une notion de raréfaction d'un ensemble de mesure nulle," *Ann. Sc. Ec. Norm. Sup.* **83** (1966), 191–200.

F. Normant & C. Tricot, "Fractal simplification of lines using convex hulls," *Geographical Analysis* (1992).

G. Peano 1, *Rend. Lincei* (4) vol.VI (1890), 54.

G. Peano 2, "Sur une courbe qui remplit une aire plane," *Mathematische Annalen* **36** (1890), 157–160.

M.J. Pelling, "Formulae for the arc–length of a curve in $\mathcal{R}^n$," *Ann. Math. Monthly* **84** (1977), 465–467.

J. Perkal, "Sur les ensembles ϵ–convexes," *Colloquium Mathematicum* **4** (1956), 1–10. And "On the length of empirical curves," *Michigan Inter–University Community of Mathematical Geographers* **10** (1966).

L. Pontrjagin & L. Schnirelmann, "Sur une propriété métrique de la dimension," *Ann. of Math.* **33** (1932), 156–162.

L. Richardson, "The problem of contiguity: an appendix of statistics for deadly quarrels," *General Systems Yearbook* **6** (1961), 139–187.

L. Santaló, *Integral Geometry and Geometric Probability* (1976), Encyclopedia of Mathematics, Addison–Wesley.

S. Sherman, "A comparison of linear measures in the plane," *Duke Math. J.* **9** (1942), 1–9.

H. Steinhaus 1, "Sur la portée pratique et théorique de quelques théorèmes sur la mesure des ensembles de droites," *C. R. du premier congrès des mathématiciens des pays slaves* (1930), 353–354.

H. Steinhaus 2, "Zur Praxis der Rectifikation und zum Längenbegrife," *Ber. Acad. Wiss.* **82** (1930), 120–130.

H. Steinhaus 3, "Length, Shape and Area," *Colloq. Math.* **3** (1954), 1–13.

T. Tagaki, "A simple example of continuous function without derivative," *Proc. Phys. Math. Soc. Japan* **1** (1903), 176–177. See *The collected papers of Teiji Tagaki*, Iwanami Shoten Publ. (1973), Tokyo.

C. Tricot, *Sur la notion de densité* (1973), Cahiers du Dept. d'Econométrie de l'U. de Genève.

C. Tricot 1, *Sur la Classification des Ensembles Boréliens de Mesure de Lebesgue Nulle* (1979), thèse de doctorat, U. de Genève.

C. Tricot 2, "Douze définitions de la densité logarithmique," *C. R. Acad. Sc. Paris* **293** (1981), 549–552.

C. Tricot 3, "Two definitions of fractional dimension," *Math. Proc. Camb. Phil. Soc.* **91** (1982), 57–74.

C. Tricot 4, "Metric properties of compact sets of measure 0 in $\mathcal{R}^2$," in *Mesures et dimensions* (1983), doctoral thesis, U. de Paris XI; and "Porous surfaces," in *Constructive approximations* **5** (1989), 117–136.

C. Tricot 5, "The geometry of the complement of a fractal set," *Phys. Lett. A* **114** (1986), 430–434.

C. Tricot 6, "Dimension fractale et spectre," *J. Chimie Phys.* **85** (1988), 379–384.

C. Tricot 7, "Local convex hulls of a curve, and the value of its fractal dimension," *Real Anal. Exchange* **5** (1990), 675–693.

C. Tricot, J.F. Quiniou, D. Wehbi, C. Roques–Carmes & B. Dubuc, "Evaluation de la dimension fractale d'un graphe," *Rev. Phys. Appl.* **23** (1988), 111–124.

J.B. Wilker, "Sizing up a solid packing," *Per. Math. Hung.* **8** (1977), 117–134.

M. Yamaguti & M. Hata, "Weierstrass's function and chaos," *Hokkaido Math. J.* **12** (1983), 333–342.

W.H. & G. Young, *The theory of sets of points* (1906), Cambridge University Press.

Index